Aircraft Valuation

"This book fills a niche in the market. It not only delivers detailed information on contemporary aircraft leasing methodologies and markets but also provides a comprehensive review of the impacts on the aviation industry from the COVID-19. This is the best book for both academia and practitioners in the aircraft leasing field."

—Chris C. Hsu, Ph.D., Professor of Finance, *Director of CUNY Aviation Institute, York College, The City University of New York*

David Yu

Aircraft Valuation

Airplane Investments as an Asset Class

David Yu, PhD. CFA, Senior ISTAT Certified Aviation Appraiser
New York University Shanghai
Shanghai, China

ISBN 978-981-15-6742-1 ISBN 978-981-15-6743-8 (eBook)
https://doi.org/10.1007/978-981-15-6743-8

This Palgrave Macmillan imprint is published by the registered company Springer Nature Singapore Pte Ltd.
The registered company address is: 152 Beach Road, #21-01/04 Gateway East, Singapore 189721, Singapore

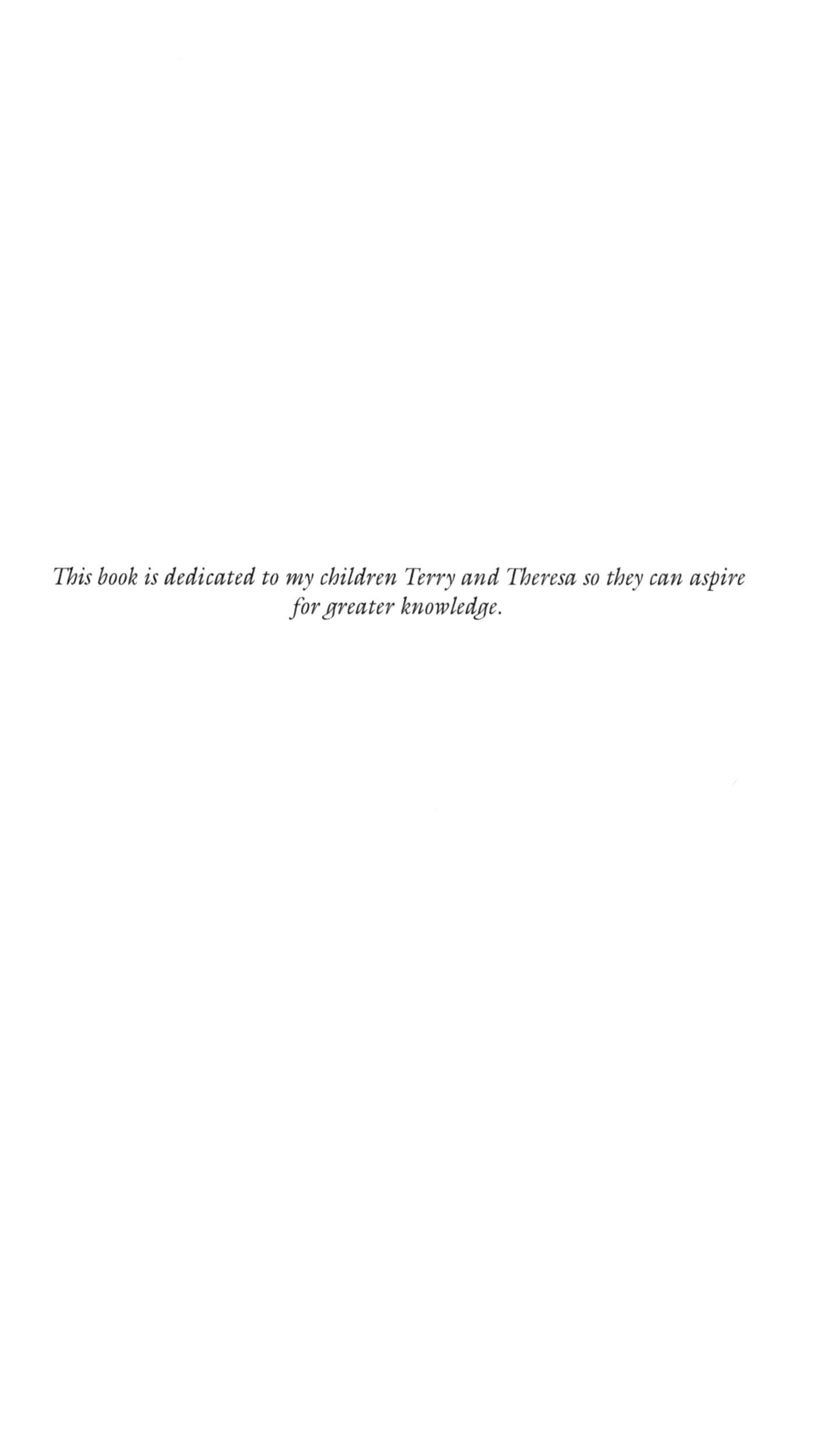

This book is dedicated to my children Terry and Theresa so they can aspire for greater knowledge.

Preface: In the News Today - On-going Significance of the COVID-19 Impacts on the Aviation Industry

Today, the world is in the midst of a vastly different environment than we have known in recent times and it seems every day there is news that has significant impact on the aviation industry, aircraft values and the global economy. While there are a few exogenous economic shocks to compare against, COVID-19 seems to have even worse effects than the most recent cases of economic shocks including SARS and the great financial crisis.

Initially, at the start of the coronavirus outbreak in January 2020, it was hoped by many that it might be contained within a few months but this has proved to be overly optimistic as it has created even more uncertainty globally. As of this publication, the author continues to be primarily at home indoors in Shanghai, China for over 90+ days since late January 2020 and has been a keen observer and commentator while the entire world has reacted to this ever evolving epidemic.

In the beginning of the viral epidemic in January and early February 2020, before the proliferation of the virus globally, the problem was framed by many as a China or Asia market issue and the direct effects were concentrated in those specific regional markets. To stop the spread of the virus, the Chinese government imposed heavy restrictions on the city of Wuhan on January 23, 2020 and then Hubei Province in its entirety on January 27, 2020. From that point on, many provinces and major cities like Beijing and Shanghai around China declared level 1 health emergencies and neighborhoods put up restrictions of movement and travel (Caixin 2020). As of February 12, 2020, there were nearly 45,000 global cases that have been confirmed and global fatalities have reached 1116, according to Johns Hopkins University CSSE (2020).

The onset of the virus was in the middle of the Chinese New Year holiday which normally sees a large increase in travel as over 300 million people move around the country and abroad to visit family and for leisure (China Daily 2020). These quickly implemented restrictions of movement also prevented some people from returning to their normal abode after visiting their familial hometowns or taking vacations during the extended New Year holiday. In addition, all tourist attractions and many public spaces were shut down to prevent further outbreak and to contain the risk of further spreading of COVID-19.

For the economy, and in turn the aviation industry, implications of the outbreak quickly reached far and wide. Businesses and production facilities were shut down due to either preventative measures, lack of workers or disruptions in supply chain inputs. This not only heavily impacted the Chinese economy, but it also affected other countries and the global economy as supply chains are integrated with the drivers of air travel supply and demand.

The current aviation production manufacturers located in China are the Airbus A320 and A330 completion facilities in Tianjin and the Boeing 737 MAX completion facility in Zhoushan in Zhejiang Province. Even during normal times, production is slow the New Year holiday but the virus has affected the normal ramp up back to full operations afterwards. Airbus' Tianjin completion facilities supplying about 10% of aircraft were temporarily shut down and did not restart until February 11, 2020 (Alcock 2020) and were able to fully resume pre-virus production levels on March 25, 2020 (The Paper 2020). Meanwhile, Boeing's 737 MAX completion Zhoushan facilities were already shut prior to the outbreak due to the grounding of the 737 MAX aircraft production.

Initial Responses to Covid-19: Phase I Virus Life Cycle (Impact on China and Interlinked Countries and Industries)

Initially, at the beginning of the response, economists slashed their forecast of China's 2020 GDP growth to as low as 4.5% from the ~6.0% previously. This is despite the People's Bank of China quickly initiating a large fiscal and monetary stimulus along with other government aid packages to boost the economy. The effects of the virus already rattled investors with China's main stock indices plunging more than 8% in the initial response on the reopening of the exchange following a 10-day break on February 3, 2020.

On the demand side, the extent of the virus' effects first became apparent in late January when Chinese airlines' domestic traffic fell 6.8% in January, reflecting the impact of flight cancellations and travel restrictions related to COVID-19. On an on-going basis, China's Ministry of Transport reported an 80% annualized fall in volumes in late January and early February. Airlines took out capacity by 0.2% and passenger load factor plunged 5.4 percentage points to 76.7% (International Air Transport Association 2020a).

International capacity was also initially slashed as the Chinese Big 3 airlines (Air China, China Eastern and China Southern) slashed international capacity between 80% and 90% in mid-February. These figures were also widely reflected by international airlines serving the Chinese market as well. This international squeeze became more acute as the domestic markets also fell, with 60% of Chinese airlines' fleets being grounded in mid-February. Domestic traffic fell 46.1% in February compared to the previous year (Civil Aviation Administration of China 2020). The Chinese aviation industry is expected to lose at least US$12.8 billion in revenue as a result of flight cancellations and weak demand among travelers in February according to the earliest estimates by International Air Transport Association on February 20, 2020 ("IATA") (International Air Transport Association 2020b).

With losses estimated to be around $3 billion in February alone, HNA Group was in trouble even before the crisis with Beijing already promising bail-outs to make up for their losses (The Economist 2020). They subsequently became the second implied airline causality of the virus stricken environment after Flybe of the UK. The control of the firm effectively changed to the newly appointed senior managers by the Hainan Province (HNA 2020).

The losses of Chinese-based traffic reverberated throughout Asia with IATA announcing a potential 13% full-year loss of passenger demand for carriers in the Asia-Pacific region which translates into a $27.8 billion revenue loss in Asia-Pacific for the full year 2020 (International Air Transport Association 2020c). Hong Kong's airlines took the brunt of the mainland shutdown and saw more than 75% drop in passenger movement. The corresponding figures for South Korea are 46%, Taiwan 38%, Thailand 33% and Singapore 32.8% in February (Heng and Yong 2020).

Global airlines that were exposed to routes with China directly or indirectly through transit, tourist or business traffic were also affected but not as severely at first. The earliest estimates by IATA were announced on February 20, 2020, estimating airline revenue losses in China at $12.8

billion, Asia at $27.8 billion and globally at $29.3 billion. This was the first of four revisions by IATA as the estimates became direr with the ramifications ever more apparent and global.

A side development which normally has a significant effect on aviation is oil prices. In mid-March, Saudi Arabia, the world's largest oil exporter, announced its intention to significantly raise the production of oil as well as lower the official selling price starting April 2020. This, in turn, caused oil prices to plummet as it foreshadowed the flood of supply coming into the market. Oil prices fell around 24%, the biggest weekly decline since December 2008 and also represented a 48% fall from the peak in early January, the start of the COVID-19 virus effects.

Despite the lower oil costs, no airline is celebrating these reduced costs. First, lower oil prices are more apparently reflected in the operations when there is higher utilization and oil is consumed. Second, many airlines had oil hedges utilizing instruments such as swaps and options in place, which were meant to lock in the price of the oil for the quantity they expected to be consumed. These airlines had to mark to market the losses on the hedges as the oil prices fell from the referenced price. This also meant that as less oil was consumed, the hedges did not serve their function to counter the price of the oil used but rather became a speculative instrument (Fig. 1).

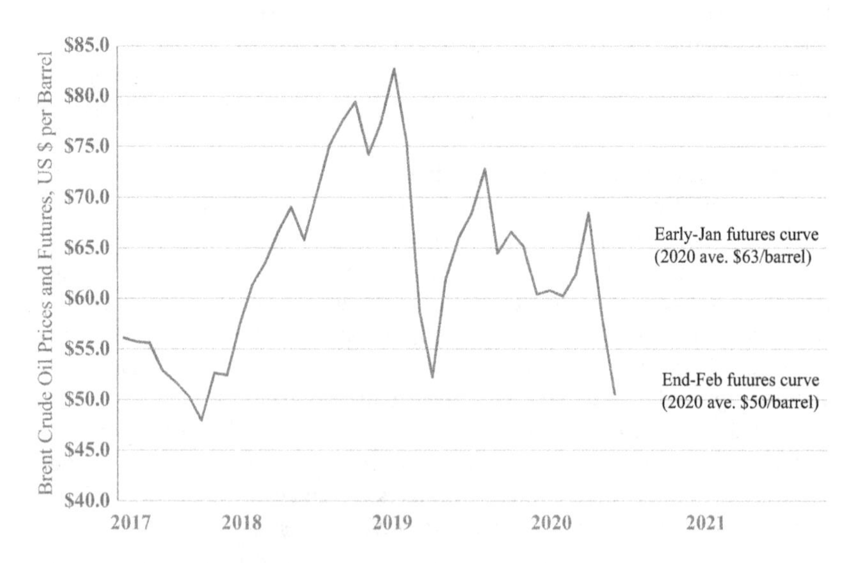

Fig. 1 Fall in oil prices. (Source: Bloomberg 2020)

Current Situation: Phase II Virus Life Cycle (More Severe Global Impact)

After the initial alarm in China and Asia, the infection rates in Europe and then the US reached significant numbers. Many governments have closed borders and restricted movement inside countries by instituting mandatory or recommended social distancing programs and even more stringent lockdown policies to stop the spread of the virus. Where borders are not actually closed, governments have implemented many restrictions on their borders, limiting international movement. Even with these measures, the virus infection rates have continued to spread. This has resulted in all airlines grounding a large portion of their capacity and heavily reducing scheduled flights.

The stock markets responded with aggressively downward movement since the start of the year. The lessor index is down 65.4% and the airline index is down 59.1% while the S&P 500 and MSCI World benchmarks are down 13.9% and 15.7%, respectively, representing the changes from the beginning of the year to April 24th, 2020 (Fig. 2).

As the virus became more widespread and the impacts more apparent, IATA issued its second revised loss figures on March 5, 2020, which estimated 2020 global airline revenue losses for the passenger business of between $63 billion (in a scenario where COVID-19 is contained in current markets with over 100 cases as of March 2) and $113 billion (in a scenario with a broader spread of COVID-19). Asia Pacific airline revenue losses were increased by 69.1% in the limited spread scenario with additional losses of 21.9% for the more extensive spread scenario (International Air Transport Association 2020d).

These estimates were subsequently updated on March 23, 2020, when the global revenue losses were estimated at $252 billion (losses increased by 123.0%) comprising of $88 billion revenue losses in Asia Pacific, $76 billion revenue losses in Europe, and $50 billion revenue losses in North America and the remaining amount from rest of the world. Asia Pacific region losses were further increased by 53.6% (Fig. 3).

On April 14, IATA issued the fourth estimate with global airline revenue losses at $314 billion (losses increased by 24.6%) comprising of $113 billion revenue losses in Asia Pacific, $89 billion revenue losses in Europe, and $64 billion revenue losses in North America and the remainder from the rest of the world. Asia Pacific losses were further increased by 28.4% (Fig. 3).

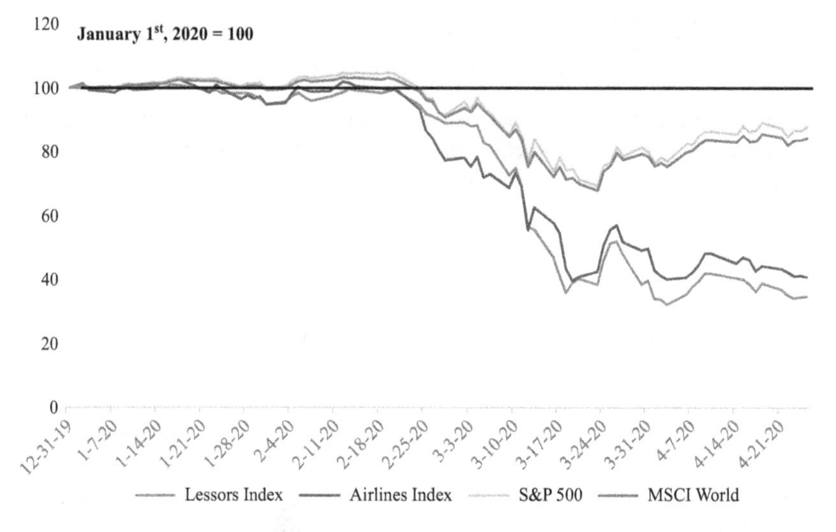

Fig. 2 Airlines and lessors indexes stock performance compared with MSCI World and S&P 500 Benchmarks since Jan 1, 2020. (Source: Bloomberg 2020. Note: Lessor Index = Average of Amedeo Air Four Plus; Aerocentury Corporation; AerCap; Aircastle; ALAFCO Aviation; Air Lease Corporation; Avation; AviaAM Leasing; Avolon; BOC Aviation; China Aviation Leasing Group; CDB Leasing; DP Aircraft 1; Doric Nimrod Air One; Doric Nimrod Air Two; Doric Nimrod Air Three; FLY Leasing; Genesis Lease; and Willis Lease Finance Corporation. Airlines Index = S&P 500 Airlines Total Return Index. See more details in Chap. 6)

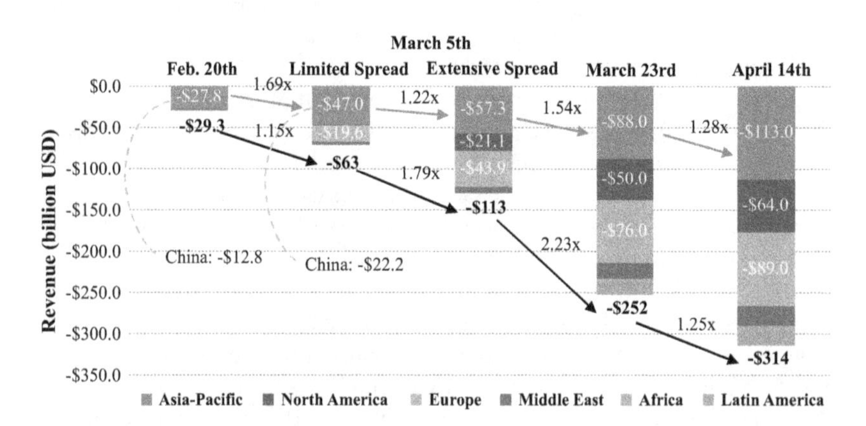

Fig. 3 IATA estimated COVID-19 impact on 2020 Airline revenue. (Source: International Air Transport Association 2020c, 2020d, 2020e, 2020f. Note: Asia-Pacific includes revenue losses from China)

Many airlines in the US and Europe also heavily reduced their capacity. American Airlines said it would slash international flying by 70% until May. International Airlines Group, the owner of British Airways and other airlines, said it was cutting capacity by 75% in April and May. Lufthansa subsidiary Austrian Airlines indefinitely halted all flight operations while the overall group reduced long-haul capacity by 90% and short-haul routes by 80%. With these reductions on a sustained basis, airline businesses are not sustainable. Even in normal times, a rough guide is that around 70% load factor per aircraft is required to sustain breakeven profitability.

IATA and other industry groups and associations are advocating for country support for the industry. The US airlines received $58 billion in aid comprising of $29 billion in payroll grants and $29 billion in loans from the $2 trillion stimulus bill, Coronavirus Aid, Relief, and Economic Security Act, which was signed into law on March 27, 2020 (United States Senate 2020). This economic stimulus bill was more than double the amount of the previous highest stimulus which came during the financial crisis and is now the largest in the history of the US. It is also the largest percentage of global relief by countries to airlines thus far.

Asia has thus far committed state support of $12.7 billion making up 15.1% of global committed state support totaling $84.6 billion. On an uncommitted state support basis, Asia has thus far $2.7 billion committed state support comprising of 10.6% of global uncommitted state support amounting to $25.4 billion. Combined, both committed and uncommitted support, Asia represents 14.0% or $15.4 billion of state support based on an overall global state support amounting to $110.1 billion (Figs. 4–6).

China, on March 4, initiated a funding scheme to incentivize the restoration and ongoing performance of flight services, including a CNY0.0176 per seat kilometer reward for flights on routes served by multiple airlines, and a CNY0.0528 per seat kilometer reward for a route served by a sole operator. The incentives are available to both domestic and international airlines serving these routes and are given on a cash grant basis through June 30 (Civil Aviation Administration of China & Ministry of Finance of the People's Republic of China 2020) (Figs. 4–6).

In China, some initial signs of the air traffic bounce back have already started to occur with strong demand from vacationers and visitors to their ancestral villages who were initially stranded by the restrictions. Following this initial increase, demand has subsequently slightly weakened and changes continue in a yo-yo manner. Domestic and international scheduled flight traffic was off the most at -70% year over year ("YoY") on the

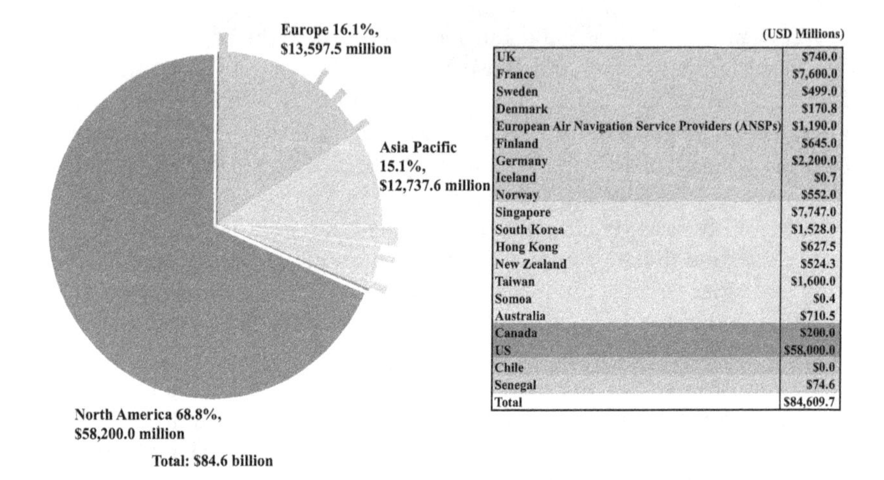

(USD Millions)

UK	$740.0
France	$7,600.0
Sweden	$499.0
Denmark	$170.8
European Air Navigation Service Providers (ANSPs)	$1,190.0
Finland	$645.0
Germany	$2,200.0
Iceland	$0.7
Norway	$552.0
Singapore	$7,747.0
South Korea	$1,528.0
Hong Kong	$627.5
New Zealand	$524.3
Taiwan	$1,600.0
Somoa	$0.4
Australia	$710.5
Canada	$200.0
US	$58,000.0
Chile	$0.0
Senegal	$74.6
Total	$84,609.7

Fig. 4 Confirmed state support for Airlines as of April 26, 2020

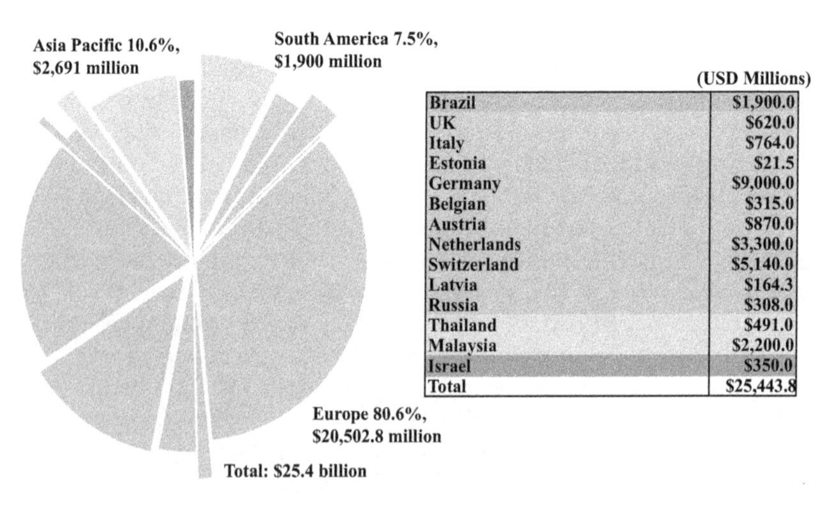

(USD Millions)

Brazil	$1,900.0
UK	$620.0
Italy	$764.0
Estonia	$21.5
Germany	$9,000.0
Belgian	$315.0
Austria	$870.0
Netherlands	$3,300.0
Switzerland	$5,140.0
Latvia	$164.3
Russia	$308.0
Thailand	$491.0
Malaysia	$2,200.0
Israel	$350.0
Total	$25,443.8

Fig. 5 Unconfirmed state support for Airlines as of April 26, 2020

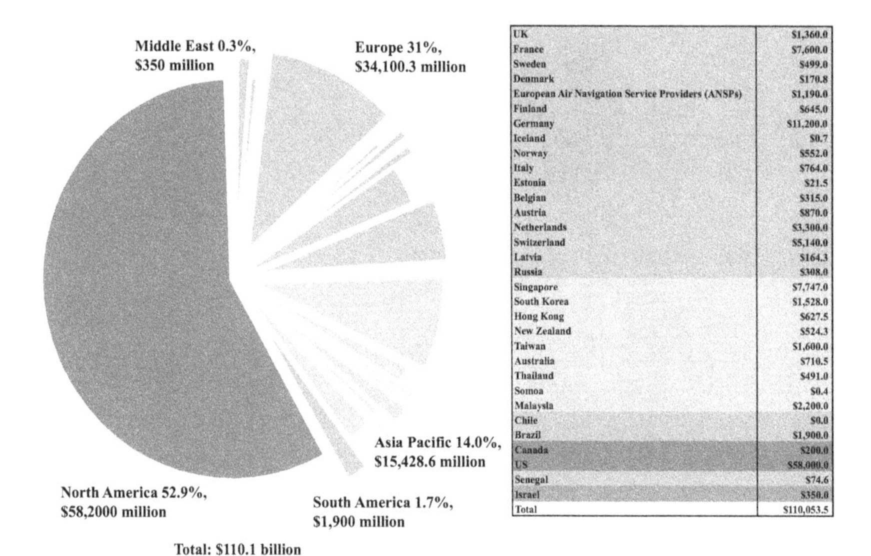

UK	$1,360.0
France	$7,600.0
Sweden	$499.0
Denmark	$170.8
European Air Navigation Service Providers (ANSPs)	$1,190.0
Finland	$645.0
Germany	$11,200.0
Iceland	$0.7
Norway	$552.0
Italy	$764.0
Estonia	$21.5
Belgian	$315.0
Austria	$870.0
Netherlands	$3,300.0
Switzerland	$5,140.0
Latvia	$164.3
Russia	$308.0
Singapore	$7,747.0
South Korea	$1,528.0
Hong Kong	$627.5
New Zealand	$524.3
Taiwan	$1,600.0
Australia	$710.5
Thailand	$491.0
Somoa	$0.4
Malaysia	$2,200.0
Chile	$0.0
Brazil	$1,900.0
Canada	$200.0
US	$58,000.0
Senegal	$74.6
Israel	$350.0
Total	$110,053.5

Fig. 6 Confirmed and unconfirmed state support for Airlines as of April 26, 2020

week of February 17 and subsequently bounced back with the traffic on the week of March 23 being -36% YoY. From that date, the figures regressed and further weakened with the week of April 20 figures at -42% YoY (OAG 2020). The trend of global traffic and the other countries specifically has been steadily downward as the crisis has gone on (Fig. 7).

For domestic scheduled flights in China, the market had a low in scheduled flight traffic at -70% YoY in the week of February 17 and subsequent high on the week of March 23 at -31% YoY. The current figures week of April 20 are -36% YoY (OAG 2020). Volatility in demand is expected to continue as the economy finds its footing and opens up more. Other than South Korea with a current -24% YoY change figure and Malaysia with -46% YoY for the week of April 20, the rest of the world and specific countries' domestic scheduled flights are at their lowest point since the beginning of January in terms of YoY change figures (Fig. 8).

For international scheduled flights in China, the market has steadily worsened with the current -95% YoY on the week of April 20 in scheduled flight traffic. This is the same for other all major countries and globally as well with YoY change figures for the week of April 20 at -90%. These

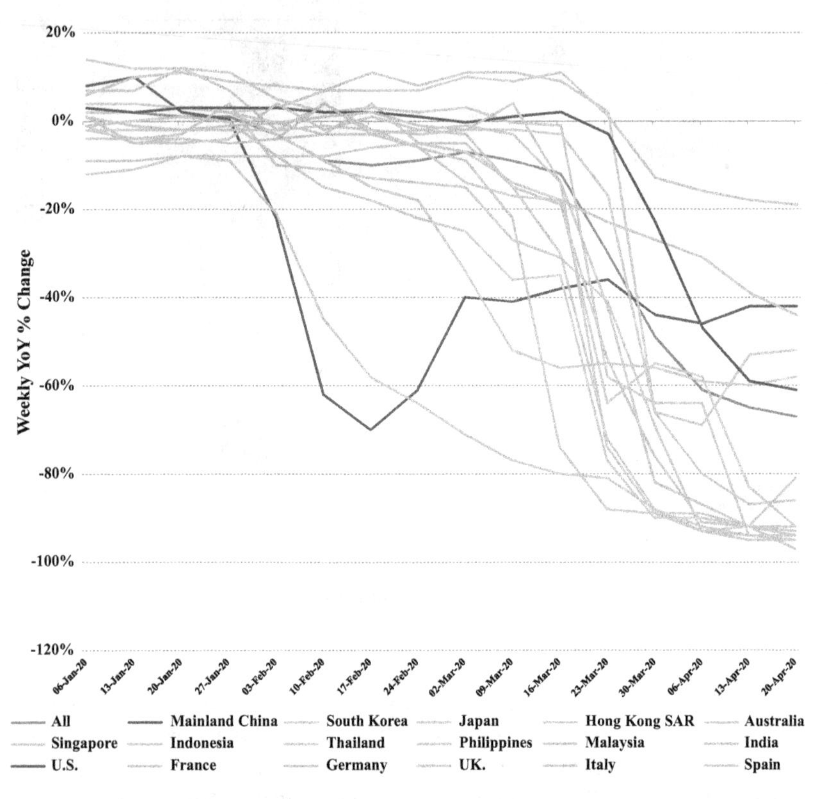

Fig. 7 International and domestic scheduled flights YoY changes by country. (Source: OAG 2020)

figures are not anticipated to rebound until border restrictions start to be lifted (Fig. 9).

Severe Acute Respiratory Syndrome ("SARS") 2002/2003

Many comparisons have been made to the most recent global outbreak, Severe Acute Respiratory Syndrome ("SARS"). SARS is a coronavirus that first appeared in China and killed nearly 800 people in 2002 and 2003 and this outbreak initially served as a reference point for the economic and

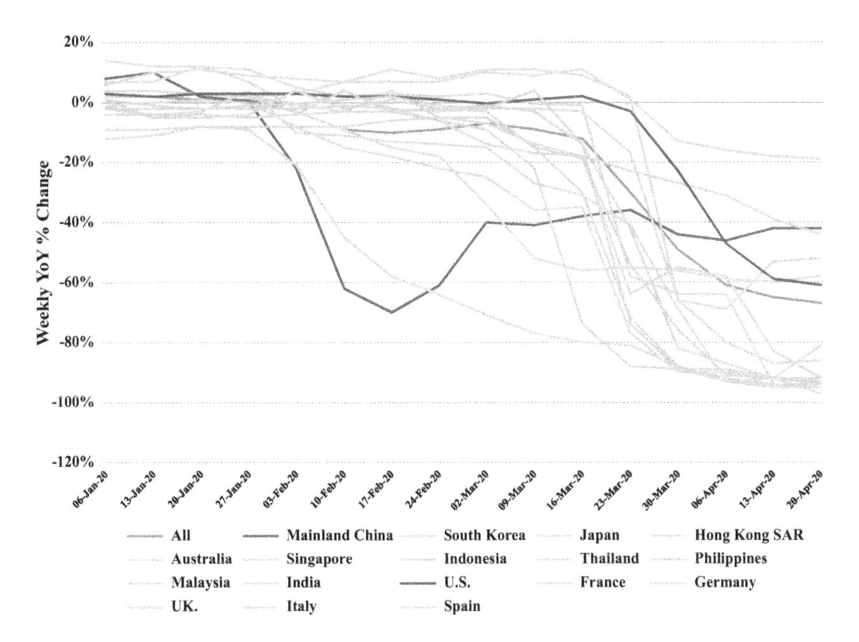

Fig. 8 Domestic scheduled flights YoY changes by country. (Source: OAG 2020)

cross-border impacts. As the current virus became a global pandemic, this would become weaker as a direct historical case comparison.

SARS contributed to an estimated 3 percentage point decrease in China's real GDP growth of 5% in its worst-affected quarter. Historically, the Chinese economy was more driven by investment and trade but has gradually rebalanced to a model led by domestic consumption and services. During the 15 years following SARS, the share of the service sector compared to China's total output rose to 52% from 40%, according to the Asian Development Bank (2020).

During the SARS outbreak in 2003, foreign direct investment ("FDI") in China actually rose from March to June, but then dropped 57% to approximately $32 million in July, the lowest monthly tally for the year. The trend quickly reversed, however, once the outbreak was brought under control. FDI inflows to China gradually picked up from December 2003 and quickly exceeded the pre-outbreak monthly high of $55 million at the beginning of 2004, according to the Asian Development Bank (2020).

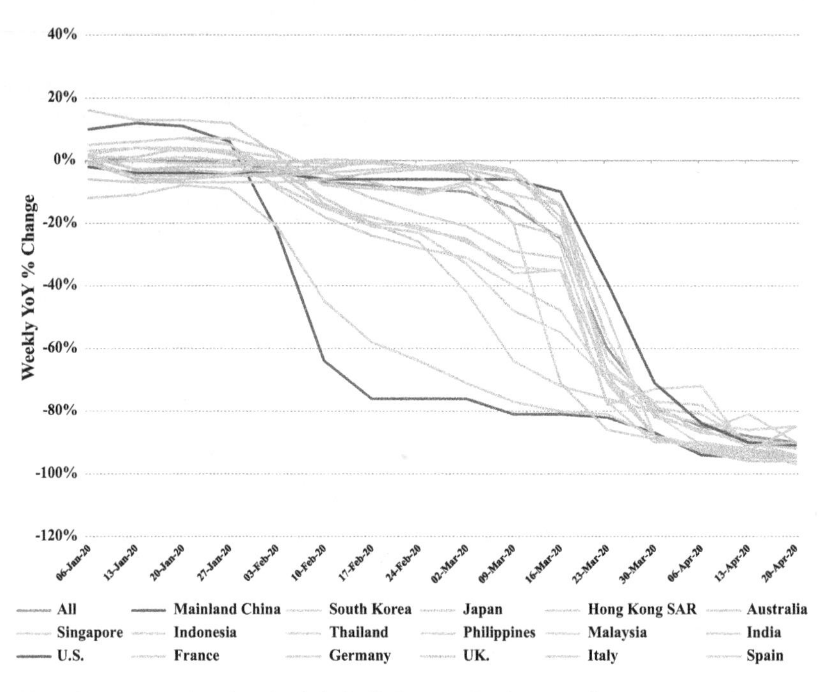

Fig. 9 International scheduled flights YoY changes by country. (Source: OAG 2020)

Globally, SARS cost airlines more than $10 billion in revenue and caused a loss of 39 billion revenue passenger kilometers ("RPK"). In North America, the loss to airlines was around $1 billion and RPK dropped 12.8 billion, or 3.7% of total international traffic, according to the International Air Transport Association (2003). In May 2003, there was a 21% YoY drop in global passenger traffic and overall capacity, expressed as available seat kilometers ("ASKs"), dropped 12.6% YoY. The global load factor showed signs of improvement, rising to 64.6% from 63.5% recorded one month prior in April. It is interesting to note that North American airlines were also hit hard during SARS but the effects were minimal for European airlines.

In terms of connectivity during SARS, airlines worldwide saw a decline in traffic with those operating out of the Asia-Pacific region losing as much as 8% on an annual basis. Asia saw similar decreased results as in China but not as severe. At the height of the outbreak (May 2003), monthly RPKs

Table 1 Global Airline performance by region

	May 2003(Percentage change over May 2002)				*Year to date results(Jan–May 2003 over Jan–May 2002)*			
Carriers	*RPK%*	*ASK%*	*FTK%*	*ATK%*	*RPK%*	*ASK%*	*FTK%*	*ATK%*
Europe	-5.5	-4.1	1.4	-2.8	-1.1	2.0	3.3	1.8
North America	-20.6	-19.0	-5.9	-10.7	-10.8	-6.0	11.4	-2.2
South America	2.0	1.9	2.8	1.6	9.1	4.9	12.0	6.4
Asia Pacific	-50.8	-30.7	0.9	-16.0	-11.9	2.3	10.0	6.9
Middle East	-1.6	-11.6	20.9	15.6	4.6	13.4	17.9	15.8
Africa	-0.1	7.7	15.0	8.5	0.2	6.3	11.1	8.4
Overall	-21.0	-12.6	3.0	-7.7	-6.4	1.2	8.7	3.4

Source: International Air Transport Association 2003

Note: RPK = Revenue Passenger Kilometers, ASK = Available Seat Kilometers, FTK = Freight Ton Kilometers, ATK = Available Ton Kilometers

of Asia-Pacific airlines were 50.8% lower than their pre-crisis levels (see Table 1). Overall SARS in 2003 cost Asia-Pacific airlines $6 billion of revenues and annual RPKs losses of 8% (International Air Transport Association 2020b). This translates to a 50.8% YoY drop in passenger traffic in RPK terms. June 2003 was the turning point when the YoY decrease reversed (35.8% YoY drop in passenger traffic) and capacity cuts were -27.2% (International Air Transport Association 2003) (Fig. 10).

In May 2003, passenger traffic in China dropped by 78% YoY while in the second quarter it dropped by 48.9% YoY (Civil Aviation Administration of China 2004). The Chinese major airlines showed similar changes but recovered from June 2003 onward, which resulted in a 1.9% YoY growth for the entire year (Civil Aviation Administration of China 2004).

What are the differences between this time and SARS? The current COVID-19 outbreak has occurred in a different environment. China's current highly indebted economy had started to show signs of weakness in 2019, as it marked the slowest GDP expansion in nearly 30 years, standing at 6.1%. These global uncertainties have already dampened Chinese outbound investment by 9.8% to $118 billion in 2019 (EY 2020).

Today, the Chinese economy accounts for 15% of global GDP in 2019, compared to 4% in 2003. As the world's largest trading nation, China represents 11.4% of global goods trade. This economic slowdown creates

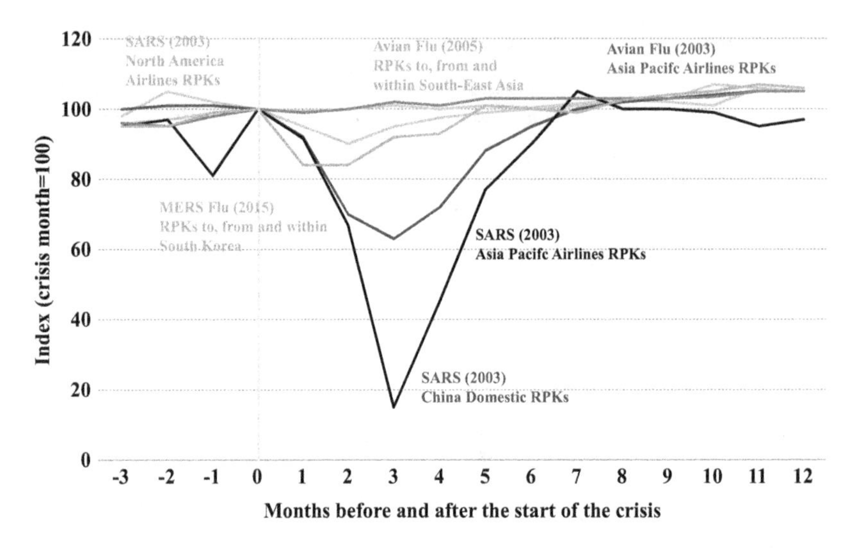

Fig. 10 Impact of past disease outbreaks on aviation. (Source: International Air Transport Association 2020b)

more drag in the face of more uncertain global growth, while the domestic economy continues to be dampened by declining consumer consumption which is now a much higher portion of GDP. China is now the source of 38% of global household consumption growth from 2010 to 2016, according to the World Bank (McKinsey & Company 2019).

The Chinese economy is also more integrated into the global economy in 2019 than it was in 2003. Increased global connectivity has made the Chinese tourism industry the main engine of growth for the global travel market in recent years. During SARS, China's tourist volume was merely 7 million, less than 10% compared to 2019 figures (Goldman Sachs 2015). Meanwhile, the number of inbound visitors from the U.S. was around 160,000, compared to 3 million in 2018 (Statista 2020). The UN World Tourism Organization estimated that Chinese tourists spent $277.3 billion overseas in 2018, up from around $10 billion in 2000 (RTE 2020) and are spending more on average than any other nationality on their trips, around $1850 per person per year (Smith 2019). China accounts for 12% of total worldwide available seat kilometers versus only 5% 15 years ago (RTE 2020). International travel in and out of China is now more than 10 times what it was in 2003, with 150 million foreign trips in 2018 (RTE 2020).

Immigration curbs and flight cancellations from COVID-19 will have a significant impact on global growth. More than 60 countries have some form of restrictions or quarantine protocols. The regions most exposed to China airline traffic by available seat kilometers are Asia's carriers with 30%, while North America is at 4%.

The SARS epidemic is the latest example of a major health outbreak. While there were more parallels in the beginning to the current COVID-19 outbreak given its similarities in geographies, this is no longer as relevant as a historical data point given the current global effects of the virus. SARS only lasted for about 4 to 6 months. One positive outcome in the aftermath of the SARS epidemic was that it was only a temporary exogenous shock to aviation, the economy and cross-border investments. It did not have long-lasting impacts and the recovery was relatively fast with the total outbound FDI recording a meteoric rise from $3 billion to $118 billion between 2002 and 2019.

Great Financial Crisis ("GFC") 2008/2009

Another historical example of an exogenous shock that is compared with COVID-19 is the Great Financial Crisis ("GFC") in 2008/2009. The GFC had large international economic impacts and was triggered by the collapse of Lehman Brothers investment bank in September 2008. This event reverberated around the world as governments and policymakers raced to secure the confidence in the global financial markets.

By September 2008, passenger travel demand had turned negative and by December, international RPKs were down 4.6% YoY. This decrease in business confidence caused business travel to fall even more and by December 2008, premium traffic was down by 13% YoY. Full-service carriers geared towards premium level customers were hit especially hard and resulted in an overall net loss of $10.4 billion for the whole year (International Air Transport Association 2009). From the early 2008 peak to the early 2009 trough, premium travel fell 25% while economy travel fell relatively less at 9% (International Air Transport Association 2010).

From mid-2009, air travel began to turn upward, boosted by the massive fiscal and monetary stimulus measures taken by governments. By the end of 2009, premium travel had recovered by 11%, while economy travel also rebounded 7%. This was accompanied by a reduction in passenger capacity in international markets of 5% and freight capacity decreasing 10%. For the full year 2009, global airlines' net losses were $9.9 billion (International Air Transport Association 2010).

In China, the 2008 passenger traffic increased by 3.3% but resulted in net loss of RMB28 billion (Civil Aviation Administration of China 2009). Asian airlines had net losses of $3.9 billion in 2008 and $2.7 billion in 2009. From Oct. 2019, RPK YoY growth was 0.9% and by September 2019 YoY growth was 2.1% (CAPA 2009). It is interesting to note that oil prices also suffered a large decrease along with the global economy. The oil bubble burst in late July 2008 and oil had fallen to $40 a barrel by the end of 2008. This was less than half of the price at the start of the year and represented a decrease of over 70% from the price's peak in July.

While this is also one of the most recent global economic shocks, it is not a direct comparison. The causes are man-made, a financial crisis of confidence while the current COVID-19 is a health related shock. While the GFC is more global in effect than during SARS, the global effects are not as significant as COVID-19.

Preliminary Outlook Going Forward: Virus Lifecycle Phase III (Global Peak and Virus Regression and Beyond)

Going in time, there are not many direct historical cases that relate directly to COVID-19. The most recent global shock that has the effects akin to what is experienced today is the 1973 oil shocks. Like GFC, that was also a manmade event. In that case, OPEC cartel in 1973 embargoed oil sales to US, UK, Canada and Netherlands which brought life to a standstill much akin to today. The most recent widespread global health pandemic is the 1918 Flu Pandemic. This lasted more than 2 years and infected an estimated 500 million people of whom 50 million were killed (Center for Disease Control and Prevention 2019). This did not have much as an effect on the budding but nascent aviation industry though, as the Wright brothers' first flight was just fifteen years before this time in 1903.

Today, as of April 26, 2020, there are over 2.9 million global cases that have been confirmed and global fatalities have reached more than two hundred thousand, according to Johns Hopkins University CSSE (2020). This is clearly a global problem, with infection rates accelerating in Europe, US and the rest of the world. With the initial outbreak shown in China, today there are signs of early recovery. Many of the restrictions on the hardest hit provinces have been lifted; Hubei Province's lifted restrictions on March 15 while Wuhan lifted its restrictions on April 8, 2020.

This is still an ongoing story that is being played out. There will be long-lasting impacts and the recovery lag period might be longer than other recent economic shocks. As with other health pandemics, the first priority is to control the spread of the virus and then keep an eye out for recovery. The quicker the virus is controlled, the quicker people can return to their normal routines and start the recovery process.

Shanghai, China
April 26, 2020

David Yu, PhD. CFA,
Senior ISTAT Certified Aviation Appraiser

Bibliography

Alcock, C. (2020). Airbus Restarts A320 Assembly in China as Virus Hits Airlines. *Aviation International News*. Retrieved from: https://www.ainonline.com/aviation-news/air-transport/2020-02-12/airbus-restarts-a320-assembly-china-virus-hits-airlines

Asian Development Bank. (2020). The Economic Impact of the COVID-19 Outbreak on Developing Asia. *ADB Briefs*. Retrieved from: https://www.adb.org/sites/default/files/publication/571536/adb-brief-128-economic-impact-covid19-developing-asia.pdf

Bloomberg. (2020). *Economic, FX, Fuel, and Funding Data*. Retrieved from: www.bloomberg.com

Caixin. (2020). Beijing, Shanghai and Other Cities Enact Highest Level Emergency in Response to Coronavirus. Retrieved from: https://www.caixinglobal.com/2020-01-25/beijing-shanghai-and-other-cities-enact-highest-level-emergency-in-response-to-coronavirus-101508199.html

CAPA. (2009). Global economic crisis has cost the aviation industry two years of growth: IATA. Retrieved from: https://centreforaviation.com/analysis/reports/global-economic-crisis-has-cost-the-aviation-industry-two-years-of-growth-iata-15921

Center for Disease Control and Prevention. (2019). 1918 Pandemic. Retrieved from: https://www.cdc.gov/flu/pandemic-resources/1918-pandemic-h1n1.html

China Daily. (2020). Over 300 mln train tickets sold for Spring Festival travel rush. Retrieved from: https://global.chinadaily.com.cn/a/202001/04/WS5e1039b7a310cf3e3558274b.html

Civil Aviation Administration of China. (2004). Monthly Statistics of Civil Aviation Industry. Retrieved from: http://www.caac.gov.cn/

Civil Aviation Administration of China. (2009). 2008 Civil Aviation Industry Annual Report. Retrieved from: http://www.caac.gov.cn/

Civil Aviation Administration of China. (2020). China Aviation Industry February Major Production Statistics. Retrieved from: http://www.caac.gov.cn/XXGK/XXGK/TJSJ/202004/P020200420522649148513.pdf

Civil Aviation Administration of China, & The Ministry of Finance of the People's Republic of China. (2020). Announcement of Corona Crisis Financial Subsidy for Aviation Industry. Retrieved from: http://www.caac.gov.cn/XXGK/XXGK/TZTG/202003/t20200304_201269.html

EY. (2020). Overview of China Outbound Investment in 2019. Retrieved from: https://www.ey.com/Publication/vwLUAssets/ey-overview-of-china-outbound-investment-in-2019-en/$FILE/ey-overview-of-china-outbound-investment-in-2019-en.pdf

Goldman Sachs. (2015). The Chinese Tourist Boom. *The Asian Consumer*. Retrieved from: https://www.goldmansachs.com/insights/pages/macroeconomic-insights-folder/chinese-tourist-boom/report.pdf

Heng, M. & Yong, C. (2020). Singapore Airlines cuts more flights; Changi sees 33% drop in passengers. *The Jakarta Post*. Retrieved from: https://www.thejakartapost.com/travel/2020/03/14/sia-cuts-more-flights-changi-sees-33-drop-in-passengers.html

International Air Transport Association. (2003). Industry Recovery Starts June Shows Signs of Improvement. Retrieved from: https://www.iata.org/en/pressroom/pr/2003-08-04-01/

International Air Transport Association. (2010). Annual Report 2010. Retrieved from: https://www.iata.org/contentassets/c81222d96c9a4e0bb4ff6ced0126f0bb/iataannualreport2010.pdf

International Air Transport Association. (2020a). COVID-19 Hits January Passenger Demand. Retrieved from: https://www.iata.org/en/pressroom/pr/2020-03-04-03/

International Air Transport Association. (2020b). Initial impact* assessment of the novel Coronavirus. IATA Economics. Retrieved from: https://www.iata.org/en/iata-repository/publications/economic-reports/coronavirus-initial-impact-assessment/

International Air Transport Association. (2020c). COVID-19 Cuts Demand and Revenues. Retrieved from: https://www.iata.org/en/pressroom/pr/2020-02-20-01/

International Air Transport Association. (2020d). IATA Updates COVID-19 Financial Impacts—Relief Measures Needed. Retrieved from: https://www.iata.org/en/pressroom/pr/2020-03-05-01/

International Air Transport Association. (2020e). Updated impact* assessment of the novel Coronavirus. *IATA Economics*. Retrieved from: https://www.iata.org/en/iata-repository/publications/economic-reports/coronavirus-updated-impact-assessment/

International Air Transport Association. (2020f). COVID-19 Updated Impact Assessment. *IATA Economics*. Retrieved from: https://www.iata.org/en/iata-repository/publications/economic-reports/covid-fourth-impact-assessment/

Johns Hopkins University CSSE. (2020). *COVID-19 Map*. Retrieved from: https://coronavirus.jhu.edu/map.html

McKinsey & Company. (2019). China and the world: Inside the dynamics of a changing relationship. *McKinsey Global Institute*. Retrieved from: https://www.mckinsey.com/featured-insights/china/china-and-the-world-inside-the-dynamics-of-a-changing-relationship

OAG. (2020). Schedules Analyser. Retrieved from: https://www.oag.com/

RTE. (2020). Chinese tourism—the main engine of global travel. Retrieved from: https://www.rte.ie/news/business/2020/0203/1112823-chinese-tourism/

Smith, O. (2019). The unstoppable rise of the Chinese traveller—where are they going and what does it mean for overtourism? *The Telegraph*. Retrieved from: https://www.telegraph.co.uk/travel/comment/rise-of-the-chinese-tourist/

Statista. (2020). Number of visitors to the United States from China from 2003 to 2024. Retrieved from: https://www.statista.com/statistics/214813/number-of-visitors-to-the-us-from-china/

The Economist. (2020). Coronavirus is grounding the world's airlines. Retrieved from: https://www.economist.com/business/2020/03/15/coronavirus-is-grounding-the-worlds-airlines

The Paper. (2020). Aviation Supply Chain Back to Normal: Airbus Tianjin Facilities Fully Resume Production. Retrieved from: https://www.thepaper.cn/newsDetail_forward_6683802

United States Senate. (2020). S.3578—COVID-19 Funding Accountability Act of 2020. Retrieved from: https://www.congress.gov/bill/116th-congress/senate-bill/3578?s=2&r=10

Acknowledgments

This book is the culmination of many years of work and experience. There are many people I would like to acknowledge in getting this book completed. First, I would like to express my gratitude to my Ph.D. thesis advisors, Prof. Xiaoquan Liu and Prof. Weimin Liu who believed in my vision and provided their words of encouragement throughout the Ph.D. journey which is the basis of this work. I would like to thank my close friends, Prof. Steve Hanke and Tasos Michael, who have provided many years of mentorship and friendship. I am also eternally grateful to the many people who have provided their comments, suggestions or good debate as we all continue to further the collective knowledge base. The list is too numerous but a subset of these individuals include Cherry Zhang, Rodrigo Zeidan, Brian Healy, Nils Hallerstrom, Stuart Hatcher, Phil Seymour, Owen Geach, Gueric Dechavanne, Killian Croke, Bryson Monteleone, Graham Deitz, Neil Whitehorse, Peter Morrell, Bijan Vasigh, Sunder Raghavan, Vitaly Guzhva, Tom Conlon and Lili Zhou. I would also like to thank the companies and their managers who provided the data which form the basis of parts of these studies, and without which many of the analyses would not have been possible. Particularly, these include the IBA Group, Collateral Verifications LLC and BK Associates, Inc. Lastly, I would like to also thank all of the numerous student research assistants who I have mentored and worked with over the years, especially Franklin Xu and Handong Xu who have gone above and beyond. I wish them all the best in their young, budding careers.

Contents

About the Author

David Yu, PhD, CFA, Senior ISTAT Certified Aviation Appraiser Prof. David Yu is a finance professor at New York University Shanghai and Stern School as well as Chairman of Asia Aviation Valuation Advisors (CAVA) and Asia Aviation Valuation Advisors (AAVA). Globally, he is one of about 20 Senior ISTAT (International Society of Transport Aircraft Trading) Certified Aviation Appraiser and is the only one in North Asia & China. He is a recognized expert in cross border finance, investing and valuation and is a frequently invited speaker at conferences. His op-eds and analysis articles have been published by *Forbes, Nikkei Asian Review, Airline Economics, Business Traveller, Airfinance Journal*, among others. He is also regularly interviewed by various top global TV, radio and print media as a business and economics commentator. He is an advisor and is on the board of directors of several companies and investment funds.

Prof. Yu is the External Thought Leader for Aviation Finance and Leasing for KPMG Ireland. He is a Director of Inception Aviation, a family investment office and previously was the Managing Director and Head of Asia for IBA Group, a UK based aviation valuation advisory firm. He was the China Chief Representative, VP Asia (Head of Asia) and Executive Committee member at Libra Group, a large Greek family investment conglomerate, where he was responsible for all of Libra's Asian interests including aviation, shipping, energy, hotels and real estate. Prior to Libra, Prof. Yu worked in investment banking with Bank of America Merrill Lynch's Global Industries Group, where he focused on M&A, debt and equity transactions in transportation, aerospace and defense, and diversified industries.

Prof. Yu is a CFA Charter holder and has a double major B.A. (full honors) and an M.S. from Johns Hopkins University, where he is a Fellow of the Applied Economics Institute. He has also studied at Peking University and National University of Singapore. He has an M.B.A. from New York University's Stern School of Business and a Ph.D. in Finance from University of Nottingham Business School.

List of Figures

List of Tables

CHAPTER 1

Introduction

The aircraft and aviation finance industry has seen large growth along with the overall aviation industry since the 1950s. It has developed from US- and European-centric origins to a more global and dynamic industry with many factors affecting the overall development. With regard to global growth, emerging markets stand out especially those of China and Asia. There has been tremendous growth in China, which has seen significant increases in all aspects of aviation financings, cross-border investment, and leasing activity. Overall, aviation finance is now a significant industry where global new aircraft deliveries worth well in excess of $126 billion are being invested annually in addition to investments in the secondary markets (Boeing Capital Corporation 2017).

Leasing, especially operating leasing, is a major driver in aircraft pricing and the asset class given it accounts for 43% of all global Airbus and Boeing and McDonnell Douglas aircraft as of 2017 (Ascend 2017). In Chap. 2, the core terms, concepts, and differences are defined including operating, finance leases, and their respective accounting and tax treatments under US GAAP (Generally Accepted Accounting Principles) and international accounting standards. The key difference between finance and operating leases is who retains the risk of ownership and residual value of the asset. There are further discussions of an example of operating lease cash flow structure. Why leasing is better compared to owning and its advantages and disadvantages from different perspectives and from practices are also discussed. There are upcoming changes and amendments in the

D. Yu, *Aircraft Valuation*,
https://doi.org/10.1007/978-981-15-6743-8_1

accounting standards and treatments to leasing. These along with the effects on leasing are discussed.

The academic literature for leasing is quite extensive since the start of the academic discussions in the late 1950s and this is the basis of understanding for aircraft leasing and pricing. Academic arguments are based on the broader buy or lease question with additional variations and forms of the analysis including effects of leases, debt, and taxation on asset pricing, the firm, and much broader economic implications. There are gaps in the argument when it comes to an understanding of the driving factors determining aircraft pricing and leasing especially where it relates to characteristics of the global market and its segments. These characteristics describe the numerous drivers of the aircraft market and its effects on pricing. While numerous articles focus on smaller sections, this chapter extends the literature and ties together these different arguments.

While there are some early disagreements on theory and empirical investigations, eventually a general consensus is reached on the lease versus buy question, of which the effects of tax- and other non-tax-based influences are the main drivers, as shown by a long progression of article arguments (Wyman 1973; Lewellen et al. 1976; Roenfeldt and Osteryoung 1973; etc.). The other line of conversation is whether leasing and debt are complements or substitutes. While there are empirical findings supporting both, the latest papers show that they are more partial complements as there are circumstances that favor leasing (Bowman 1980; Ang and Peterson 1984; Finucane 1988; Lewis and Schallheim 1992; Yan 2006). While leasing is viewed initially as finance leases, there is a later trend to address it through the operating leasing context but at times the term is more ambiguously defined.

Asymmetric information has effects on leasing through tax differences and residual value knowledge (Krahan and Meran 1987; Lease et al. 1990; Edwards and Mayer 1991; Graham et al. 1998). Also, asymmetric information costs drive preference to leasing and this preference for leasing occurs if companies have better ratings and pay more dividends (Sharpe and Nguyen 1995). Also, the sale and leasebacks case is tax benefit driven and a method for companies to raise capital. Studies show that there are abnormally positive returns and this is due to the lowered tax expectation created by sale and leasebacks (SLBs) (Slovin et al. 1990; Handa 1991; Ezzell and Vora 2001). Operating lease SLB also supports the notion of expanded credit capacity (Schallheim et al. 2013).

Non-tax incentives for leasing are also investigated such as contractual provisions, increasing the firm's debt capacity, managing credit ratings, and bankruptcy costs (Smith and Wakeman 1985; Vasigh et al. 2014; Eisfeldt and Rampini 2008; Lim et al. 2017; Krishnan and Moyer 1994). Real options and mathematical programming techniques are used to address different facets of leases and valuation. These include investigating residual value guarantees, asymmetry relationships, and valuation of operating leases (Schallheim and McConnell 1985; Sharpe and Nguyen 1995). There have been extensions of valuation methods for lease contracts and aircraft fleet decision-making (Stonier 1998, 1999, 2001; Bellalah et al. 2002; Clarke et al. 2003; etc.). Utilizing mathematical programming methods, airline decision-making in regard to buy, sale, and lease decisions showed the use of operating leases specifically for different types of airline business models (Hsu et al. 2011; Bazargan and Hartman 2012; Chen et al. 2018).

Chapter 3 reviews the dynamics, drivers, and outlook of the aviation finance and leasing landscape with respect to the global market. The overall drivers affecting the industry and aircraft pricing include demand, supply, and business model changes. Demand drivers include economic factors, business cycles, exogenous shocks, fuel prices, and traffic flows along with population demographics. Supply drivers include aircraft manufacturers, parked or retired aircraft, operating leases, secondary trading of aircraft, the financing environment along with current trends and segments such as commercial banks, capital markets, and export credit financing.

In addition, an analysis of the drivers affecting cross-border mergers and acquisitions in the industry is developed along with the characteristics of the increased use of leasing, specifically aircraft operating leasing. There are further discussions on the resurgence of sidecar joint ventures. The major global jurisdictions of aircraft leasing including Ireland, Singapore, Hong Kong, are also discussed. China is discussed in greater depth in Chap. 4. Tax and government incentives are shown as major drivers as well as other non-tax factors for leasing.

Chapter 4 is a more in-depth review of the Chinese market. This includes a discussion about the various drivers of demand and supply specifically with regard to the Chinese market. The differences and similarities between China and the other global jurisdictions are also discussed.

Chapter 5 focuses on the empirical data and the historical market characteristics and analysis of aircraft asset pricing over different time segments and across multiple cycles. The main question addressed is determining the market characteristics of the aircraft asset class in terms of its returns, volatility, and trends. This chapter fills in the gap in the academic literature by examining empirical analysis and looks into the economic shocks through aviation asset pricing. This hand-collected dataset includes time series of specific aircraft type valuation data from a collection of major aircraft appraisers over a 21-year period from 1996 to 2017 representing the entire range of the large commercial aircraft asset class.

The aviation asset class is then segmented into five different major aircraft type groupings with different weighting construction effects including all aircraft types, all narrowbody aircraft, all widebody aircraft, all narrowbody classic aircraft, and all narrowbody next generation aircraft. These groups are analyzed under four time segments that look at the overall picture and periods to see the effects of the great financial crisis.

After the establishment of the aircraft asset class and the market segmentations, further comparative analyses are conducted in Chap. 6 to deduce the implications to the other real asset classes and major investable benchmarks. The main hypothesis is that the aircraft asset class has lower stand-alone risk in terms of standard deviation and variance over the time segments. This extends the academic conversation by expanding on empirical portfolio theory and the effects of economic shocks.

The aircraft asset class characteristics are compared to the 20 other asset classes including publicly listed aircraft lessors, infrastructure, shipping, real estate, transportation, commodities, precious metals, agricultural land, and timberland. Other comparables include common liquid benchmarks and indices including US Treasuries 1-3 M, specific commodities including crude oil and gold, and other interest rate indicators such as US dollar-denominated floating to fixed swap instruments 10 Years and USD three-month LIBOR. LIBOR (London Inter-bank Offered Rate) is the most commonly used reference rate for short term floating rates. In addition to return, correlation, and covariance testing, regression and significance testing are conducted to assess the aircraft asset class with the other comparable asset classes to extend the academic knowledge.

Bibliography

Ang, J., & Peterson, P. P. (1984). The leasing puzzle. *The Journal of Finance, 39*(4), 1055–1065.

Ascend. (2017). *Yearly Operating Lease Percentage as of September 14*. Retrieved from: www.flightglobal.com

Bazargan, M., & Hartman, J. (2012). Aircraft replacement strategy: Model and analysis. *Journal of Air Transport Management, 25*, 26–29.

Bellalah, M., THEMA, & CEREG. (2002). Valuing lease contracts under incomplete information: A real-options approach. *The Engineering Economist, 47*(2), 194–212.

Boeing Capital Corporation. (2017). *Current Market Finance Outlook 2017.* Retrieved from Seattle, WA: www.boeing.com/resources/boeingdotcom/company/capital/pdf/2017_BCC_market_report.pdf

Bowman, R. G. (1980). The debt equivalence of leases: An empirical investigation. *Accounting Review, 55*, 237–253.

Chen, W.-T., Huang, K., & Ardiansyah, M. N. (2018). A mathematical programming model for aircraft leasing decisions. *Journal of Air Transport Management, 69*, 15–25.

Clarke, J.-P., Miller, B., & Protz, J. (2003). *An options-based analysis of the large aircraft market.* Paper presented at the AIAA/ICAS International Air and Space Symposium and Exposition, Daytona Ohio.

Edwards, J. S., & Mayer, C. P. (1991). Leasing, taxes, and the cost of capital. *Journal of Public Economics, 44*(2), 173–197.

Eisfeldt, A. L., & Rampini, A. A. (2008). Leasing, ability to repossess, and debt capacity. *The Review of Financial Studies, 22*(4), 1621–1657.

Ezzell, J. R., & Vora, P. P. (2001). Leasing versus purchasing: Direct evidence on a corporation's motivations for leasing and consequences of leasing. *The Quarterly Review of Economics and Finance, 41*(1), 33-47.

Finucane, T. J. (1988). Some empirical evidence on the use of financial leases. *Journal of Financial Research, 11*(4), 321–333.

Graham, J. R., Lemmon, M. L., & Schallheim, J. S. (1998). Debt, leases, taxes, and the endogeneity of corporate tax status. *The Journal of Finance, 53*(1), 131–162.

Handa, P. (1991). An economic analysis of leasebacks. *Review of Quantitative Finance and Accounting, 1*(2), 177–189.

Hsu, C.-I., Li, H.-C., Liu, S.-M., & Chao, C.-C. (2011). Aircraft replacement scheduling: a dynamic programming approach. *Transportation research part E: logistics and transportation review, 47*(1), 41–60.

Krahan, J. P., & Meran, G. (1987). *Why leasing. An Introduction to Comparative Contractual Analysis.* Berlin: Bamberg, G. & Spremann, K., eds.

Krishnan, V. S., & Moyer, R. C. (1994). Bankruptcy costs and the financial leasing decision. *Financial Management*, 31–42.

Lease, R. C., McConnell, J. J., & Schallheim, J. S. (1990). Realized returns and the default and prepayment experience of financial leasing contracts. *Financial Management*, 11–20.

Lewellen, W. G., Long, M. S., & McConnell, J. J. (1976). Asset leasing in competitive capital markets. *The Journal of Finance, 31*(3), 787–798.

Lewis, C. M., & Schallheim, J. S. (1992). Are debt and leases substitutes? *Journal of Financial and Quantitative Analysis, 27*(4), 497–511.

Lim, S. C., Mann, S. C., & Mihov, V. T. (2017). Do operating leases expand credit capacity? Evidence from borrowing costs and credit ratings. *Journal of Corporate Finance, 42*, 100–114.

Roenfeldt, R. L., & Osteryoung, J. S. (1973). Analysis of financial leases. *Financial Management*, 74–87.

Schallheim, J., Wells, K., & Whitby, R. J. (2013). Do leases expand debt capacity? *Journal of Corporate Finance, 23*, 368–381.

Schallheim, J. S., & McConnell, J. (1985). A model for the determination of "fair" premiums on lease cancellation insurance policies. *The Journal of Finance, 40*(5), 1439–1457.

Sharpe, S. A., & Nguyen, H. H. (1995). Capital market imperfections and the incentive to lease. *Journal of Financial Economics, 39*(2), 271–294.

Slovin, M. B., Sushka, M. E., & Polonchek, J. A. (1990). Corporate Sale and Leasebacks and Shareholder Wealth. *The Journal of Finance, 45*(1), 289–299.

Smith, C. W., & Wakeman, L. (1985). Determinants of corporate leasing policy. *The Journal of Finance, 40*(3), 895–908.

Stonier, J. (1999). Airline long-term planning under uncertainty. *Real Options and Business Strategy, Lenos Trigeorgis ed., Risk Books.*

Stonier, J. (2001). Airline fleet planning, financing, and hedging decisions under conditions of uncertainty. *Handbook of Airline Strategy*, McGraw-Hill.

Stonier, J. (1998). Marketing From A Manufacturer's Perspective: Issues In Quantifying The Economic Benefits Of New Aircraft, And Making The Correct Financing Decision. *Handbook Of Airline Marketing.*

Vasigh, B., Fleming, K., & Humphreys, B. (2014). *Foundations of airline finance: Methodology and practice*: Routledge.

Wyman, H. E. (1973). Financial lease evaluation under conditions of uncertainty. *The Accounting Review, 48*(3), 489–493.

Yan, A. (2006). Leasing and debt financing: substitutes or complements? *Journal of Financial and Quantitative Analysis, 41*(3), 709–731.

CHAPTER 2

Background Concepts and Definitions

1 Accounting and Tax Treatments

To understand aircraft and aviation valuation and leasing, one must understand the dynamics between the ecosystem, namely, the owner, financier, and user. The owner can be the airline or lessor and lessors account for 43% of all Airbus and Boeing and McDonnell Douglas aircraft in the world as of 2017 (Ascend 2017). An airline or operator is the ultimate user of the aircraft and one of the main stakeholders in the aircraft leasing industry. To operate as an airline, it needs to buy or obtain the use of one or multiple aircraft and other large fixed assets in addition to many smaller items. The operator has a variety of options that are assessed to complete this objective. After going through the various analyses and processes in choosing the type of aircraft, the classic question of how to fund the acquisition comes into play, where there are also a multitude of options at one's disposal.

Most conservatively and simply, the airline can buy the aircraft for all cash or equity. In addition, they can raise debt financing in different forms such as on a senior secured basis, where the aircraft asset is collateral. Another option is to finance the acquisition by debt financing that is unsecured or company-secured financing facility, which may or may not be backed by specific pledged collateral from the company. The

D. Yu, *Aircraft Valuation*,
https://doi.org/10.1007/978-981-15-6743-8_2

other financing options can be funding the acquisition through mezzanine debt, capital markets, export financing, or manufacturer's support, also known as seller financing. Instead of debt financing, the other option is to lease the aircraft either by completing a sale and leaseback or directly through the leasing company. Directly, this aircraft lease can be either a finance (or capital lease) or an operating lease depending on different classifications and can have significant differences in tax treatments and financial ramifications for the company. Under a sale and leaseback, the company buys the asset and then simultaneously sells the asset to the leasing company but also leases it back for a certain period and terms.

To fully understand leasing, one must understand the various tax and accounting definitions and differences that exist for aircraft leasing. Currently, the industry relies heavily on lease accounting guidelines as governed by International Accounting Standards (IAS) 17 Leases, maintained by the International Accounting Standards Board (IASB). The objective of the standard is to "prescribe, for lessees and lessors, the appropriate accounting policies and disclosures to apply in relation to leases" (IFRS 2003). In aircraft leasing, the lessor is defined as the company that owns or has title to the aircraft asset and grants the use of such assets or aircraft to another party under a lease agreement. The party that obtains the use of the aircraft asset is referred to as the lessee. While the "lease is an agreement whereby the lessor conveys to the lessee in return for a payment or series of payments the right to use an asset for an agreed period of time … A finance lease is a lease that transfers substantially all the risks and rewards incidental to ownership of an asset. The title may or may not eventually be transferred. An operating lease is a lease other than a finance lease" (IFRS 2003). The main idea or the concept behind the classification is "based on the extent to which risks and rewards incidental to ownership of a leased asset lie with the lessor or the lessee" (IFRS 2003). The lease itself can be quite flexible or standard as it is a negotiated agreement between each of the two parties.

There are multiple, slightly different definitions of operating and financial leases depending on the oversight body. It is important to understand that the IASB, through the IAS and the International Financial Reporting Standards (IFRS), only advises and influences each country's GAAP (Generally Accepted Accounting Principles) but does not have the power to set the rules, which lie with the respective country bodies. For example, in the US, the Financial Accounting Standards Board (FASB) makes up the rules and regulations which become US GAAP. The current leasing treatment standard under US GAAP is FASB Number 13 or Accounting Standards Codification (ASC) Topic 840. FASB Number 13 has been in existence since 1976. Currently, under US GAAP (FASB 13 or ASC 840) and IFRS (IAS 17), the treatment of leases is generally similar (Fig. 2.1).

- The lease transfers ownership of the asset to the lessee by the end of the lease term
- The lessee has the option to purchase the asset at a price which is expected to be sufficiently lower than fair value at the date the option becomes exercisable that, at the inception of the lease, it is reasonably certain that the option will be exercised
- The lease term is for the major part of the economic life of the asset, even if title is not transferred
- At the inception of the lease, the present value of the minimum lease payments amounts to at least substantially all of the fair value of the leased asset
- The lease assets are of a specialized nature such that only the lessee can use them without major modifications being made

Other situations that might also lead to classification as a finance lease are:

- If the lessee is entitled to cancel the lease, the lessor's losses associated with the cancellation are borne by the lessee
- Gains or losses from fluctuations in the fair value of the residual fall to the lessee (for example, by means of a rebate of lease payments)
- The lessee has the ability to continue to lease for a secondary period at a rent that is substantially lower than market rent

Fig. 2.1 IAS 17. (IFRS 2003)

- The lease transfers ownership of the asset to the lessee by the end of the lease term
- The lessee has the option to purchase the asset at a price which is expected to be sufficiently lower than fair value at the date the option becomes exercisable that, at the inception of the lease, it is reasonably certain that the option will be exercised
- The lease term is for the major part of the economic life of the asset, even if the title is not transferred
- At the inception of the lease, the present value of the minimum lease payments amounts to at least substantially all of the fair value of the leased asset
- The lease assets are of a specialized nature such that only the lessee can use them without major modifications being made

Other situations that might also lead to classification as a finance lease are:

- If the lessee is entitled to cancel the lease, the lessor's losses associated with the cancellation are borne by the lessee
- Gains or losses from fluctuations in the fair value of the residual fall to the lessee (e.g., by means of a rebate of lease payments)
- The lessee has the ability to continue to lease for a secondary period at a rent that is substantially lower than the market rent

The US GAAP is more specific in guidance than the IFRS. While the IFRS refers to finance leases, the US GAAP defines this as a capital lease. Leveraged leases treatment is defined in US GAAP while it does not exist in IFRS.

2 Operating and Finance Lease

US GAAP under ASC 840 has ruled that a lease should be treated as a finance or capital lease if it meets any one of the following four conditions:

(a) "if the lease life exceeds 75% of the life of the asset;
(b) if there is a transfer of ownership to the lessee at the end of the lease term;
(c) if there is an option to purchase the asset at a 'bargain price' at the end of the lease term;

(d) if the present value of the lease payments, discounted at an appropriate discount rate, exceeds 90% of the fair market value of the asset." (Financial Accounting Standards Board 1976)

Every lease that falls outside these criteria is considered an operating lease, where the ownership of the asset is retained by the lessor, during and after the period of the lease, there is no bargain price option by the lessee, or the lease period is less than 75% of the life of the asset and the present value of the discounted asset at an appropriate discount rate lease payment is less than 90% of the fair market value of the asset. For accounting purposes, under an operating lease for the lessee, the lease payments are considered an operational expense and flows through the income statement while the asset does not appear on the lessee's balance sheet nor can the respective depreciation be claimed by the lessee.

For the lessor, the asset appears on its balance sheet along with any debt finance associated with the asset. In the income statement, the lease payments made by the lessee are considered the lessor's revenue, while depreciation can be claimed and interest expense associated with the financing, if any, can be deducted. The lessor retains the risk of ownership and in aviation terms retains the residual value risks of the aircraft.

The opposite holds true for a finance lease where the ownership of the asset might be retained by the lessor during but transferred to the lessee after the period of the lease or there exists a bargain price option for the lessee. The lease period can be greater than 75% of the life of the asset and the present value of the lease payments, discounted at an appropriate discount rate, is greater than 90% of the fair market value of the asset. For accounting purposes under a finance lease, for the lessee, the lease is considered a loan and the lessee is considered the owner of the asset. As such, the lessee's present value of the lease payments is treated as debt on the balance sheet and the interest expense is calculated from this debt amount, which flows through the income statement and its depreciation can be claimed.

For the lessor, the present value of the future cash flows is recognized as a revenue asset and appears on its balance sheet along with any debt finance associated with the asset. In the income statement, the lease payments by the lessee is considered revenue, while depreciation can be claim, and interest expense associated with the financing, if any, can be deducted (Damodaran 2017).

3 Typical Operating Lease Structure

Some of the characteristics of a typical operating lease are that the lessor acquires the aircraft directly from the manufacturer or by way of the airline known as an SLB (sale and leaseback). If ordered directly from the OEM, the lessor pays any deposits or advance payments such as pre-delivery payments (PDPs) and the lessor arranges for the delivery of the aircraft and its financing such as senior secured bank debt. In the case of an SLB, the airline would be responsible for any deposits and PDPs to the OEM but the lessor would need to put up a form of security, such as a deposit, to the airline. The lessee will sign the lease to the aircraft to commence at the delivery of the aircraft. The typical lease term is 5–12 years for a lease with narrowbody aircraft having a shorter tenor while widebody aircraft have a longer tenor.

In the interim of the lease, the airline is responsible for rent and any maintenance reserves if required. Ultimately, the airline is responsible for any maintenance upkeep costs for the aircraft but qualified maintenance tasks can be reimbursed through the maintenance reserves paid if any (described in more detail below). After the lease ends, the lessee returns the aircraft to the lessor in a technical condition that is pre-negotiated. The return technical condition can be a minimum physical redelivery condition with a monetary adjustment to a delivered technical condition or another predetermined benchmark. Usually, redeliveries are timed to involve a heavy maintenance overhaul event for the aircraft. At the end of the lease term, the lessee may have a lease embedded option to renew the lease at a set price and term or the purchase of the aircraft at a predetermined price. Practically, these options usually exist but would need to be negotiated between the parties near the end of the lease.

For a typical operating lease contract, the cash flows to the lessor can be broken down into three main components—the actual lease rate costs, maintenance reserves, and security deposits. Historically, the lease rate is approximately 1.0% lease rate factor (LRF) or 1% of the purchase price of an aircraft cost per month but this is dependent on a variety of factors including aircraft supply and demand, among others. Maintenance reserves are paid by the lessee to the lessor as the aircraft is used and wear-down is accrued. This is usually accounted for by the actual time or flight usage on pre-agreed metrics. As the cost of overhauls, especially major ones, are expensive, these maintenance reserve funds are a support for the lessor and lessee to ensure that the lessee has the funds to ensure adequate repairs. The lessee is able to be reimbursed from this fund for completed qualified maintenance. This is very similar to the sinking fund provisions for corporate bonds and other financings.

Security deposits are typically three months of lease cost but this can vary depending on the lessee and market conditions. Also, maintenance reserves and security deposits can be in the form of cash, letters of credit, or a combination of both. These items are all the results of the final negotiated document between the two parties and as such there are cases where this differs from the norm where no maintenance reserves or security deposits are required but this is usually reserved for companies with the highest creditworthiness or is supplemented by other credit enhancements such as other forms of security that the lessee can provide (Fig. 2.2).

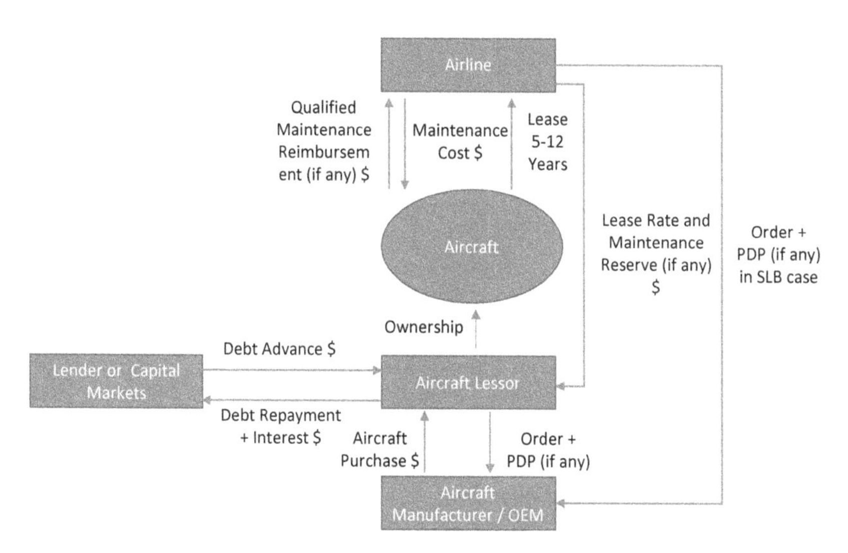

Fig. 2.2 Operating lease cash flow structure diagram

4 Why Lease?: Advantages and Disadvantages to Operating and Finance Leasing from Practice

Operating leases are where the owner or lessor has retained the residual risk of the asset and the investment nature of such versus a finance lease, which is characterized more as a loan. Also, most of the world's aircraft leasing companies are focused on only operating leases while the minority are involved with finance leases or other types of assets as well.

There are many advantages to operating leases from the user's or lessee's point of view. Operating leases provide flexibility to the lessee especially those that need to update or replace their equipment frequently. The flexibility though can go both ways, especially in jurisdictions with US Chapter 11 bankruptcy style situations where the aircraft can be handed back. As the lessee has no residual risk in the asset, they are protected from the potential drop in value as well as the risk of obsolesce of the asset. As the asset is not on the lessee's balance sheet, the lease payments flow through the income statement, which is fully tax-deductible and accounting is simpler. In this case, this is considered off-balance sheet financing, which is one of the aims of new accountings discussed in later sections. In terms of overall metrics, as there is no asset on the balance sheet, the return on asset (ROA) metric is higher compared to a finance lease. While airlines can be either scheduled operators or non-scheduled charter carriers, there is a trend to introduce newer aircraft as it brings prestige, creating more demand for the companies' products. These trends favor operating leases.

The main advantage of a finance lease for the lessee is that the expenses can be recognized faster and front weighted. The lessee can deduct the depreciation of the asset along with the interest component of the lease payment. Balance sheet-wise, the present value of the lease payments is an asset that depreciates while there is a liability recognized as the debt and this is reduced by the interest expense. All of the risks and rewards of the asset lie with the lessee while the lessor holds on to the title and ownership of the asset until the purchase option at the end.

One of the other considerations is that with operating leases, the lessee may be exposed to lease rate fluctuation especially if it is on a floating rate basis or on the renewal of the lease at the conclusion of the term. Another drawback is that without exposure to the residual value of the asset, the lessee does not gain equity with the appreciation of the asset but, at the same time, the lessee is not taking on the risk of the fluctuating value of the aircraft.

With regard to taxes, the lessee may have smaller tax benefits than in the case of purchase or finance lease as depreciation cannot be claimed, which is the major value of the asset in accelerated depreciation terms especially at the beginning of the lease. There are extra costs that the lessee must meet to fulfill lease contract requirements such as maintenance, reporting, and administration. In addition, there may be restrictive covenants placed on the lessee by the lessor in relation to the use of the aircraft for certain types of activities such as for Haj flights or to different jurisdictions, for example. The contract can also restrict sub-leasing arrangements when the lessee temporarily has no use for the asset or restricts how the asset is technically maintained in regard to procedures and usage of alternatives to the OEM produced parts such as parts manufacturer authorized (PMA).[1] Also, the contract can restrict the maximum control of the aircraft's configuration by the lessee as to when the asset is redelivered. In most leases, the lessee pays for any additional withholding or other additional taxes relating to the rental payment can be especially acute for cross-jurisdictional payments not covered by country-to-country double tax treaties.

The disadvantages to a finance lease are that the lessee retains the risks and the rewards of the asset, which could mean losses due to a decrease in the value of the asset. Like an operating lease, the lessee is responsible for the repairs and maintenance of the asset. As the lessee retains the risks and the rewards of the asset, there could be losses due to a decrease in the value of the asset. The finance lease while increasing the assets on the balance sheet will also increase the liabilities and may affect how analysts analyze the lessee's financial situation such as credit ratings. Again, compared to an outright acquisition, the lessor will still have some control and say over what can be done with the asset and hence the associated costs. Due to these implications, the lessee's ROA ratio will be lower than under an operating lease.

One of the major advantages of operating leases for the lessee is that the amount of capital for security deposit is very little relative to the asset value. In this respect, it can be smaller than both finance and outright loans. In finance leases, the lessee is required to put up typically 10–20%

[1] The use of PMA parts versus OEM parts is controversial in the industry. The use of PMA parts is frowned upon by OEMs and lessors due to the potential difficulties in transitioning the aircraft, affecting values, among others; however, airlines, tend to view it more favorably given their decreased operating costs.

of the amount of the total asset price. Younger and aggressively growing airlines tend to be not as well capitalized, thus utilizing operating leases compared with better-funded mature airlines. This preference for operating leases might also be the case when the aircraft is different than the existing fleet due to starting new business concepts and areas. In this case, the operator is unsure about the cost characteristics and longevity of usage of the aircraft and prefers to start with the lease option instead of purchasing the aircraft outright.

There are also instances where the manufacturer's order books are at capacity in a given time and the only way to get the desired aircraft is to lease it from a lessor with an existing order book slot or buy the asset secondhand. This is especially acute today when the manufacturer order books stretch ten-plus years out for popular aircraft. The lease payments can also be structured in fixed or floating terms depending on the lessee's requirements and lessor's preferences.

5 Lease Versus Own or Debt from Practice

The airline's selection of leasing or owning the aircraft will be driven by a variety of drivers such as the legal considerations, tax considerations, availability of funding, cost of capital, flexibility of asset usage, its outlook, and its own return projections. There is a lot of financial literature on this topic based on various scenario analyses looking at questions focused on the "lease versus own" or "lease versus buy" approach. More about the historical literature can be found in the Literature Review in Sect. 2.2.

There are disadvantages to acquiring with debt and owning the asset outright compared with operating leases. It requires a larger investment and initial outlay such as a down payment or minimum equity required. Like a finance lease, ownership entails the owner to retain residual value risk and the rewards associated with this. The financing institution might also place restrictions that are included in the loan covenants. It requires the company to arrange for the financing. Ownership entails less fleet planning flexibility than an operating lease. Instead of just handing the aircraft back at the end of the lease, selling the aircraft that is owned entails additional overhead on the company and might not create optimal solutions given the need to dispose of assets that are not a core competency of the airline.

Like all types of leases, valuation depends on a variety of factors, including supply demand of aircraft, interest swap rates, how long one plans to use the aircraft, leverage, and so on. These viewpoints change depending on the long- and short-term point of view. Generally, the rule of thumb is if the plan is to utilize the aircraft for a long time (i.e., longer than ~12 years, the longest length for a typical aircraft operating lease), then owning the aircraft is more beneficial. Like long-dated assets such as real estate, buying is cheaper than renting or leasing in the long term. So perhaps given the popularity of aircraft leasing for airlines and its growth and profitability as a business, this might be a reflection of how short term the airline business has become.

6 Upcoming Amendments to Accounting Standards for Leases and Its Effects

An update of the international accounting standards such as IAS 17 will be replaced by IFRS 16 which comes into effect in January 2019. As noted, as a result of the 2008 global financial crisis (GFC), there is a speedier passage of reforms (Bertomeu and Magee 2011; Kothari and Lester 2012). The aims are put on the balance sheet recognition of lease liabilities by lessees, and reduce opportunities for structuring and improve information about lessor exposure to the retained risks in the underlying asset. For lessees, there is now only one choice for all lease classifications with two exceptions, low value assets and short leases.

The situation before incorporation of IAS 17 can potentially hide leverage through off-balance sheet footnote disclosures and result in the lack of direct comparability. While this is true at face value, it has been common for research analysts to adjust the balance of these companies to capitalize these operating leases and bring them on the balance sheet for easier comparison (Damodaran 2017). This has become quite popular as "[off balance sheet] financing increased 745% as a proposition of total debt from 1980 to 2007" (Cornaggia et al. 2013). This new accounting rule effectively does the job for researchers. The effect of these changes is $3 trillion (Burgess et al. 2016).

For lessees, this change includes calculating and recognizing a "right of use" (ROU) asset by a present value of the lease payments and this is depreciated like before. As this ROU is an asset there is a corresponding liability using the discount rate determined at lease start on balance sheet and this reduces as payments are made. The effect of the new rules is that interest and depreciation costs will replace lease costs as if the assets are actually on the balance sheet and owned. The net effects on the income statement is that expenses are generally front-loaded (bigger early recognition). There are minimal effects to lessors.

One of the stated goals of the IFRS is for the convergence of US GAAP and the global standard, but this has not fully materialized. One of the main criticisms of the IFRS is that it is not adopted in the US, the largest economy having outsized influence. The SEC (US Securities And Exchange Commission) provided its own critic of the IFRS in 2012 and the IASB responded by saying "while acknowledging the challenges, the analysis conducted by the IFRS Foundation staff shows that there are no insurmountable obstacles for adoption of IFRSs by the United States and that the US is well placed to achieve a successful transition to IFRSs, thus completing the objective repeatedly confirmed by the G20 leaders" (US Securities And Exchange Commission 2012; IFRS 2012).

Comparing the US GAAP and the IFRS, there are many differences between IAS 17 and US GAAP's ASC 840 that apply to aircraft assets. In terms of lease classification, the US GAAP classification of a lease depends on whether the lease meets certain criteria while the IAS classification of a lease depends on the substance of the transaction while specific indicators and case examples are provided. Another is the lessee's implicit rate in the lease to discount minimum lease payments under the IAS but US GAAP uses this implicit rate if it's known and lower than the incremental borrowing rate. Another is how and when to recognize a gain or loss on a sale and leaseback transaction.

Given the large number and percentage of aircraft leasing community based in Ireland, the Irish GAAP needs to be acknowledged. If Irish companies have debt or equity listed on a regulated market of any European Economic Area State, then they are required to prepare their group annual financial statements using IFRS but an Irish company can elect for either IFRS or Irish GAAP mainly under the Statement of Standard Accounting Practice (SSAP) 21. IAS 17 and SSAP 21 have much in common but, specifically for aviation, SSAP 21 includes lease and hire purchase contracts while the former does not. SSAP 21 like US GAAP has a quantitative test for a finance lease while only IAS 17 provides additional guidance. The treatment of lease incentives accounting can differ due to the timing and how it's accounted for. The accounting of lease income by lessees under a finance lease is a difference as SSAP 21 is more flexible on the methods on how to recognize income. The last major difference is that the required disclosures under IAS 17 are more detailed than under SSAP 21.

When analyzing tax issues, especially for cross-border aircraft leasing, one needs to consider and understand the domicile and the tax home for both the lessor and the lessee. Some other things to consider include taxation treatment of income for lessor, deductibility of the expenses for lessee, tax depreciation eligibility, any withholding of taxes (if any), any tax incentives offered by various jurisdictions (if any), exposures for "taxable presence" (if any), coverage by the double tax treaty network, and any indirect taxes such as VAT, GST, and so on.

Under US GAAP, FASB issued a revision in February 2016 known as Lease Accounting Revision or ASC 842. The last effective date for implementation is the fiscal years beginning December 15, 2018, for public companies and fiscal years beginning after December 15, 2019, for private companies (Financial Accounting Standards Board 2016). The aim of the update is to more accurately reflect off-balance sheet operating leases onto companies' balance sheets and be more closely aligned with the IFRS.

Differences include clean-up such as renaming capital leases to finance leases, which is to conform to IFRS terminology. Other costs that used to be re-billed by the lessor such as taxes and insurance are now capitalized as they qualify to be excluded. The test conditions which determine whether the lease is a finance or operating lease remain pretty much unchanged. One change in the addition of a criterion for finance lease classification is that "the underlying asset is of such a specialized nature that it is expected to have no alternative use to the lessor at the end of the lease term" (Financial Accounting Standards Board 2016). In real-world situations, this already meets the present value criterion test. Another change is that the sale and leaseback accounting is not allowed anymore if the lessee still has a "continuing right of control" to the asset. This is in the case of the repurchase clause of the asset at a specific fixed price, which is not the specific bargain price criteria used before. Otherwise, this is treated as a financing transaction. Leveraged leases as a classification will also be eliminated.

7 Brief Academic Literature Review: Leasing

The academic literature for aircraft leasing is quite extensive since the start of the academic discussions in the late 1950s and this is the basis of understanding for aircraft pricing. Academic arguments are based on the broader buy or lease question with additional variations and forms of the analysis including effects of leases, debt, and taxation on asset pricing, the firm and much broader economic implications. There are gaps in the argument when it comes to an understanding of the driving factors determining aircraft pricing and leasing especially where it relates to characteristics of the global market and specifically the Chinese market. These characteristics describe the numerous drivers of the aircraft market and its effects on pricing. While there are numerous articles that focus on smaller sections, this chapter extends the literature and ties together these different arguments.

While there are some early disagreements on theory and empirical investigations, eventually a general consensus is reached on the lease versus buy question of which the effects of tax- and other non-tax-based influences are the main drivers as shown by a long progression of article arguments (Wyman 1973; Lewellen et al. 1976; Roenfeldt and Osteryoung 1973; etc.). The other line of conversation is whether leasing and debt are complements or substitutes. While there are empirical findings for both, the latest papers show that they are more partial complements as there are circumstances that favor leasing (Bowman 1980; Ang and Peterson 1984; Finucane 1988; Lewis and Schallheim 1992; Yan 2006). While leasing is viewed initially as finance leases, there was a later trend to address it through the operating leasing context but at times the term is more ambiguously defined.

Asymmetric information has effects on leasing through tax differences and residual value knowledge (Krahan and Meran 1987; Lease et al. 1990; Edwards and Mayer 1991; Graham et al. 1998). Also, asymmetric information costs drive preference to leasing and this preference for leasing occurs if companies have better ratings and pay more dividends (Sharpe and Nguyen 1995). Also, the sale and leasebacks case is tax benefit driven and a method for companies to raise capital. Studies show that there are abnormally positive returns and this is due to the lowered tax expectation created by sale and leasebacks (SLBs) (Slovin et al. 1990; Handa 1991; Ezzell and Vora 2001) Operating lease SLBs also support the notion of expanded credit capacity (Schallheim et al. 2013).

Non-tax incentives for leasing are also investigated such as contractual provisions, increasing the firm's debt capacity, managing credit ratings and bankruptcy costs (Smith and Wakeman 1985; Vasigh et al. 2014; Eisfeldt and Rampini 2008; Lim et al. 2017; Krishnan and Moyer 1994). Real options and mathematical programming techniques are used to address different facets of leases and valuation. Those include investigating residual value guarantees, asymmetry relationships, and valuation of operating leases (Schallheim and McConnell 1985; Sharpe and Nguyen 1995). There have been extensions of valuation methods for lease contracts and aircraft fleet decision-making (Stonier 1998, 1999, 2001; Bellalah et al. 2002; Clarke et al. 2003; etc.). Utilizing mathematical programming methods, airline decision-making with regard to buy, sale, and lease decisions showed the use of operating leases specifically for different types of airline business models (Hsu et al. 2011; Bazargan and Hartman 2012; Chen et al. 2018).

There are also extensions of the literature, by directly examining empirical analysis in regard to segmentation and economic shock timing in aviation through the analysis of 21 years of the aircraft asset class dataset from 1996 to 2007. There is also construction of five different major aircraft type groupings with different weighting construction effects from this dataset which differentiates the categories of aircraft. In addition to the entire dataset, empirical tests are carried out on economic shocks by examining the effects and characteristics before and after the 2008 global financial crisis. Also, Chap. 5 expands on the academic conversation and analysis to contribute to further understanding of the aviation asset class under the portfolio theory framework and the effects of economic shocks.

8 Aircraft Values, Liquidity, and Operating Leasing

Starting in 1998, more papers were published concerning observations on the pricing of commercial aircraft and airlines, especially with regard to the US market, along with different elements of the aircraft operating lease industry. "The question of valuing commercial aircraft has not been explored in any great depth by researchers in academia" (Gorjidooz and Vasigh 2010). With regard to airlines, Pulvino (1998) found that fire sales do exist as these aircraft are sold at a 14% discount to the average market price by distressed airlines that have more asset sales than under normal situations. Goolsbee (1998) concluded from his empirical studies of the airline's decision rationale for the sale or retirement of an aircraft type specifically common at the time, the Boeing 707 type. The study shows that the decision on a particular aircraft type is based on the airlines' financials, its cost of capital, and the overall business cycle. This is not a broad enough sample set and also this only focuses on the airline itself and not aircraft traders or lessors who commonly are owners of aircraft. Continuing the asset sales studies in distressed situations, Pulvino (1999) studied the pricing effects caused by Chap. 11 reorganization or Chap. 7 liquidation, both under US bankruptcy laws. The study found that under both bankruptcy types, the asset prices are lower than those for assets sold by non-distressed airlines, and there is no difference in obtaining higher prices or limiting the number of discounted aircraft sold with either bankruptcy type. Looking at the determinates of the liquidation value of assets, it is noted that in a liquidity event, an asset is likely to be below the value of its best use as industry peers at that time are also likely to be experiencing problems (Williamson 1988; Shleifer and Vishny 1992).

Comparing types of leasing firms, Habib and Johnsen (1999) concluded in their study that lenders with specialist knowledge have a greater advantage compared to generalists. Given this advantage, they postulate that generalists lend less and have a higher default rate than specialized companies such as those who have "fully integrated redevelopment [functions] including asset valuation, monitoring, repossessions and resale" (Habib and Johnsen 1999). They have an advantage in the management and redeployment of specific assets. This summary statement echoes the prevailing thoughts in the industry today. Eisfeldt and Rampini (2008)'s empirical study shows that the leasing segment is the largest external finance source especially for smaller, capital-constrained companies.

Flipping the question around, Oum et al. (2000a) looked at the airline or end user's point of view of the optimal ratio of operating leases for airlines. Employing a two-step model, first, the airline profit maximization function based on uncertain demand or revenues accounting for variable costs and the different weighting of owned or long-term capital leases and shorter operating leases to find the *k* capital constraint.

Given the *k* capital constraint, we get Eq. (2.1), which demonstrates that one expects to have a higher risk premium on operating leases compared to longer-term capital leases:

$$\int (w_s - w_k) f(\tau) d\tau = E(w_s) - w_k > 0 \tag{2.1}$$

R = revenue	*y(τ) = demand*	*Z = capacity of the airline = K+S*	*K = capital stock leased for longer term (capital lease)*
S = capital stock leased for shorter time (operating lease)	τ = future state of nature	w_k = costs of long-term capital	Max $\pi = R[y(\tau), Z] - V[y(\tau), Z] - w_k K - w_s(\tau)S$

w_s = costs of short-term capital

This model was tested on a dataset of 23 major international airlines between 1986 and 1993 and the empirical results showed that 40–60% is the optimal ratio of operating leases for airlines. This chapter is an extension of Gritta (1974) and Gritta et al. (1994), which looks at the effects of leases for airlines especially in their debt ratio and capital structure.

An updated empirical study by Bourjade et al. (2017) looked at the use of only operating leases (not other types) and whether that improves the global airline's financial performance. The dataset was public data from 73 global airlines from 1996 to 2011 and the results showed that the impact of operating leasing on the airline's operating margin is concave. It also shows that low-cost carriers by business model differentiation are more affected by the benefits of leasing than full service carriers and that younger companies are more profitable utilizing operating leasing. It was also shown that the optimal percentage of leasing that maximizes the operating profit is 53.4% with a 5% confidence interval of [0.468,0.601], which is consistent with the empirical findings of Oum et al. (2000a).

Not all papers advocate operating leasing in aviation. One of the earliest papers for leasing in aviation is Parrish (1970) and Gritta and Lynagh (1973)'s "Aircraft Leasing—Panacea or Problem," in which the authors advocated for the cautious view of operating leasing in aviation and a warning of its hidden charms but this view seems out of date compared to the newer research.

In addition, more discussions spoke about liquidity and various characteristics affecting aircraft valuation such as Vasigh and Erfani (2004). Gavazza (2010) empirically studied aircraft liquidity specifically in the secondary markets to address the question of "leasing or owning". He argues that for commercial aircraft, more liquid assets are more likely to be operating leases instead of finance leases. His findings also suggested that these more liquid assets also tend to have shorter-term duration as operating leases and longer-term duration as finance leases. In an extension of the research, Gavazza (2011)'s empirical study on aircraft transactions in the secondary market found that an aircraft that is leased is traded more often and the usage is higher than in the case of owned aircraft. In addition, lessors, as capital reallocation intermediaries, have advantages with regard to transaction costs and the redeployment of capital. Gorjidooz and Vasigh (2010) studied liquid aircraft, especially Airbus A320-200, Airbus A330, Boeing 737-700LR, Boeing 767-400, by observing FAA Form 41 quarterly data from 1995 to 2005. The authors created a model based on the passenger and cargo revenues of the aircraft and also used Monte Carlo simulations to test it.

In summary, even after numerous academic articles and studies that have been written on leasing theory, its characteristics, its effects on the capital structure or as a potential substitute for debt, different scenarios of trading, and so on, there are currently still questions on why leasing is a good option and also what are the drivers and characteristics of the market. This research aims to fill in the holes for both the business economic cycles, and it's how the class fits in the portfolio theory for other asset classes. As is noted, more literature needs to specifically address the aviation industry and the operating leasing.

9 Economic and Business Cycles

There are many papers that have studied cyclicality in the economy and various industries (Smith 1776; Robertson 1915; Meadows 1971). The main general conclusion is that these cycles are caused by the failure to grasp the delay feedback effects from managing the existing inventory and product acquisition among other resources. There have been cycles in the aviation sector in the US since the 1978 deregulation and Jiang and Hansman (2004) shows that the average one full cycle period is approximately ten years with a long-run mean very close to zero. There have been discussions of the cyclicality found in the aviation industry, which has been addressed by Liehr et al. (2001) and Lyneis (2000) in system dynamics literature. This has been expanded upon by Pierson and Sterman (2013) to include an endogenous account of feedbacks omitted from some earlier work, including price setting, wages, and air travel demand.

Another research article addressing the question of whether airline capacity investment is efficient is Wojahn (2012), who examines the causes of overinvestment in aircraft capacity. He empirically tests the subject through a sample set of 69 global publically listed airlines. He concludes that airlines in all regions over the last 30 years did not cover their cost of capital and the returns have structurally declined, with airlines all the while continuing to add to capacity. Also, the findings show that causes of this include agency problems such as myopia and empire building, shifting business models toward low-cost and existing Asian carriers, coupled with remnants of capital in legacy airlines and economies of scale are all associated with overinvestment. While this chapter focuses on airlines, one important area that is missing for adding aircraft capacity is the oligopoly of aircraft manufacturers. This is addressed broadly in Brander and Lewis (1986) where they argue that there are linkages of product markets and financial market linkages, and limited liability effect of the corporation can produce more debt-induced aggressive output, which leads to overcapacity. Oum et al. (2000b) also support this argument that oligopolistic markets do not minimize the total social costs and thus lead to excess capacity. Through their empirical studies of ten US airlines, they found that excess debt load appears to lead to excess capacity in the relationship between an air carrier's debt and capacity.

There has also been an examination of the global financial crisis in relation to aviation. Franke and John (2011) described the differences between the 2008 and 2001/2003 economic shocks and how this resulted in different effects by airline types, which further enhanced different and new business models and operations such as the low-cost airline model in continental Europe. Pearce (2012) looks at the state of air transport markets and the airline industry post the 2008 global financial crisis. He finds that demand rebounded from the lows in 18 months and is still robust after repeated exogenous shocks while the industry is still highly leveraged and fragile. He finds that airlines cut capacity by underutilization of aircraft, which was challenging given they had committed for aircraft deliveries ordered at the peak of the previous cycle, which implies excess invested capital.

Bjelicic (2012) focuses on the aviation finance market after the 2008 global financial crisis. He finds that capital became very scarce for airlines with a greater emphasis on export credit agencies along with a decrease in bank loans, and aircraft lessors became an important source of funding for new airlines. He also notes that the availability of finance may be a barrier to market entry. Mann (2009) agrees with Bjelicic with regard to the retreat of capital availability in traditional areas of aviation finance after the 2008 global financial crisis and the expansion of export credit agencies in financing.

10 Real Assets Portfolio Theory

There has been much discussion of portfolio theory. Aviation is one of the subsegments of the real assets asset class which also includes real estate, infrastructure, agricultural land, timberland, commodities, and shipping. Most academic literature specifically relates to real estate, commodities, timberland, and its relation to portfolio theory. These asset classes are considered the "traditional" real asset classes by institutional investors who have been investing for decades while "new" real asset classes include infrastructure, agricultural land, and so on (Martin 2010). Martin does not consider other differentiated classes and considers shipping and aviation as part of the real asset category. With the other subclasses, this can be best summed up by this statement "little empirical work has been done on infrastructure as an asset class despite increased allocations by institutional investors" (Bird et al. 2014). There are no discussions of aviation in relation to portfolio theory and this is a gap in the academic literature that is aimed to be addressed in Chap. 5.

According to a review of the literature on infrastructure, based on the initial arguments by researchers, infrastructure has characteristics similar to those of real estate, with a long life, illiquidity, and the ability to hedge inflation (Dechant and Finkenzeller 2010; Newell and Peng 2007, 2008). Froot (1995) by empirically using a variety of real assets subclasses shows that real assets are good hedging tools for portfolios with their negative inflation correlation characteristics.

There has been more argument on infrastructure as a separate asset class with its own characteristics, namely, natural monopolistic or oligopolistic elements due to restrictions of use or ownership and economies of scale (Finkenzeller et al. 2010; Inderst 2009, 2010). Infrastructure which exhibits these natural monopolistic characteristics is often government regulated and a premium is needed for this regulatory risk.

Rothballer and Kaserer (2012) find that the empirical evidence on infrastructure's risk characteristics is still limited and they empirically tested 1400 publicly listed firms worldwide and found that infrastructure exhibit significantly lower market risk than MSCI World equities, which confirms the portfolio diversification benefits. The article notes that the class has idiosyncratic risks such as construction risks, operating leverage, the exposure to regulatory changes, and the lack of product diversification.

Bibliography

Ang, J., & Peterson, P. P. (1984). The leasing puzzle. *The Journal of Finance, 39*(4), 1055–1065.

Ascend. (2017). *Yearly Operating Lease Percentage as of September 14*. Retrieved from: www.flightglobal.com

Bazargan, M., & Hartman, J. (2012). Aircraft replacement strategy: Model and analysis. *Journal of Air Transport Management, 25*, 26–29.

Bellalah, M., THEMA, & CEREG. (2002). Valuing lease contracts under incomplete information: A real-options approach. *The Engineering Economist, 47*(2), 194–212.

Bertomeu, J., & Magee, R. P. (2011). From low-quality reporting to financial crises: Politics of disclosure regulation along the economic cycle. *Journal of Accounting and Economics, 52*(2), 209–227.

Bird, R., Liem, H., & Thorp, S. (2014). Infrastructure: Real assets and real returns. *European Financial Management, 20*(4), 802–824.

Bjelicic, B. (2012). Financing airlines in the wake of the financial markets crisis. *Journal of Air Transport Management, 21*, 10–16.

Bourjade, S., Huc, R., & Muller-Vibes, C. (2017). Leasing and profitability: Empirical evidence from the airline industry. *Transportation Research Part A: Policy and Practice, 97*, 30–46.

Bowman, R. G. (1980). The debt equivalence of leases: An empirical investigation. *Accounting Review, 55*, 237–253.

Brander, J. A., & Lewis, T. R. (1986). Oligopoly and financial structure: The limited liability effect. *The American Economic Review*, 956–970.

Burgess, K., Agnew, H., & Daneshkhu, S. (2016, January 13). Reporting rule adds $3tn of leases to balance sheets globally. *Financial Times*.

Chen, W.-T., Huang, K., & Ardiansyah, M. N. (2018). A mathematical programming model for aircraft leasing decisions. *Journal of Air Transport Management, 69*, 15–25.

Clarke, J.-P., Miller, B., & Protz, J. (2003). *An options-based analysis of the large aircraft market*. Paper presented at the AIAA/ICAS International Air and Space Symposium and Exposition, Daytona Ohio.

Cornaggia, K. J., Franzen, L. A., & Simin, T. T. (2013). Bringing leased assets onto the balance sheet. *Journal of Corporate Finance, 22*, 345–360.

Damodaran, A. (2017). Operating versus Capital Leases. Retrieved from http://pages.stern.nyu.edu/~adamodar/New_Home_Page/AccPrimer/lease.htm

Dechant, T., & Finkenzeller, K. (2010). Real estate or infrastructure? Evidence from conditional asset allocation.

Edwards, J. S., & Mayer, C. P. (1991). Leasing, taxes, and the cost of capital. *Journal of Public Economics, 44*(2), 173–197.

Eisfeldt, A. L., & Rampini, A. A. (2008). Leasing, ability to repossess, and debt capacity. *The Review of Financial Studies, 22*(4), 1621–1657.

Ezzell, J. R., & Vora, P. P. (2001). Leasing versus purchasing: Direct evidence on a corporation's motivations for leasing and consequences of leasing. *The Quarterly Review of Economics and Finance, 41*(1), 33-47.

Financial Accounting Standards Board. (1976). Statement of Financial Accounting Standards No. 13. Retrieved from www.fasb.org/jsp/FASB/Document_C/DocumentPage?cid=1218220124481&acceptedDisclaimer=true

Financial Accounting Standards Board. (2016). Accounting Standards Update: Leases (Topic 842). Retrieved from www.fasb.org/jsp/FASB/Document_C/DocumentPage?cid=1176167901010&acceptedDisclaimer=true

Finkenzeller, K., Dechant, T., & Schäfers, W. (2010). Infrastructure: a new dimension of real estate? An asset allocation analysis. *Journal of Property Investment & Finance, 28*(4), 263–274.

Finucane, T. J. (1988). Some empirical evidence on the use of financial leases. *Journal of Financial Research, 11*(4), 321–333.

Franke, M., & John, F. (2011). What comes next after recession?–Airline industry scenarios and potential end games. *Journal of Air Transport Management, 17*(1), 19–26.

Froot, K. A. (1995). Hedging portfolios with real assets. *Journal of Portfolio Management, 21*(4), 60.

Gavazza, A. (2010). Asset liquidity and financial contracts: Evidence from aircraft leases. *Journal of Financial Economics, 95*(1), 62–84.

Gavazza, A. (2011). Leasing and secondary markets: Theory and evidence from commercial aircraft. *Journal of Political Economy, 119*(2), 325–377.

Goolsbee, A. (1998). The business cycle, financial performance, and the retirement of capital goods. *Review of Economic Dynamics, 1*(2), 474-496.

Gorjidooz, J., & Vasigh, B. (2010). Aircraft valuation in dynamic air transport industry. *Journal of Business & Economics Research (JBER), 8*(7).

Graham, J. R., Lemmon, M. L., & Schallheim, J. S. (1998). Debt, leases, taxes, and the endogeneity of corporate tax status. *The Journal of Finance, 53*(1), 131–162.

Gritta, R., & Lynagh, P. (1973). Aircraft Leasing-Panacea or Problem. *Transp. LJ, 5*, 9.

Gritta, R. D. (1974). The Impact of the Capitalization of Leases on Financial Analysis. *Financial Analysts Journal, 30*(2), 47–52.

Gritta, R. D., Lippman, E., & Chow, G. (1994). The impact of the capitalization of leases on airline financial analysis: An issue revisited. *Logistics and Transportation Review, 30*(2), 189.

Habib, M. A., & Johnsen, D. B. (1999). The financing and redeployment of specific assets. *The Journal of Finance, 54*(2), 693–720.

Handa, P. (1991). An economic analysis of leasebacks. *Review of Quantitative Finance and Accounting, 1*(2), 177–189.

Hsu, C.-I., Li, H.-C., Liu, S.-M., & Chao, C.-C. (2011). Aircraft replacement scheduling: a dynamic programming approach. *Transportation research part E: logistics and transportation review, 47*(1), 41–60.

IFRS. (2003). IAS 17. Retrieved from www.ifrs.org/Documents/IAS17.pdf

IFRS. (2012). IFRS Foundation Staff Analysis of SEC Final Staff Report on IFRS. Retrieved from www.ifrs.org/Alerts/PressRelease/Pages/IFRS-Foundation-Staff-Analysis-of-SEC-Final-Staff-Report-on-IFRS.aspx

Inderst, G. (2009). Pension fund investment in infrastructure.

Inderst, G. (2010). Infrastructure as an asset class. *EIB papers, 15*(1), 70–105.

Jiang, H., & Hansman, R. (2004). *An analysis of profit cycles in the airline industry.* Thesis (SM), Department of Aeronautics and Astronautics, Massachusetts.

Kothari, S., & Lester, R. (2012). The role of accounting in the financial crisis: Lessons for the future. *Accounting Horizons, 26*(2), 335–351.

Krahan, J. P., & Meran, G. (1987). Why leasing. *An Introduction to Comparative Contractual Analysis.* Berlin: Bamberg, G. & Spremann, K., eds.

Krishnan, V. S., & Moyer, R. C. (1994). Bankruptcy costs and the financial leasing decision. *Financial Management*, 31–42.

Lease, R. C., McConnell, J. J., & Schallheim, J. S. (1990). Realized returns and the default and prepayment experience of financial leasing contracts. *Financial Management*, 11–20.

Lewellen, W. G., Long, M. S., & McConnell, J. J. (1976). Asset leasing in competitive capital markets. *The Journal of Finance, 31*(3), 787–798.

Lewis, C. M., & Schallheim, J. S. (1992). Are debt and leases substitutes? *Journal of Financial and Quantitative Analysis, 27*(4), 497–511.

Liehr, M., Größler, A., Klein, M., & Milling, P. M. (2001). Cycles in the sky: understanding and managing business cycles in the airline market. *System Dynamics Review, 17*(4), 311–332.

Lim, S. C., Mann, S. C., & Mihov, V. T. (2017). Do operating leases expand credit capacity? Evidence from borrowing costs and credit ratings. *Journal of Corporate Finance, 42*, 100–114.

Lyneis, J. M. (2000). System dynamics for market forecasting and structural analysis. *System Dynamics Review: The Journal of the System Dynamics Society, 16*(1), 3–25.

Mann, E. D. (2009). Aviation finance: An overview. *Journal of Structured Finance, 15*(1), 109.

Martin, G. A. (2010). The long-horizon benefits of traditional and new real assets in the institutional portfolio. *The Journal of Alternative Investments, 13*(1), 6.

Meadows, D. L. (1971). *Dynamics of commodity production cycles.*

Newell, G., & Peng, H. W. (2007). *The significance of infrastructure in investment portfolios.* Paper presented at the Pacific Rim Real Estate Conference, Fremantle.

Newell, G., & Peng, H. W. (2008). The role of US infrastructure in investment portfolios. *Journal of Real Estate Portfolio Management, 14*(1), 21–34.
Oum, T. H., Zhang, A., & Zhang, Y. (2000a). Optimal demand for operating lease of aircraft. *Transportation Research Part B: Methodological, 34*(1), 17–29.
Oum, T. H., Zhang, A., & Zhang, Y. (2000b). Socially optimal capacity and capital structure in oligopoly: the case of the airline industry. *Journal of Transport Economics and Policy*, 55–68.
Parrish, R. (1970). Aircraft Leasing. *Airline Management and Marketing, II (June 1970), 50.*
Pearce, B. (2012). The state of air transport markets and the airline industry after the great recession. *Journal of Air Transport Management, 21*, 3–9.
Pierson, K., and Sterman, J. D. (2013). Cyclical dynamics of airline industry earnings. System Dynamics Review, 29(3), 129–156.
Pulvino, T. C. (1998). Do asset fire sales exist? An empirical investigation of commercial aircraft transactions. *The Journal of Finance, 53*(3), 939–978.
Pulvino, T. C. (1999). Effects of bankruptcy court protection on asset sales. *Journal of Financial Economics, 52*(2), 151–186.
Robertson, D. H. (1915). *A study of industrial fluctuation: an enquiry into the character and causes of the so-called cyclical movements of trade*: London: PS King.
Roenfeldt, R. L., & Osteryoung, J. S. (1973). Analysis of financial leases. *Financial Management*, 74–87.
Rothballer, C., & Kaserer, C. (2012). The risk profile of infrastructure investments: Challenging conventional wisdom. *Journal of Structured Finance, 18*(2), 95.
Schallheim, J., Wells, K., & Whitby, R. J. (2013). Do leases expand debt capacity? *Journal of Corporate Finance, 23*, 368–381.
Schallheim, J. S., & McConnell, J. (1985). A model for the determination of "fair" premiums on lease cancellation insurance policies. *The Journal of Finance, 40*(5), 1439–1457.
Sharpe, S. A., & Nguyen, H. H. (1995). Capital market imperfections and the incentive to lease. *Journal of Financial Economics, 39*(2), 271–294.
Shleifer, A., & Vishny, R. W. (1992). Liquidation values and debt capacity: A market equilibrium approach. *The Journal of Finance, 47*(4), 1343–1366.
Slovin, M. B., Sushka, M. E., & Polonchek, J. A. (1990). Corporate Sale and Leasebacks and Shareholder Wealth. *The Journal of Finance, 45*(1), 289–299.
Smith, A. (1776). *The wealth of nations.*
Smith, C. W., & Wakeman, L. (1985). Determinants of corporate leasing policy. *The Journal of Finance, 40*(3), 895–908.
Stonier, J. (1999). Airline long-term planning under uncertainty. *Real Options and Business Strategy, Lenos Trigeorgis ed., Risk Books.*
Stonier, J. (2001). Airline fleet planning, financing, and hedging decisions under conditions of uncertainty. *Handbook of Airline Strategy*, McGraw-Hill.

Stonier, J. (1998). Marketing From A Manufacturer's Perspective: Issues In Quantifying The Economic Benefits Of New Aircraft, And Making The Correct Financing Decision. *Handbook Of Airline Marketing.*

US Securities And Exchange Commission. (2012). *Work Plan for the Consideration of Incorporating International Financial Reporting Standards into the Financial Reporting System for U.S. Issuers.* Retrieved from www.sec.gov/spotlight/globalaccountingstandards/ifrs-work-plan-final-report.pdf

Vasigh, B., & Erfani, G. (2004). Aircraft value, global economy and volatility. *Airlines Magazine, 28.*

Vasigh, B., Fleming, K., & Humphreys, B. (2014). *Foundations of airline finance: Methodology and practice*: Routledge.

Williamson, O. E. (1988). Corporate finance and corporate governance. *The Journal of Finance, 43*(3), 567–591.

Wojahn, O. W. (2012). Why does the airline industry over-invest? *Journal of Air Transport Management, 19*, 1–8.

Wyman, H. E. (1973). Financial lease evaluation under conditions of uncertainty. *The Accounting Review, 48*(3), 489–493.

Yan, A. (2006). Leasing and debt financing: substitutes or complements? *Journal of Financial and Quantitative Analysis, 41*(3), 709–731.

CHAPTER 3

Global Aircraft Leasing Industry Characteristics

1 Background and History of the Aircraft Leasing Industry

Leasing has been a key driver of the aircraft trading market. Lessors, especially relating to aviation, are both a source of financing and also one of the market participants for aircraft trading that has increased in significance over time. As noted in Chap. 2, there are finance and operating leases depending on the classification of the lease. Finance leasing in aviation has been in existence for a long time and the earliest mention is in the "Report on Air Transportation of Aviation Securities Committee of Investment Bankers Association of America" in 1949. It is not until the late 1960s that aircraft operating leasing started to develop. The aircraft operating leasing industry can be traced to the manufacturer McDonnell Douglas Corporation (now part of the Boeing Company) and the founding of its wholly owned subsidiary, McDonnell Douglas Finance Corporation (MDFC) in 1968 (renamed Boeing Capital Corporation in 1997 after the merger). MDFC was set up for the explicit purpose of assisting with the financing of new and used aircraft for the company and industry. MDFC began by leasing two DC-9s to Air West and three DC-8-63CFs to Flying Tiger Line. It was initially capitalized by a $200 million revolving credit from ten banks that also financed its parent

D. Yu, *Aircraft Valuation*,
https://doi.org/10.1007/978-981-15-6743-8_3

company (Smith 1968). This first wave of aircraft leasing were more of finance leases than operating leases as the tenor of the leases ran usually between 15 and 18 years and focused on US carriers by the manufacturers. This first active period was between 1968 and 1972 and encompassed deliveries of new DC-9s and Boeing 727s along with the first wide-bodied aircraft (Dallos 1986).

After 1968, the aircraft operating leasing industry began with its origins by the founding of International Lease Finance Corporation (ILFC) by Steve Udvar-Hazy in California, US in 1973. Not long after that in 1975, in Europe, Guinness Peat Aviation (GPA) was founded in Ireland by Aer Lingus executives lead by Tony Ryan and Guinness Peat Group. Both of the founders come from humble beginnings, are the pioneers of the industry, and have bigger than life reputations ("High-flying Irishman" 2007). The importance of these two individuals in aviation trading and leasing reflects the increasing role of operating lessors in terms of aircraft pricing. Udvar-Hazy is known as the "godfather" or "father" of the aviation leasing industry and Ryan is also a legend in the Irish commercial community and is known as the founder of GPA but is better known more universally for the biggest European low-cost airline (LCC) that he subsequently founded bearing his name—Ryanair. LCCs as an airline business model would have increased penetration of the world airlines and have large effects on aircraft types and pricing with most preferring to have single-aisle narrowbody aircraft, all in economy-seating configuration and eschewing from other types and configurations by the traditional airlines.

Udvar-Hazy founded ILFC with two partners in 1973 with $150,000 to buy a used DC-8 (Wayne 2007). ILFC started being traded over the counter in 1983 and was later acquired by American International Group, Inc., (AIG) in 1990 for $1.26 billion. At Udvar-Hazy's retirement from ILFC in February 2010, the company had an owned fleet of 933 with managed aircraft pushing over 1,000 aircraft in total with a net book value of $38.5 billion. He subsequently co-founded Air Lease Corporation which later went public on the NYSE in 2011.

GPA was formed by Guinness Peat Group and Aer Lingus, where Ryan was one of the executives in 1975. At its peak, it had a valuation of €4 billion and 400 aircraft fleet. In 1992, it tried to go public by having a public offering of €850 million which collapsed due to the lack of demand given the bearish environment at the time due to the first Gulf War ("High-flying Irishman" 2007). This failed IPO (Initial Public Offering) created a liquidity crisis for the company as it had a very large aircraft order book of 700 aircraft worth "$17 billion and not enough capital to fund them" and started the downfall of the company (Humphries and Hepher 2014).

By 1993, the liquidity crisis forced GPA to restructure and most of its assets and employees were transferred to GE Capital under the GE Capital Aviation Services (GECAS) subsidiary. GECAS built on this transaction is still one of the heavyweights of the industry and to this day has been either the top one or top two aircraft lessors globally. GECAS took over "about 60 percent of GPA's fleet and 75 percent of its people" (Humphries and Hepher 2014). It also included an option to buy a large portion of the company.

After the GECAS transaction, GPA continued to have a large fleet and own a substantial fleet and to innovate. In March 1996, GPA sold 229 aircrafts for $4.048 billion, which was the second-largest securitization at that time. This was used to refinance its debts (Bowers 1998). In November 1998, Texas Pacific Group (TPG) acquired 62% of the company and renamed it AerFi Group plc. They also restructured GECAS's option down to 23%. AerFi acquired Indigo Aviation, a Swedish aircraft lessor in December 1999 and in November 2000 AerFi was acquired for $750 million by debis AirFinance (owned by DaimlerChrysler AG). In March 2005, debis AirFinance was acquired by Cerberus Capital. Debis AirFinance was subsequently renamed AerCap, which later acquired ILFC from its parent American Insurance Group. By 1990, GPA had close to $17 billion order book for 700 new aircrafts spread out for the future years.

At its peak, it had over 400 employees and many subsidiaries outside of its mainline business including a helicopter leasing joint venture set up in June 1990 with CHC Helicopter called GPA Helicopters Ltd (Reuters 1990). In addition, many of GPA's employees have spread throughout the industry and helped found some of the biggest players in the industry including GECAS, CIT, AerCap (acquired ILFC and Genesis Lease), Standard Charter (acquired Pembroke Capital), SMBC Aviation Capital (renamed from RBS Aviation Capital and International Aircraft Management Group previously), Aircastle, BBAM and Fly Leasing (from

Babcock and Brown). GPA set up Ireland as the premier center of the aircraft leasing industry that it enjoys today. The existence of specialized personnel along with a relatively low corporate tax environment and the supporting services industries has allowed the country to remain the center of the aircraft leasing industry.

2 Jurisdictions for Leasing: Ireland, Singapore, Hong Kong

In the aircraft leasing industry, the main country and hub of operation for most of the industry is Ireland. Currently, Ireland is domicile to 50% of the world's aircraft fleet representing over €100 billion in assets and Ireland has the overwhelming amount of the global top 15 lessors ranked by fleet size (IDA Financial Services 2017). This is due to a couple of primary reasons, namely, government support, which results in a favorable taxation regime in terms of tax incentives effectively resulting in low taxes along with a large number of double tax agreements with other countries that lower the friction costs of moving capital across jurisdictions. Double tax treaties are treaties signed between two countries that avoid or mitigate the need to pay double tax—in both countries of domicile and country of incorporation.

These low taxes are the result of favorable governmental support in terms of various policies to support the growth of the industry. The government has published a draft national aviation policy that specifically endorses and advocates for the aircraft leasing industry. During 1980–2003, the foreign tax rate was 10% and currently, the 'trading' status firms enjoy a tax rate of 12.5% while a standard company tax rate without presence for trading status is 25%. There are established, specific safe harbor rules for the industry for what constitutes 'trading company' status. This is one of the lowest corporate tax rates in the world. Generally, there are 0% withholding taxes on inbound and outbound lease payments as well as on interest, dividends and royalties to the EU and tax treaty countries. In 2012, Ireland enacted unilateral credit relief on withholding taxes on lease payments from non-double tax treaty countries. The extensive number of double-tax treaties signed (72 have been signed) are in addition to the EU Directives.

There are stamp duty exemptions and an effective 0% VAT regime. For capital markets transactions such as asset backed securities (ABS) and enhanced equipment trust certificates (EETCs), specific legislation has been enacted to support these in addition to the existence of efficient structures for these securitizations and other financings. Another key in the favorable tax regime is the depreciation of the cost of the aircraft that can be used to offset the rental income. The depreciation policy is a straight line depreciation basis over eight years. As an EU country, the country is not considered a 'tax haven', which has been an increasing area of concern for companies and financing. The cornerstones of this favorable tax regime were first enacted in the 1950s by the launch of the International Finance Service Centre to support its growth. While the industry is large in assets, it has been criticized for paying just €23 million of Irish Corporation Tax in 2014 and €29 million in 2013. Over the past five years as of 2015, it was confirmed by the Ministry of Finance that the industry had paid only €123 million in taxes (Deegan 2015).These figures reinforce the concept of low taxation for the industry.

Another important factor is the sound legal regime with Ireland's use of common law. This type of legal system is used extensively in the world due to the spread of English law. Another important genesis is the birth and demise of GPA as the experienced veterans then moved on to different companies creating the entire value chain needed to sustain a healthy ecosystem and a hub of aviation leasing. The number and robustness of the aircraft servicing industry in the country include general corporate services such as banking, accounting, audit, and legal. As things have developed, these support services have specialized further and developed into industry-specific focused firms and professionals including specific industry consulting firms such as aircraft technical management, aircraft asset management, placement of aircraft, remarketing of new and second-hand aircraft, arrangers for various financings, third-party aircraft appraisal firms, and so on.

Knowledgeable workers are also supported by local universities with a focus on aviation and leasing that is a feedstock for new employees of these companies. Geographically, the country is situated in between continental Europe and USA which enhances its recruitment efforts. The aircraft leasing industry employs more than 1200 people directly and indirectly through professional services firms linked to the industry (IDA Financial Services 2017). It was noted by another source that the industry employs 1000 people directly and 2000 staff indirectly and spends over €135 million a year on professional services and infrastructure as of 2015.

In addition to Ireland, Singapore is another popular jurisdiction for aircraft leasing. Again, the main drivers are governmental support resulting in low taxation and numerous double tax treaties. Singapore has 76 double tax treaties including three countries that have signed but have yet to be ratified. The main government support scheme is the Aircraft Leasing Scheme (ALS) which started in 2012. It provides a low corporate tax of 5% or 10% on the profits for aircraft leasing companies that fulfill certain requirements as compared with a statutory corporate tax rate of 17%. For withholding taxes, there is an exemption on aircraft purchase and automatically on payments for financing payments with offshore lenders. For tax depreciation of the assets, approved companies under the ALS have an irrevocable option to depreciate the aircraft over any number of years from 5 to 20 years (EDB Singapore 2012). In January 2017, ALS was recently renewed for another five years along with the initial qualified party, Bank of China (BOC) owned BOC Aviation.

In addition to the favorable government support, Singapore's geographical location and climate is favorable for business. It is known as a key hub for finance and a regional beacon for Southeast Asia and Asia itself and is physically located closer in proximity to many locations in Asia giving it an advantage for some companies. It also has a sound legal environment that is also based on common law similar to Ireland. While there are many experienced professionals in finance and related industries, there is not the depth of experienced talent who are specialized in aircraft leasing as compared to Ireland. The finance center is an active hub for capital market and lending, among others. Industry-specialized and related professional service firms have grown in the past few years but are not in abundance as compared to Ireland.

Hong Kong is the new jurisdiction trying to become the leader and hub for aircraft leasing with special emphasis in capturing outbound Chinese capital flow. Similar to other jurisdictions, government support through taxation regimes is an important point along with double tax treaties. Historically, the tax laws in the 1990s were changed to clamp down on the abuse of the popular Hong Kong leveraged leasing but it also meant that there were fewer incentives for aircraft lessors to establish in Hong Kong. Recently on January 18, 2017, in the Chief Executive's policy address, he proposed in the Proposed Dedicated Tax Regime to Develop Aircraft Leasing Business in Hong Kong that the tax rate for qualified aircraft lessors and its profits will be half of the normal profits tax rate for corporations, so 8.25% reduced from the 16.5% prevailing rate (Hong Kong Legislative Council 2017). He also noted the taxable amount of lease rental payments derived to a non-Hong Kong-based lessee will be equal to 20% of the tax base which means the gross rentals less any deductible expenses. This is better than the previous gross rental taxation but did not go as far as other competitor jurisdictions by including tax depreciation for aircraft leased to non-Hong Kong-based airlines and lessors (Hong Kong Legislative Council 2016). These updated policies were finally set into law under the Inland Revenue (Amendment No. 3) Ordinance 2017 in July 2017 to mark the 20th anniversary of the handover of Hong Kong.

For double tax treaties, Hong Kong currently only has 36 signed double tax treaties. Chief Executive Mr. C.Y. Leung first announced this intention on January 14, 2015 and signed the updated double tax treaty with China on April 1, 2015 to reduce withholding tax rate on aircraft and ship leasing to 5% from 7%, which is the lowest double tax treaty signed by China. This compares to 6% withholding tax rate for Ireland- and Singapore-based companies from China (Hong Kong Inland Revenue Department 2017). These are part of the continued efforts by Hong Kong to act as the gateway for Chinese capital flowing outbound as mainland China is also fast becoming one of the main regions for aircraft and ship leasing.

Hong Kong also enjoys a sound legal system based on common law similar to Ireland and Singapore. Hong Kong is viewed as a bigger financial hub in Asia compared to Singapore but it lags even further in terms of personnel and service firms to support the aviation industry. There are current efforts to increase the government support through more favorable taxation regimes of the jurisdiction along with other characteristics that will make domiciling aircraft in Hong Kong more competitive.

3 Aircraft and Airline Market Drivers

Aircraft leasing is a complex business. There are so many moving pieces and drivers that affect valuation, usage, and others. In this discussion, aircraft technical aspects will be ignored and focus will be on the discussion of the economic and financial drivers that affect this business. At the very top level, the major stakeholders here are the original equipment manufacturers (OEMs), airlines, financiers, and governments. Each of these stakeholder's viewpoints contributes to the set and subset of drivers both on the demand and supply side that contributes to the growth trend of the aircraft leasing and its valuation.

4 Overall Demand Drivers

The global airline industry has been growing rapidly since the mid-1940s. Overall traffic growth and the most referenced proxy for the health of the industry is determined by the number of kilometers that passenger has flown expressed by revenue passenger kilometers (RPKs). See Fig. 3.1 for the tremendous world annual traffic growth from 0.4 trillion RPKs in 1965 to over 6 trillion RPKs in 2015.

This growth is resilient to temporary shocks of the various crises such as the oil crisis, Gulf crisis, Asia crisis, World Trade Center crisis, SARS, and the global financial crisis. There are many important drivers and

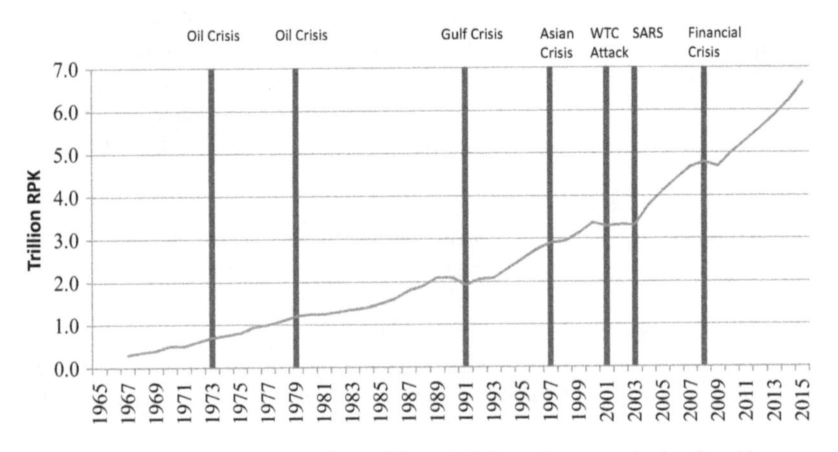

Fig. 3.1 Annual world traffic (trillion RPK) and external shocks. (Source: International Civil Aviation Organization 2016; Airbus 2016)

metrics for this industry to understand the fundamental demand drivers of the aircraft and airline market.

4.1 Economic Drivers

Generally, economics and finances of people and airlines are a major component driving demand for the airlines and thus the aircraft leasing industry. For a measure of general wealth and the economy, the most generally used metric is gross domestic product (GDP). There are several ways to view this, including nominal GDP, real GDP, and per capita GDP. Nominal GDP can be derived by the expenditure approach, which is one of the common approaches.

GDP=C (consumption) + I (investment) + G (government expenditures) + (X (export) – M (imports) (net exports)

Real GDP takes into account inflation which is subtracted from nominal GDP and per capita nominal or real GDP is the respective figures divided by the number of the population to obtain an average wealth per person figure. Per capita GDP is a good gauge for comparison of relative wealth between various countries and regions especially with adjustments for population differences.

As GDP is broken down by its components, consumption by private entities and non-government consumers is one of the main drivers for aviation. While the other inputs in this model are also important to aviation, they are more indirect and consumption is the most direct for demand. Generally, economic output drives activity in business travel and cargo services but also stimulates consumers for more leisure-type travel. Both Holloway (1997) and Lenoir (1998) described cycles in the context of the air transport industry and Lenoir (1998) found that nominal GDP and air traffic growth were positively correlated (Fig. 3.2).

These growth rates are shown to be mostly correlated to each other except for periods of crisis such as the oil crisis, Gulf crisis, WTC crisis, and SARS. The period during the latest global financial crisis showed the growth to be nearly identical compared with the other previous exogenous shocks. While this correlation has been shown over long periods historically, there have been some recent changes to this relationship. Further work by IBA Group has shown that the correlation has continued over 1985–2015 but the trends have shifted during the period of

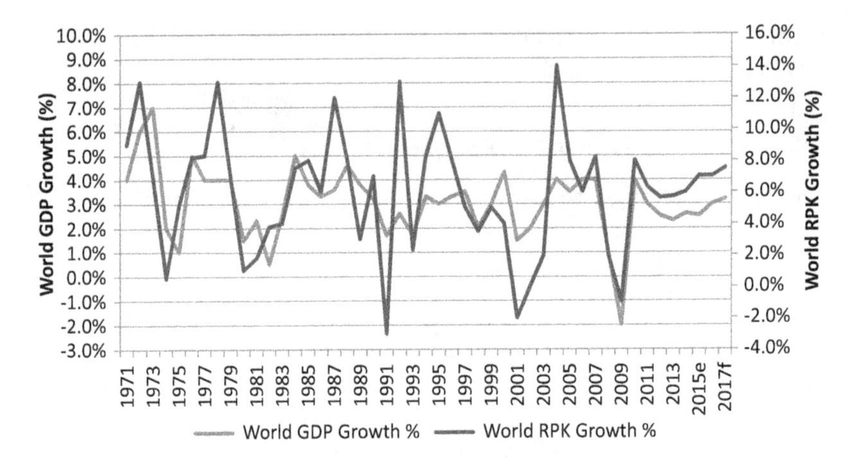

Fig. 3.2 World airline RPK growth and world GDP growth: 1971 to 2017F. (Source: International Monetary Fund 2016; Airline Monitor 2016; CAPA 2016)

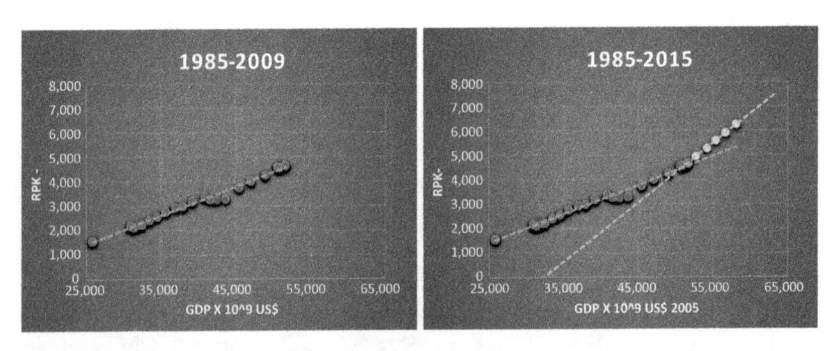

Fig. 3.3 Global traffic and nominal GDP correlation. (Source: International Air Transport Association 2016a; IBA Group 2017)

1985–2009 between RPK passenger growth and global nominal GDP as compared to the steeper, correlation found during 2009–2015.

Figure 3.3 notes a recent observation of higher growth of air traffic per increase of wealth as measured by nominal GDP. Airbus has also looked at the relationship between real GDP and global traffic. As shown in Fig. 3.4,

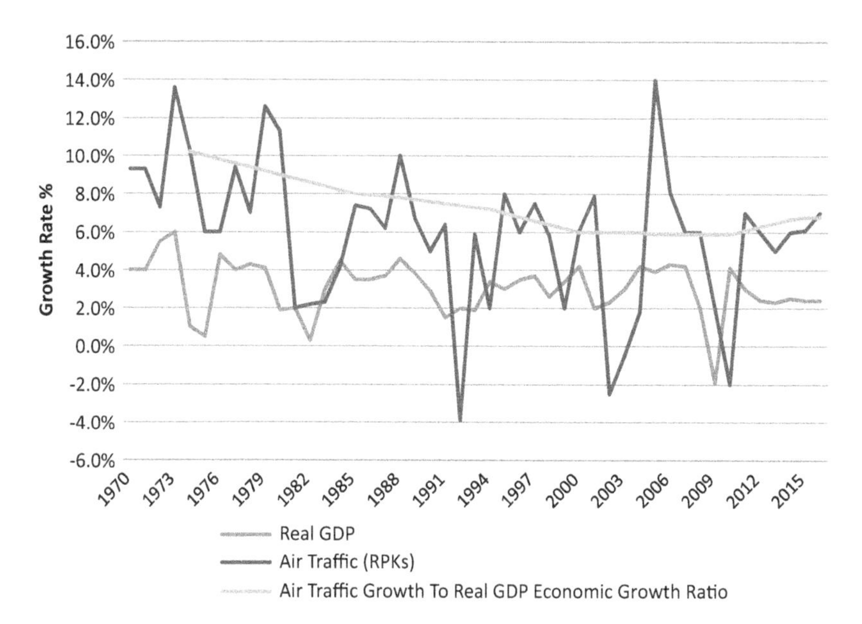

Fig. 3.4 Relationship between passenger traffic and real GDP evolving over time. (Source: International Civil Aviation Organization 2017; IHS Economics 2016; Airbus 2016)

there is a decreasing ratio of the passenger traffic to real GDP as the different decades progressed. This decreasing relationship ratio has reverted upwards back only in the latest decade.

In Fig. 3.5, emerging markets show signs of higher GDP growth compared with advanced markets and the world average, which tends to be correlated. Emerging markets do have more volatility than the other two groups. In some of these high-growth emerging economies such as China and India, Airbus has found that private consumption is more of a specific main driver than just the broader GDP. Airbus' yearly forecast models have recently been adjusted in their 2016 version to reflect this change. In the US market, while historically GDP and air traffic tracked well, domestic US passenger growth has outpaced US real GDP growth (Airbus 2016).

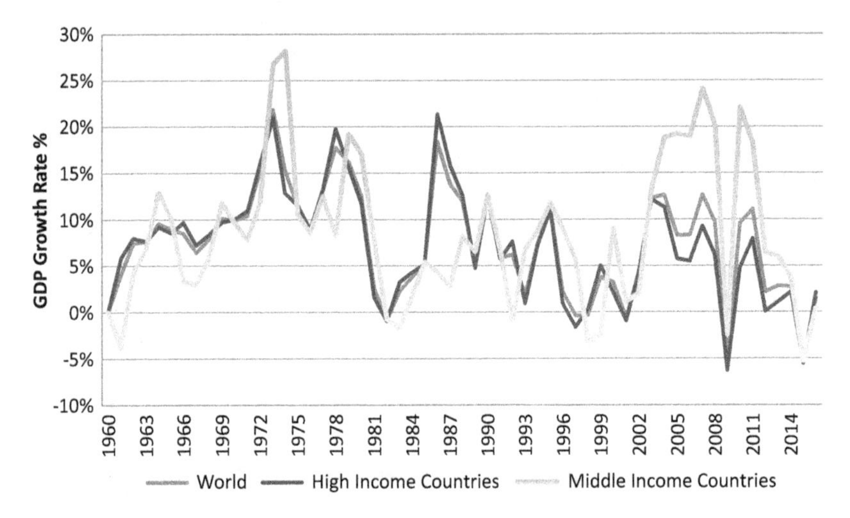

Fig. 3.5 GDP growth by country income. (Source: World Bank 2017)

4.2 *Business Cycles in Aviation and Industry Profitability*

There has been much research conducted about general business cycles since the founding of modern economics by Smith (1776) and his theories of economic growth and how it is impacted by cycles. Robertson (1915) noted that businesses have a cycle in relation to investments and are prone to over-investment and then periods of underinvestment as too many goods are in the market, thus creating a cycle. Eisfeldt and Rampini (2008) show how capital reallocation is important especially during the stages of the business cycle.

Like all types of business, the air transport industry is prone to business cycles. These cycles of boom and recession are particularly acute and common in the airline industry. Figure 3.6 shows the various parts of the aircraft and airline cycle from boom to bust and back to the growth stage.

Shearman (1992) notes that economic conditions regionally, nationally and internationally affect the air transport industry. Goolsbee (1998) showed the decision making of aircraft in business cycles. Liehr et al. (2001) and Lyneis (2000) describe the dynamics of the aviation business cycles through a systems dynamics problem. The industry is prone to under- and over-investment like other businesses as evidenced by airline

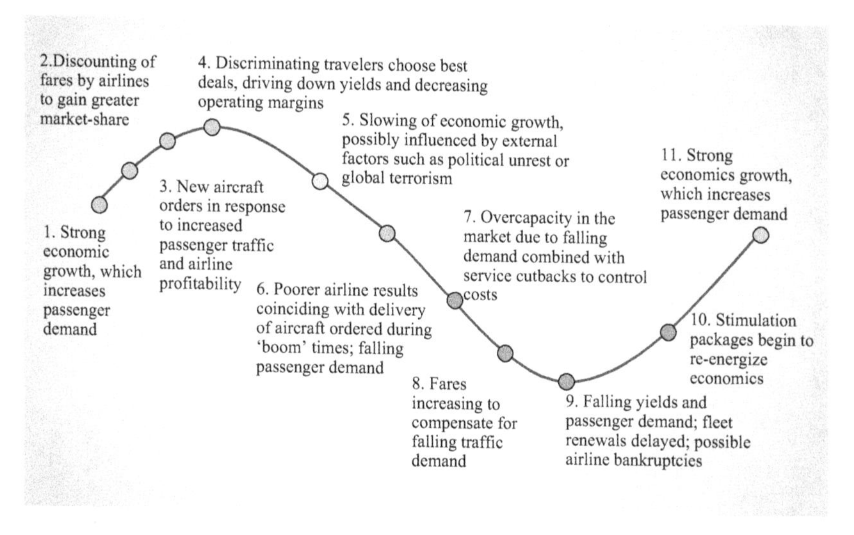

Fig. 3.6 Aircraft and airline market cycle

capital expenditures as a percentage of operating cash flows. Figure 3.7 shows there are only a few years (mid-1980s, 1990s, 2000) that the world airlines are spending less on capital expenditures than operating cash flows which is non-sustainable in the long run.

Interest rates set as part of monetary policy by the US Federal Reserve and other national institutions globally have tremendous effects upon the economy as well as the airline business due to the changes in business financing costs as well as the overall business sentiment. Depending on the surprise and expectation nature of the changes, different responses may result. Interest rates along with interest swap rates, where floating rates are fixed over a period of time of the contract, have a direct input into debt cost of financing. For aircraft leasing, interest swap rates are a main indicator of the direction the lease rates offered for aircraft on lease as it reflects an underlying cost of capital for the transaction. Figure 3.8 shows the historical trends of the various USD Swap and LIBOR rates tenors. The bands between the different rates are much tighter in the preceding years than the later years.

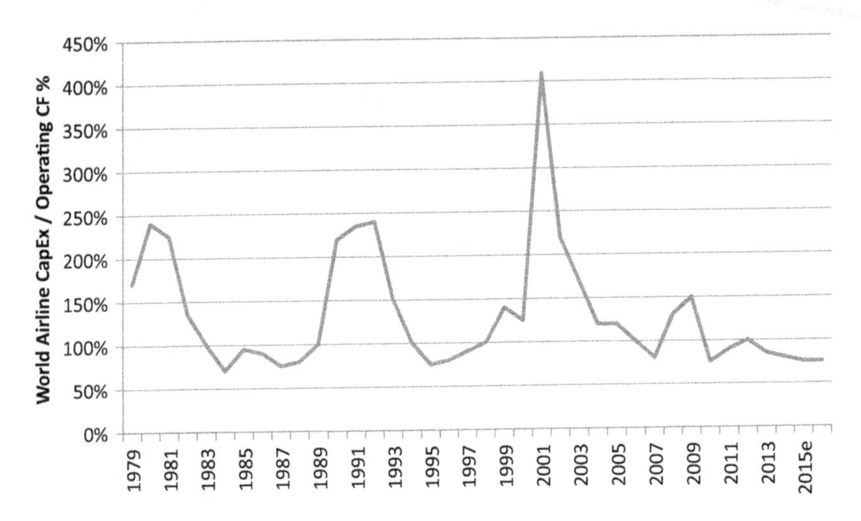

Fig. 3.7 World airline capital expenditure as a percentage of operating cash flow: 1979 to 2016E. (Source: Airline Monitor 2016; International Air Transport Association 2016b; CAPA 2016)

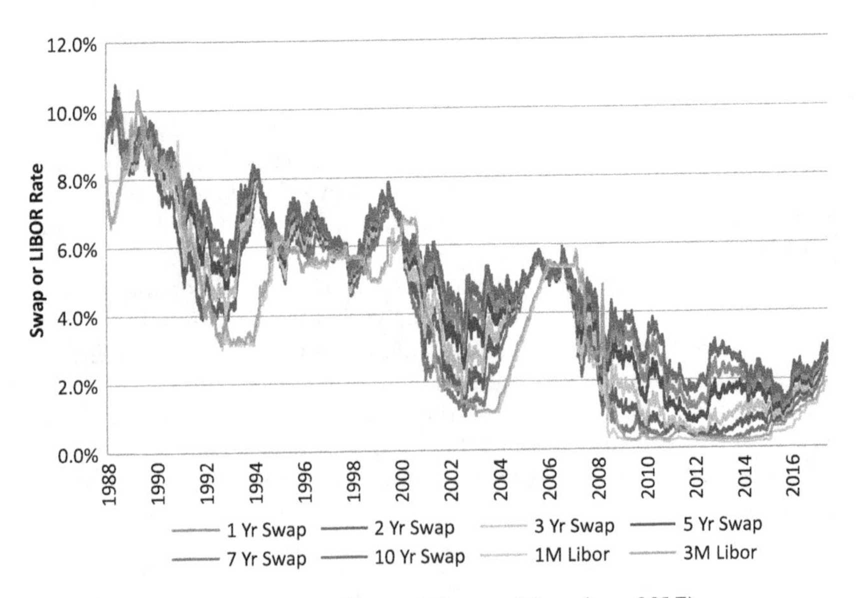

Fig. 3.8 USD swap and LIBOR rates. (Source: Bloomberg 2017)

Given the cyclical nature of the industry, cash flows and therefore profitability is a key metric of sustainability and growth of the aviation industry as well as with other industries and as shown, it tracks the global economy. There are many important input factors that affect airline cash flows and profitability and many studies have been conducted over the decades on this subject (Kasper 1988; Pryke 1987; Doganis 1985; Laprade 1981; Pearson 1976; Straszheim 1969; etc.).

As the correlation of passenger traffic and GDP growth is shown in Fig. 3.2, it is not surprising that there is a correlation between GDP growth and the airline's financials as measured by net post tax profit margin. This is seen in Fig. 3.9, which shows a weaker correlation between these metrics, with the differences being more acute after the WTC and SARS crises. On a further comparison historically, Fig. 3.10 shows the airline's operating profit or EBIT margin is more correlated to global GDP growth than the net post tax profit margin shown in Fig. 3.9. Figure 3.10 also shows the extreme variances of the EBIT margin as it is both larger and smaller than the growth in the various cycles.

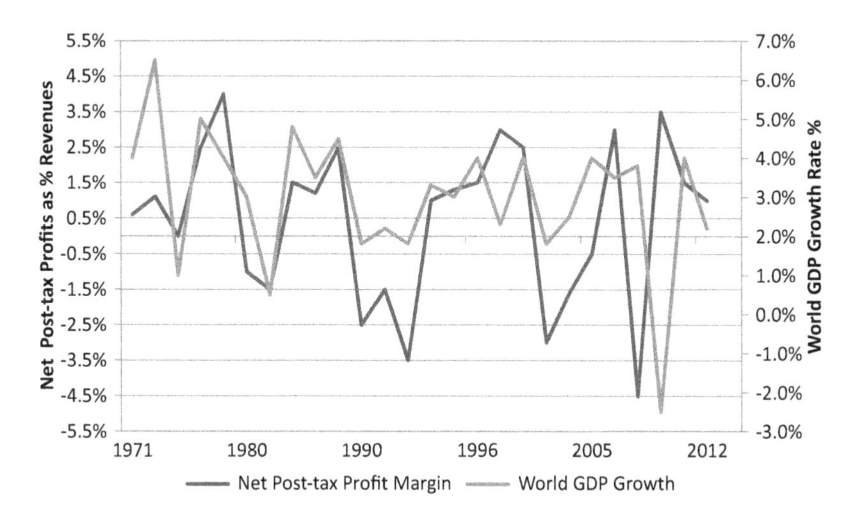

Fig. 3.9 World economic growth and airline profit margins: 1970 to 2011. (Source: International Air Transport Association 2012; International Civil Aviation Organization 2012a; Haver Analytics 2012)

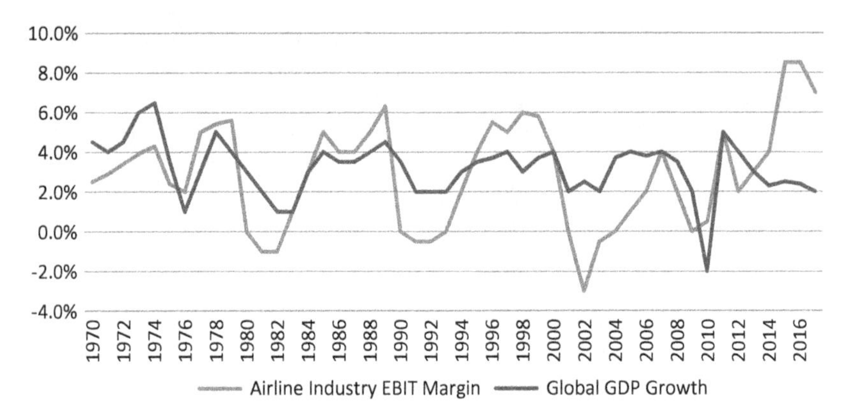

Fig. 3.10 EBIT or operating margins and global GDP growth. (Source: International Air Transport Association 2016b; International Civil Aviation Organization 2016; International Monetary Fund 2016)

4.3 *Exogenous Shocks*

In addition to normal business cycles, there have been many major exogenous shocks to the economy which, therefore, in turn, have had significant effects on the aviation market. These major shocks were usually major global economic events and have affected the aviation market in terms of the significant reduction in traffic demand, aircraft values, and the other drivers directly or indirectly inherent to aviation. These major exogenous shocks to the system include the First Oil Crisis 1973, Second Oil Crisis 1978–1979, Gulf War 1 Crisis 1990–1991, Asian Financial Crisis 1997–1998, 9/11 World Trade Center Attack 2001–2002, Avian Flu or SARS 2003–2004, and Global Financial Crisis of 2008–2009.

In addition to these very large events, there have been smaller and more regional exogenous shocks that have also affected the air transport industry and general economics—including the Swine Flu 2008, Icelandic Ash Cloud 2010, Japanese Tsunami 2011, and Hurricane Sandy 2012. Going forward, inevitability will be more of these exogenous shocks and smaller regional events, and one that is still being played out is Brexit 2016. See Fig. 3.1 showing the temporary negative effects of these exogenous shocks to the overall world traffic. Franke and John (2011), Pearce (2012), and Bjelicic (2012) all look at the state of affairs post the latest 2008 economic shocks.

4.4 *Fuel Prices*

For airline's cost breakdown, the historical rule of thumb is 50% for direct operating costs including all costs relating to flight operations like aircraft, fuel, maintenance, pilots; 30% for ground operation costs including servicing the passenger and aircraft and landing fees, and sales fees; and 20% for system operating costs including administrative, marketing, and other general costs like in-flight services and group equipment ownership (International Civil Aviation Organization 2017). Given the high cost of fuel historically, this is the number one cost category of around 30%, labor being 20%, depreciation 6%, aircraft rentals 4%, and everything else 38% according to International Air Transport Association (IATA) in 2008. There is much volatility in the jet fuel prices over time as seen in Fig. 3.11.

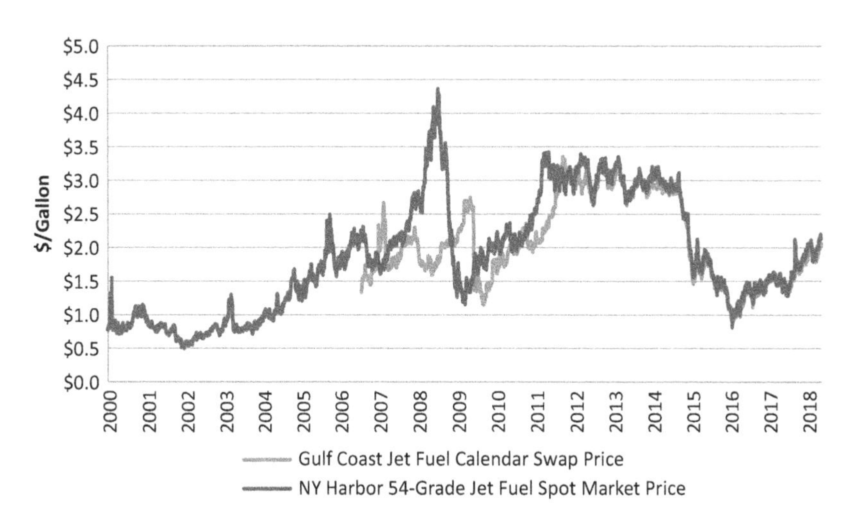

Fig. 3.11 Jet fuel prices. (Source: Bloomberg 2017)

Fuel prices have large spikes with the global financial crisis. Currently for US airlines, fuel has come down to the second largest cost component representing about 16%, labor increasing to 33%, aircraft rentals increasing to 7% of the cost (A4A 2017). Fuel cost is still an important driver of the success of all airlines and the recent run up in prices from the low point is having significant impacts on airlines. Specifically, the fuel price and the financial state of airlines has direct correlation as aircraft traffic consumes large quantities of fuel and even small cent changes in the fuel price can result in millions of dollars to the bottom line financially.

4.5 *Traffic Flows and Population Demographics*

Another significant metric is passenger yield which is a measure of the average fare paid per kilometer or mile. Yield differs due to different supply and demand and can show the differences between the market segments, seat classes, seasons, and geographies. While generally the higher the better, it should not be viewed only in isolation but in addition to the other metrics to be able to put the figures into the whole market perspective. This can help the airline develop price discrimination and differentiation strategies. There can be varied movements in the different sub-geographies and as Airbus notes, in the past year "yield has played a significant role" specifically in the US domestic market (Airbus 2016). Figure 3.12 shows the decrease of global yield based on an actual and constant USD exchange rate basis over the past few years.

Historically, besides business travel, only the upper class of the population was a consumer of aviation travel. The global rise of the middle class has become a large driver of aviation for leisure activities. According to Airbus and McKinsey, there is a large growth in the middle class globally and especially acute in the emerging economies compared with more stable projections of mature economies.

Shown in Fig. 3.13, Asia has 1284 million middle-class households (46% of global total) with a total global middle class of 2793 million (38%) compared to a global population of 7350 million. Middle-class household is defined as households with annual income between \$20,000 and \$150,000 (PPP constant 2014 prices). Another factor

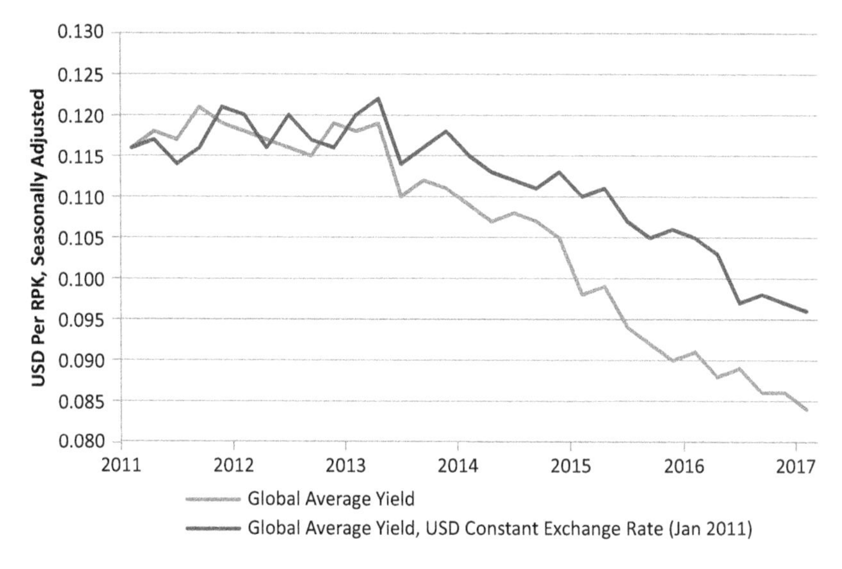

Fig. 3.12 Airline (passenger) yield. (Source: International Air Transport Association 2016b; DIIO 2017; Thomson Reuters Datastream 2017)

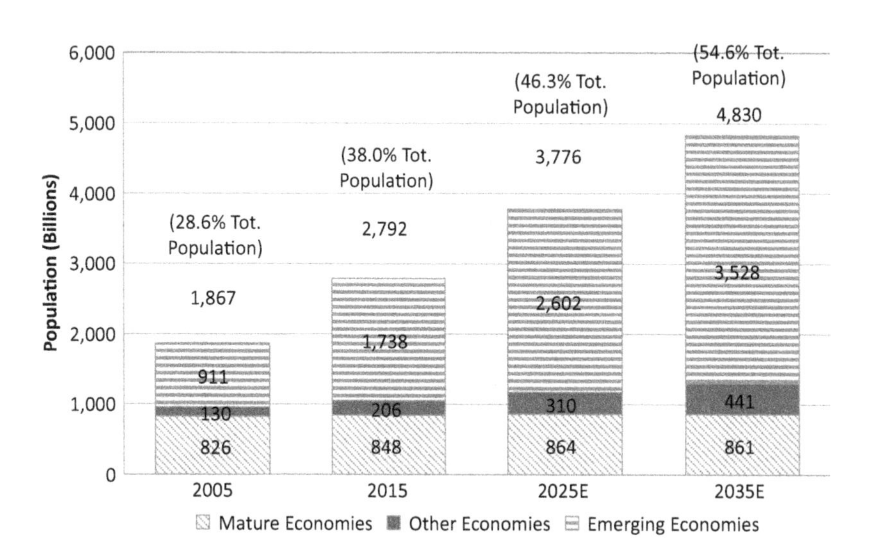

Fig. 3.13 Growth of the world's middle class. (Source: Oxford Economics 2016; Airbus 2016)

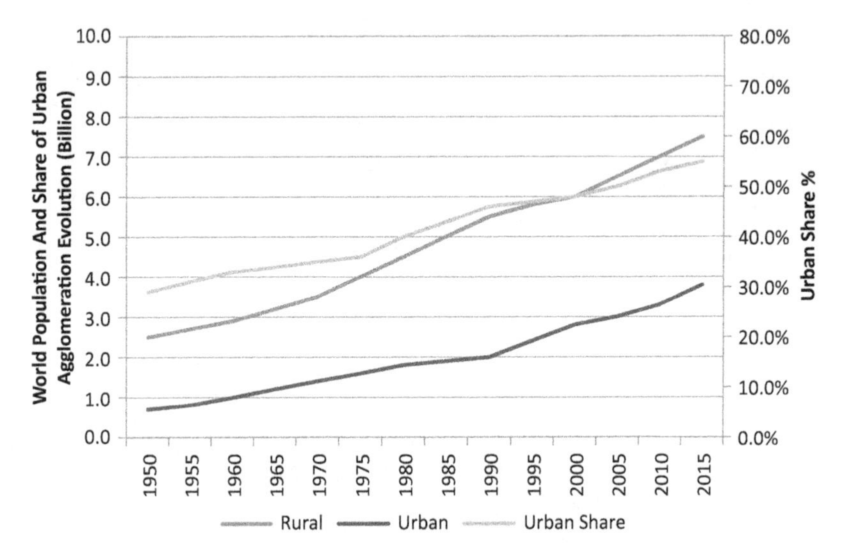

Fig. 3.14 Growth of urbanization. (Source: UN Population Division 2016; Airbus 2016)

that has contributed to increased wealth and consumption is the trend for global urbanization as shown in Fig. 3.14. Currently, the urban share of the global population is approximately 55% which is forecasted to reach almost 70% on an overall population of nearly 10 billion people by 2050.

Hand in hand in this globalization is the increased liberalization for immigration including the simplification and lengthening the number and duration of visits of visas including visa free, visa-on-arrival, and visa waiver schemes. This has been evident by recent announcements of China and USA, China and Australia visa schemes, Schengen visa-free area in Europe, and others. These increased liberalizations have helped propel both business and leisure-travel demand. Increased number of bilateral air agreements and regional treaties and associations such as ASEAN, NAFTA, the proposed TPP, and so on, have increased the demand. As shown in Fig. 3.15, there is a significant growth of bilateral air service agreements from 0 in 1945 to more than 2500 in 2010. All these trade agreements

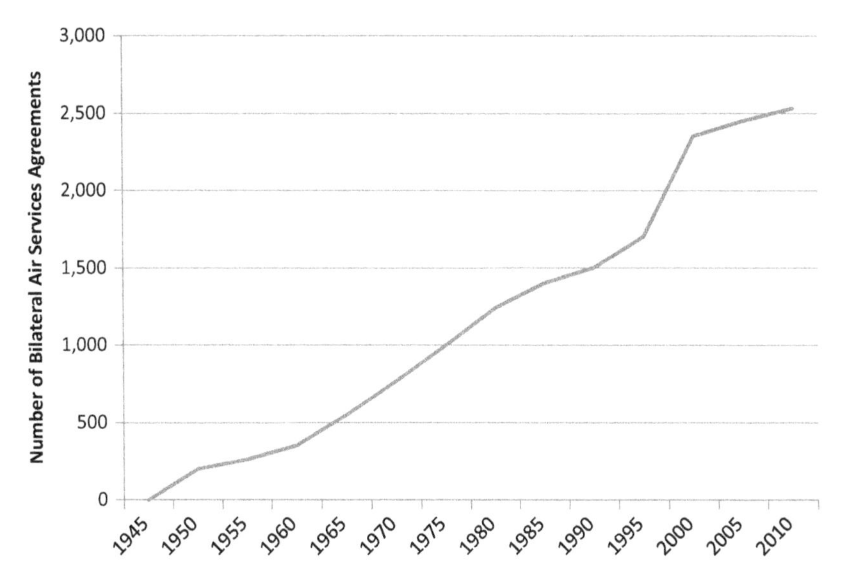

Fig. 3.15 Growth of bilateral air services agreements in the world. (Source: International Civil Aviation Organization 2012b; Airbus 2016)

and regions further stimulate demand for aviation both for passenger and cargo services.

5 Overall Supply Drivers

5.1 Manufacturers and Parked or Retired Aircraft

In addition to the demand drivers, there are a variety of supply drivers including new orders, deliveries, parked aircraft levels, retirements, secondary trading activity, and useful economic life of aircraft. Starting with the main manufacturers, there are currently two large commercial aircraft manufacturers of significance—Boeing and Airbus—who dominate the market as a roughly split duopoly. This discussion will focus on large aircraft with greater than 100 seats. In the past several years, there has been a renewed interest in new aircraft manufacturing, especially on the regional and larger single aisle or narrowbody aircraft. Most of these new entrants are focused on the regional aircraft of up to 100 seats including Canada's Bombardier CS300 and CS100, Russia's Sukhoi SSJ100, and Japan's Mitsubishi MRJ90 and MRJ70. The only narrowbodies in advanced development and testing are Russia's Irkut MC-21 and China's COMAC C919. The only widebody currently in conceptual design development is the COMAC and Russian joint venture, the C929. In addition to these manufacturers, there are also several other regional commercial and business aircraft manufacturers such as Embraer.

As there are only two main manufacturers, they control the order book and set the tone for the pricing of new aircraft. As is the norm in the industry, manufacturers set list prices per model and update them yearly. There can be discounts and other incentives depending on the buyer, geography, quantity of aircraft, and a variety of other factors including strategic or competitive aims. Traditionally, airlines stick with one manufacturer for the entire fleet or a certain purpose such as short haul until they have a big enough fleet to have multiple suppliers as changes are very sticky due to the investment of not only the asset but the associated supply chain including spare parts, crew and pilot training, certifications, and other things that need to be changed as well. These days, more startup airlines diversify earlier, utilizing both aircraft manufacturers. As seen in Figs. 3.16–3.17, the gross and net orders, deliveries, cancellations, and backlogs of Airbus and Boeing are shown. There can be large demand for popular aircraft and the delivery times can be stretched out into the future stretching out over ten years.

Most new aircraft orders include both firm and options that can be declared by the buyer. In addition, there are options to convert to another type of the same aircraft family such as Airbus A319, A320, A321 in the A320 family or Boeing 737-700, 800, 900 in the 737 NG family. New orders have a lot of customization as there are a lot of options that need to be set per order such as weight certifications, crew training, spare parts, and other various supports that can be provided. As part of these orders, there are generally pre-delivery payments (PDPs) due to the manufacturer representing roughly 30% of the total cost while the aircraft price has a referenced base year and escalation to the delivery date based on a manufacturer inflation index representing inflation and other factors.

These orders have a variety of purposes for the buyers. They can be for new routes and other new growth areas or replacement for retiring or old aircraft. In addition, as shown in Fig. 3.18, oil pricing and new orders are directionally correlated with a lag effect. Boeing and Airbus' new orders

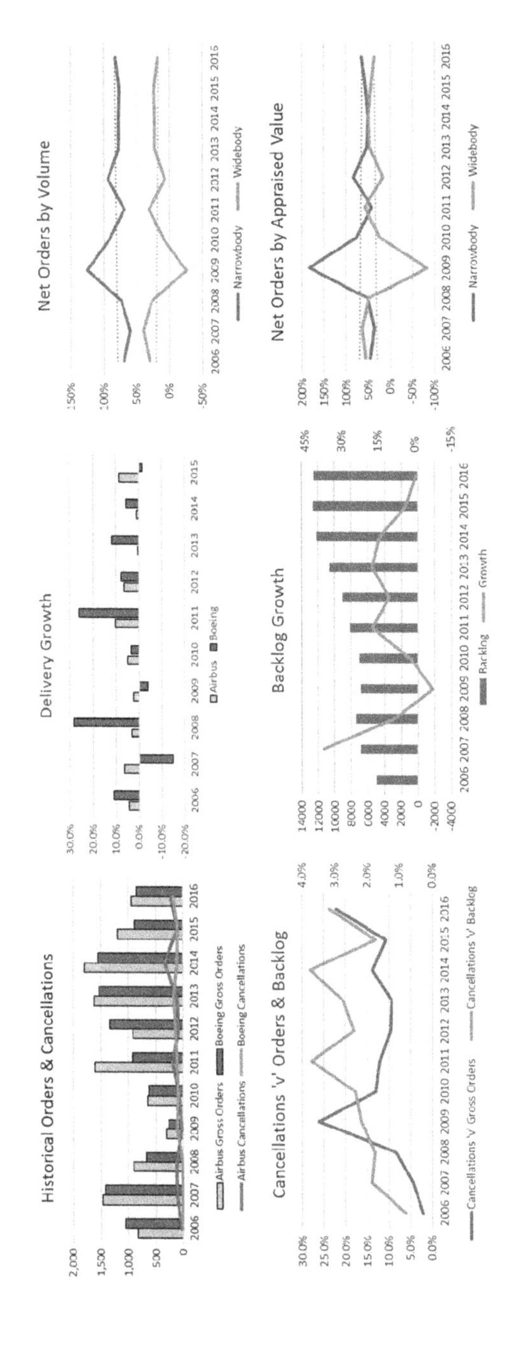

Fig. 3.16 OEM orders, cancellations, and backlog activity. (Source: Airbus 2017; Boeing Corporation 2017)

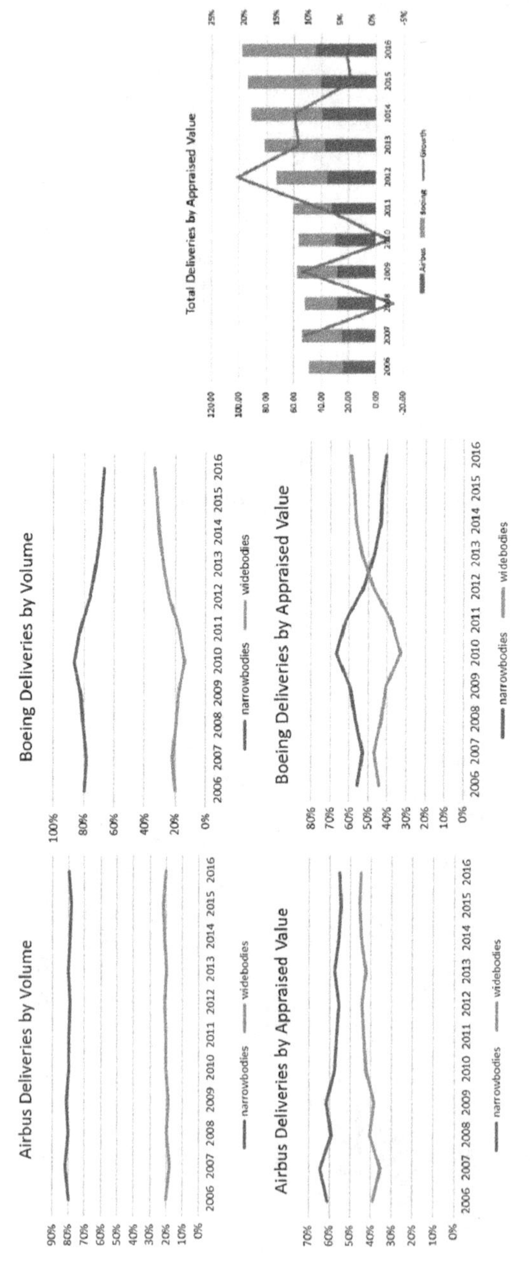

Fig. 3.17 Airbus and Boeing delivery activity. (Source: Airbus 2017; Boeing Corporation 2017)

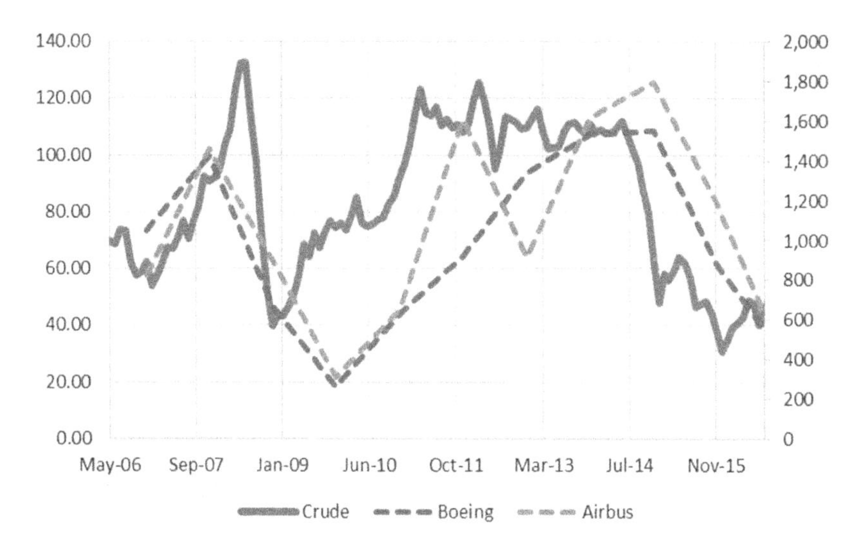

Fig. 3.18 Brent crude oil and OEM gross orders: May 2006–October 2016. (Source: Bloomberg 2017; Airbus 2017; Boeing Corporation 2017)

are similar directionally with some differences shown. This lag effect is acute with the drop in oil prices in 2008 and 2014. As order backlog can stretch for many years, the availability of the latest order slots can be an advantage for operating lessors or sellers when negotiating with airlines.

Another driver in the supply of aircraft is the number of aircraft parked or stored. Aircraft which is not wanted can be parked in areas temporarily or for longer periods in mostly desert locations such as Arizona and Europe to prevent corrosion damage. In addition, these aircraft should be put on specific short- or long-term aircraft maintenance packages for parked aircraft that monitors for corrosion, test the engines periodically, and other tasks to enable it to return to service and prevent more costly wear down. When oil prices have been high, there have been more instances of older and more heavy fuel consuming aircraft such as four engines instead of two engines that have been parked. These older aircraft and heavy fuel consuming aircraft generally have higher operating costs and less financial viability than newer more efficient aircraft and are the first to be parked or returned to the lessor.

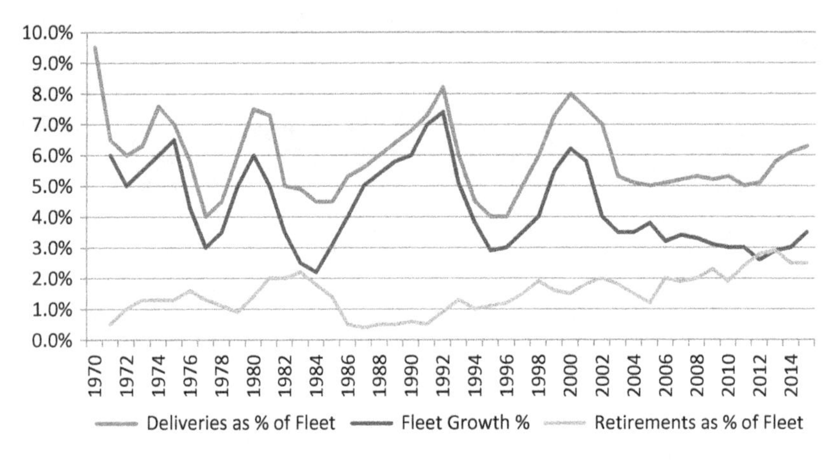

Fig. 3.19 World aircraft deliveries and retirements as percentage of fleet and fleet growth: 1971 to 2017F. (Source: Airline Monitor 2016; CAPA 2016)

Once an aircraft is deemed to be not wanted and retired, they can be parked or sold. On a global scale, retirements are a small percentage of the overall fleet while deliveries and fleet growth are highly correlated as shown in Fig. 3.19 with the drops of the delivery percentage aligned with the exogenous shocks.

The retirement and replacement decision analysis is done when the costs of operating the aircraft are more than the revenues that can be generated as compared with other aircraft types. As discussed earlier in section 4.4, fuel is a large factor in the operating costs of an airline. As the percentage of fuel costs increase for an airline, there is need for more efficient, newer aircraft to replace aging fuel inefficient aircraft and conversely this is true as well in times of low fuel prices when airlines tend to extend the age of their fleets by decreasing retirements as shown in Fig. 3.20.

There is a lag that occurs from the observation to decision and implementation. For 2017, the average overall retirement age was 23.6 years with the most commonly retired aircraft being narrowbodies as the average retirement age was above the common 25 year retirement assumption. While widebody aircraft were retired at an average age of 23.5 years, there are noticeable differences between the different widebody models (Fig. 3.21).

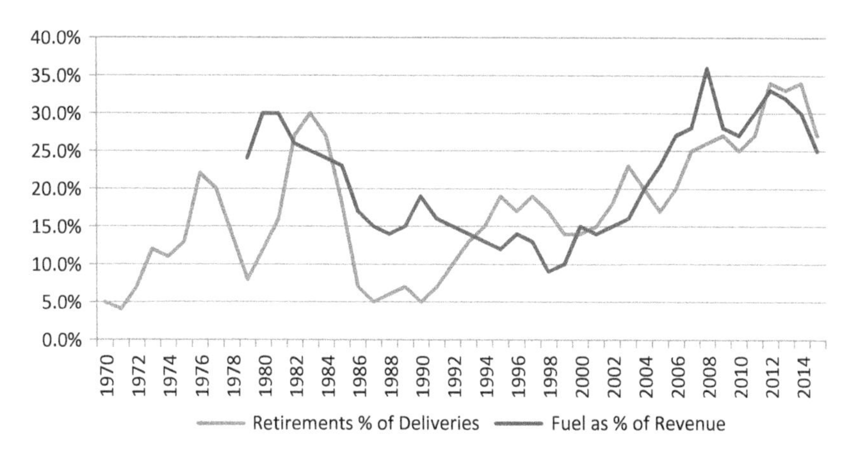

Fig. 3.20 Aircraft retirements: 1970 to 2015. (Source: Airline Monitor 2016; CAPA 2016)

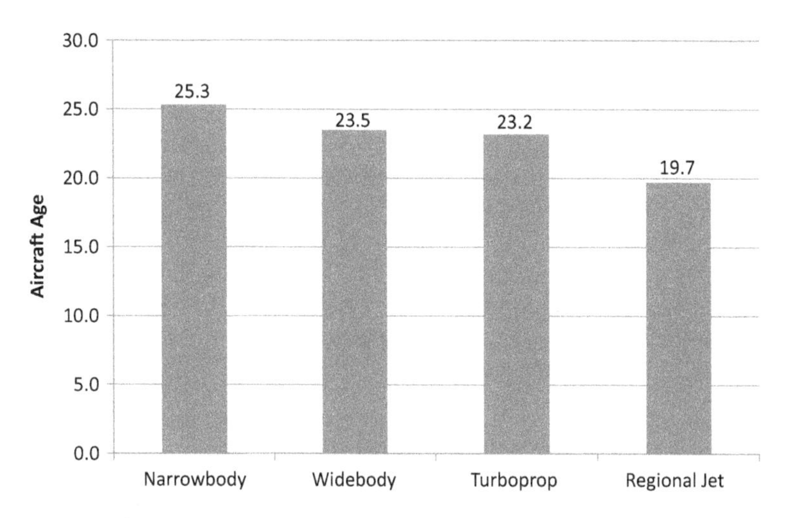

Fig. 3.21 Average age of retirements in 2017 by aircraft class. (Source: IBA Group 2017)

5.2 *Financing Environment and Current Trends*

Throughout the history of aircraft leasing, various financing trends have come and gone while the industry has grown significantly to the current $126 billion forecasted of new delivery funding requirements for 2017

(Boeing Capital Corporation 2017). In addition to the various options for funding, currently, there are more commercial banks than ever before involved in the equity side of the business along with newer capital sources such as insurance companies. These new players are additionally more diverse in geographical breakdown by their funding sources as compared between the traditional western players and newer eastern players. As 2017 further develops, this section will look at the upcoming trends especially in relation to geographical sources of the funding as well as a comparison of its characteristics.

5.3 *Global State of Aircraft Funding*

Airlines and aircraft have traditionally been financed by equity and bank financing facilities. These sources have continued to evolve with the support of the aircraft manufacturers and other innovative structures by the continued increase in demand from the users, mainly the airlines. These innovative financing structures include the use of the operating lease by capital sources such as commercial banks and leasing companies, both captive and non-captive to banks or manufacturers. The other sources of funding include cash or equity (26% of 2017 funding); export credit agencies (9% of 2017 funding); bank debt (34% of 2017 funding) by commercial banks, institutional players such as private equity and hedge funds, and tax equity; capital markets (31% of 2017 funding), including ABS and EETCs; aircraft and engine manufacturer financings (0% of 2017 funding); and insurance financings (0% of 2017 funding) with a total of $126 billion of new aircraft funding requirements in 2017 (Boeing Capital Corporation 2017). See Fig. 3.22 for a historical evolution of the percentage of the funding source for Boeing.

5.4 *Export Credit Financing*

Export credit agencies (ECAs) have been a significant source of funding, especially during times of financial distress, but this has been trending lower as capital markets financing has increased. In this case, development banks, policy banks, and other governmental institutions are all considered ECAs as they support the domestic industries. Export credit support by US Export-Import Bank (US EXIM) for domestic-built aircraft started in the 1970s. The main beneficiaries in the aircraft manufacturer space were Boeing and McDonnell Douglas. The recent low volume exhibited

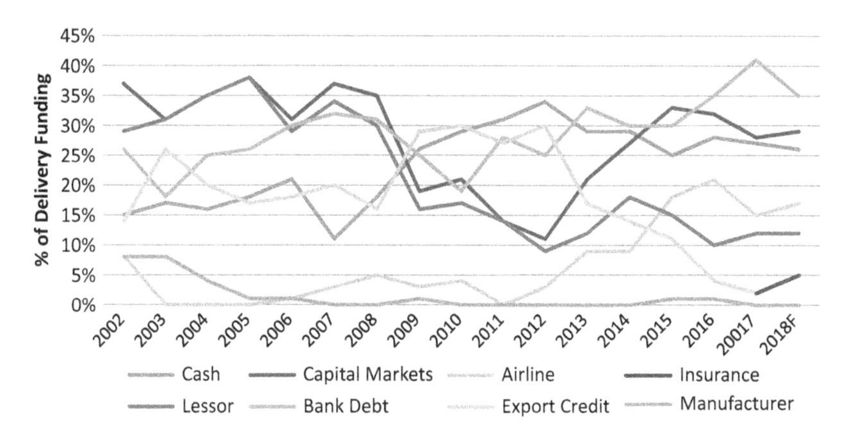

Fig. 3.22 Share of Boeing delivery funding by capital source. (Source: Boeing Capital Corporation 2010, 2017)

in 2016 has been the uncertainty of US EXIM, given its large historical support, especially to aircraft deliveries. After letting its charter expire in July 2015, the US Congress reauthorized the bank in December 2015, but it is still in a state of limbo because it is not able to conduct new business given that it awaits Senate confirmation of members to reconstitute a quorum. Only with a quorum can new funding decisions be made.

The Canadian, Brazilian, and other European export credit support organizations are active in the aircraft segment as these are the areas where active manufacturers are located. These specific ECAs include Export Development Canada, Brazil Development Bank, Export Credits Guarantee Department or UK Export Finance (UK), Federal Export Credit Guarantees managed by Euler Hermes Aktiengesellschaft (Germany), and Compagnie Francaise d'Assurance pour le Commerce Exterieur (COFACE) (France).

Since the tail end of the financial recession, capital markets have seen a significant rise in the number and magnitude of deals and it has risen to represent about one-third of all new aircraft funding as shown in Fig. 3.22. Figure 3.23 shows the elevated percentage of Boeing deliveries funded by EXIM while post the global financial crisis, the funding share has decreased significantly to 0%.These are represented by various securitization transactions, including ABSs and EETCs. During this period, the vast majority of the capital markets deals have been completed in the West but, in Asia, the

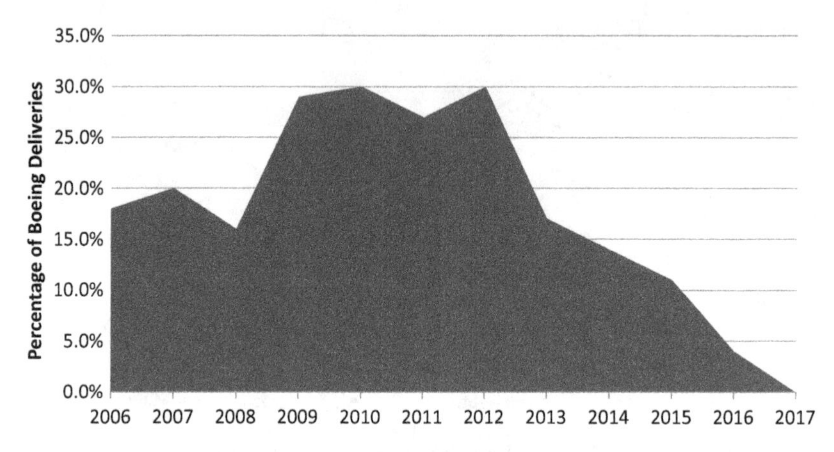

Fig. 3.23 Percentage of Boeing deliveries funded by US EXIM. (Source: Boeing Capital Corporation 2017)

market has only just started. There are more innovative financing structures now being completed in onshore China and Hong Kong.

The author views the continued trend to continue due to more capital markets deals being completed globally in 2017 but not by much in relative terms to other sources. The number of capital markets transactions in Asia will increase substantially, along with the expanding interest by financial players as described below, but as a percentage of the overall global market, this will not move the needle much in 2017. Boeing's view of historical sentiment of the different financing types is found in Fig. 3.24.

5.5 *Capital Markets*

Capital markets is a significant portion of the aircraft financing market with 30% of 2016 global funding and 31% projected for 2017. Airlines can tap the capital markets for syndicated unsecured debt or equity for the company itself. They can also issue syndicated senior secured debt that is backed by aircraft or other assets such as securitizations. In addition, they can issue EETCs which are tranche debt secured by aircraft or other aviation assets. EETCs are generally created tax reasons as they have higher ratings than the airline issuing the security, thus decreasing the cost of the borrowings. It also creates more security for the owner in the event of bankruptcy as it is secured by the equipment asset such as aircraft. These

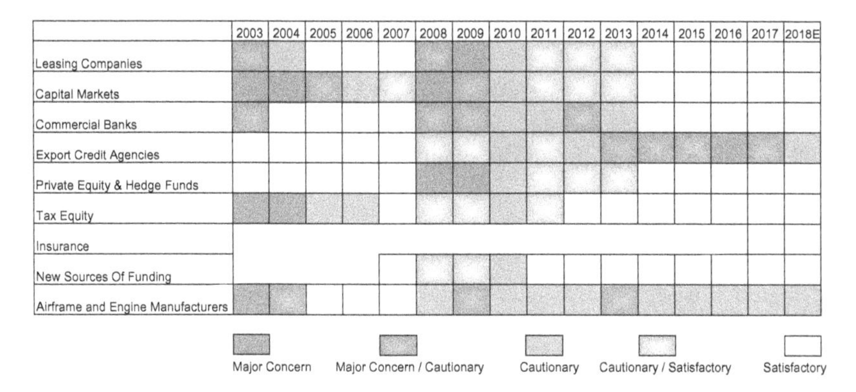

Fig. 3.24 Aircraft financing environment. (Source: Boeing Capital Corporation 2010, 2017)

were first developed in the early 1990s. The differences between an equipment trust certificate (ETC) and EETC is a liquidity facility attached to the EETC to support the credit rating and the tranched nature of an EETC similar to the securitization.

ABS is similar to EETC but is issued by a special purpose vehicle managed by a lessor or investor instead of airlines. The major difference between EETC and ABS is EETC is issued by one airline while ABS spreads the risk with many underlying airline credits. The structure was first brought to market in 1992 by GPA and Citicorp with ALPS 92-1 with 14 aircrafts valued at $521 million. ABS usually contains a spread of manufacturers, lessees, geographies, and narrowbody and widebody types of aircraft. GPA continued to use this structure to finance its aircraft until its demise through ALPS 94-1, Airplanes 96, and ALPS 96-1 (to refinance the original ALPS 92-1). The market was nascent until the relaunch of the structure by other lessors including Aercap, ACG, and Aircastle starting in 2005.

Both of these structures have tranches including, senior, various more junior tranches and equity which all have varying amounts and interest rates. This provides optimal matching of the desired credit exposure at the lowest cost. These structures also have retained outside servicers and consultant providers such as technical and appraisals to monitor the transactions. This might create an increased burden on the underlying airline lessees. These structures have been popular in USA especially the Chapter 11 bankruptcy regime, and especially Section 1110, overseeing the repossession of a bankruptcy lessee and possible replacement of the aircraft. Some of the other benefits of these capital markets transactions are the ability to move large amounts of assets off-balance sheet at lower borrowing costs, while retaining the management of the portfolio (still providing fees) or still retaining the equity tranches.

5.6 Commercial Banks: Financiers and Re-Entering as Investors

Commercial banks have always played a significant role in the aviation finance market. They have traditionally provided financing facilities, both secured and unsecured term or revolving credit facilities. Recently there have been more unsecured term loan facilities completed at the lessor company level where traditionally the commercial banks have focused on the senior secured financing of specific aircraft assets. In addition, more commercial banks have again reentered the equity investment in the industry and have acted as lessors and equity players.

One of the subtrends is in the change of the overall mix in terms of geography. In the late 1980s and early 1990s, banks such as Morgan Stanley not only arranged but also used their balance sheets to become equity investors. Dean Witter was one of the original founding shareholders of Aviation Capital Group (ACG) in 1986. The merged Morgan Stanley Dean Witter combined their aircraft portfolios into the acquired AWAS in 2000. The end of this era came when John Mack took over as chief executive officer and almost immediately sold AWAS to Terra Firma once it was determined to be a non-strategic asset.

Today in USA, Bank of America Merrill Lynch and CIT are still active as bank-owned lessors through their mainly Irish subsidiaries. CIT will soon be removed from this list when the expected closing of the sale to HNA Group is completed in the first quarter of 2017 because it too has been deemed a non-core asset. Wells Fargo too entered the space through a joint venture with Avolon in 2013.

In Europe, DVB, Santander, and Standard Chartered, through its acquisition of Pembroke in 2007, are still active as investors. It is interesting to note that Standard Chartered is now in a joint venture with an undisclosed Chinese investor for a separate aircraft leasing investment entity. The once active HSH Nordbank, through its formation of the Amentum platform, has since been sold to its management in an management buy-out (MBO).

This was because the bank has downsized as a result of the problems with its shipping portfolio.

The story of Boullioun Aviation Services, Inc. (Boullioun) is interesting because it was bought from Sumitomo Trust and Banking by Deutsche Bank in 1998 and subsequently sold to WestLB, another German bank, in 2001. WestLB, once quite active as an equity investor, also had a 35.5% shareholding in Singapore Aircraft Leasing Enterprise (SALE) until its sale to Bank of China in 2006 which later renamed it to BOC Aviation, when it decided these were noncore assets and refocused on its traditional European banking business. WestLB too had issues in its shipping portfolio and has further retrenched and this could be said about many of the European banks in the space. The shipping problem story will continue to be an impact, especially for European banks.

In Russia, VEB, Serbank, and VTB banking groups have also been active through their leasing subsidiaries. Lately, they have encountered difficulties with sanctions and currency issues. All of the above has happened as the East and the Middle East have seen large increases in activity in the sector. Japan originally had a lot of interest in aircraft leasing—for example, with Sumitomo Trust and Banking Company's acquisition of Boullioun in 1994 from the lessor's namesake founder and its subsequent sale to Deutsche Bank in 1998 as a result of the financial crisis of Japan Inc. This interest in aviation was resurrected post-2010 and was highlighted by Sumitomo Mitsui Banking Corporation (SMBC)'s acquisition of RBS Aviation Capital in 2012, among other merger and acquisition transactions by other local parties. Australia's Macquarie, Commonwealth Bank of Australia, and Investec have all been active principal investors. Middle Eastern banks have now joined in the mix including the National Bank of Abu Dhabi.

Another driver rationale for this trend is the increased implementation of higher reserve capital requirements on global banks by Basel III regulations enacted by the global financial crisis and set for implementation shortly. These proposed amendments to the final Basel III even before its implementation, unofficially Basel IV, has even more stringent requirements that standardize risk models and do away with internal risk ratings that have been discussed in depth previously. ECAs, on the other hand, have a cover effect on this standardized higher risk rating and lower the costs. These areas requiring large capital requirements such as aircraft leasing and private equity may propel banks to reexamine and restructure further or leave these investments.

In addition, the industry is seeing more insurance companies come into the space. Insurance companies with their large investment mandates have traditionally invested in public equities, capital markets, and alternative investments such as hedge funds. Through these asset classes, insurance companies have had exposure to the aircraft leasing companies through one or multiple streams. Some insurance companies have direct investments in aircraft leasing assets, such as ILFC when it was acquired by AIG in 1990. Subsequently, AIG sold its ILFC subsidiary after the global financial crisis to AerCap and took a large shareholding in the new combined entity in 2014.

Other large notable direct investments by insurance firms include Pacific Life Insurance Company in ACG in 1996. The group has continued to expand, including acquiring Boullioun in 2005 and embarking on a new joint venture with NWS Holdings Limited (NWS) in 2016. Other mid-sized insurance companies in Europe have also started to invest directly into aircraft leasing assets. Generally, these firms have invested in similar profiled investments such as infrastructure or real assets through ABS, EETC or other public and non-public equity and debt.

5.7 *New Aircraft Operating Leases and Aircraft Secondary Trading*

Another finance composition driver is the growing amount of operating leases as the percentage of all global Airbus, Boeing, and McDonnell Douglas aircraft has grown from 22% in 1997 to 43% as shown in Fig. 3.25 (Ascend 2017).

Current global operating leases are around 50%. New operating leases and the secondary trading levels of leased aircraft are strong drivers contributing to the financing. All of these are drivers that affect the number of aircraft produced and the sustainability of aircraft in use by the airlines. New operating leases increased in 2017 for the Airbus A320-200 and Boeing 737-800, by 32% and 20% respectively after a flat growth year in 2016 as shown in Fig. 3.26.

In the circa 150-seat segment, the A319-100 is outperforming the Boeing 737-700 in terms of new operating leases while new operating leases of 737-700s have declined in each of the last two years. The most popular narrowbodies—Airbus A320-200 and Boeing 737-800 models—show increased demand year on year. While there is still demand for Boeing 737 classic aircraft, they now represent a smaller share of the

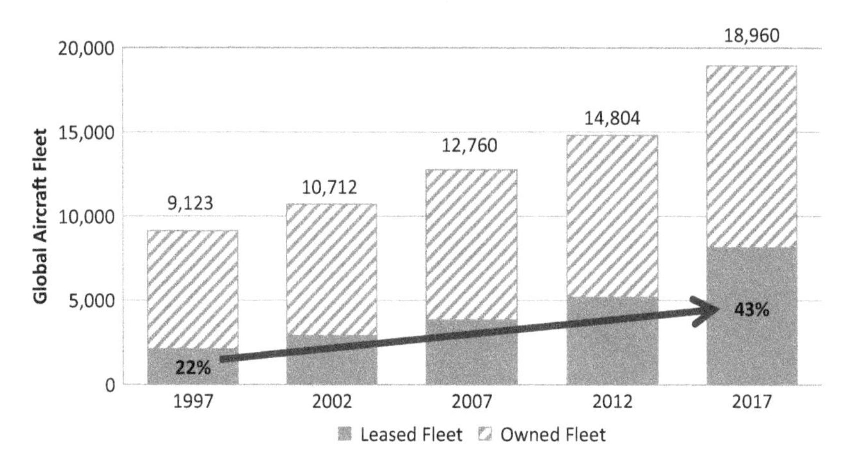

Fig. 3.25 Growth of global operating lease market share. (Source: Ascend 2017)

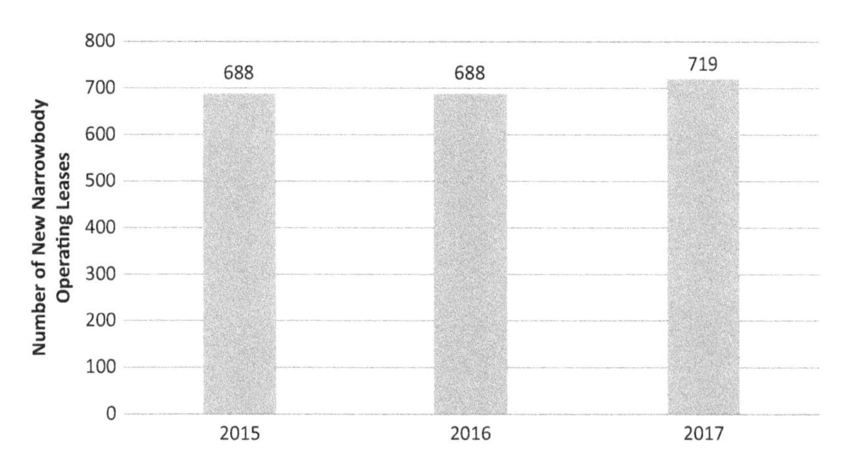

Fig. 3.26 New passenger narrowbody operating leases: 2015–2017. (Source: IBA Group 2017)

leasing market. A321 operating lease demand peaked in 2016 but showed a 19% year-on-year decrease in 2017 as shown in Fig. 3.27.

For new operating leases of widebodies, the market has been different than the narrowbodies with a growth year in 2016 and then a down year in 2017 as shown in Fig. 3.28.

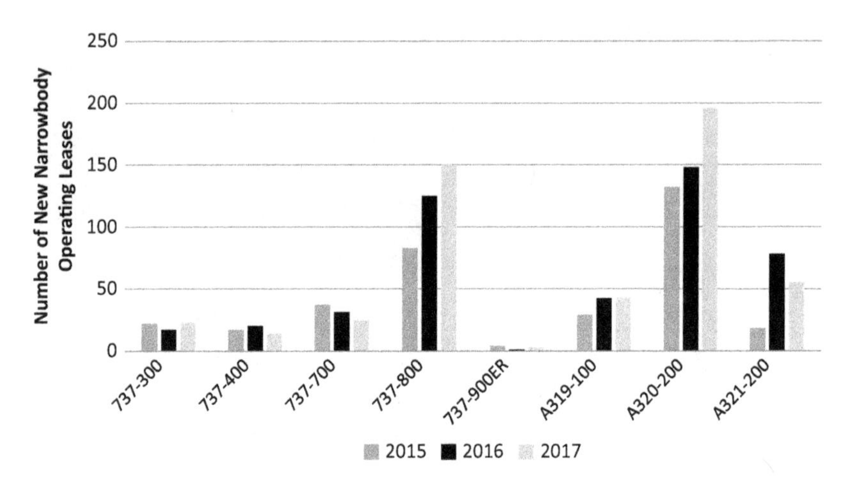

Fig. 3.27 New passenger narrowbody operating leases by type: 2015–2017. (Source: IBA Group 2017)

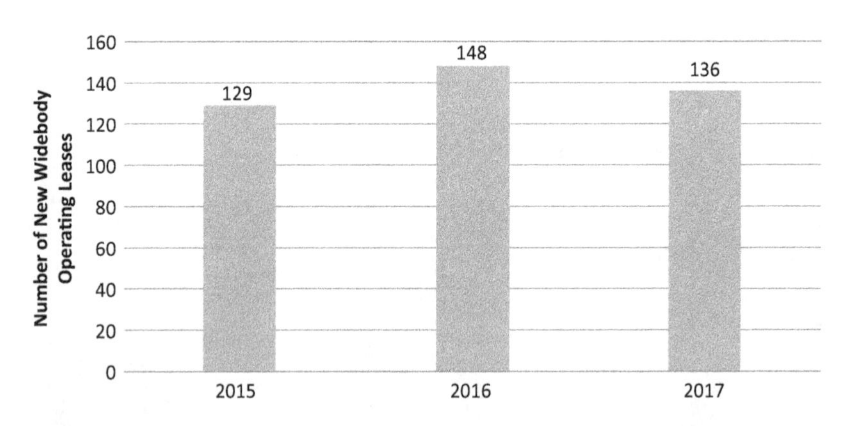

Fig. 3.28 New passenger widebody operating leases: 2015–2017. (Source: IBA Group 2017)

The Airbus A330-300 has seen an increase in the number of new leases. New Airbus A350-900 leases have increased as a result of increased deliveries and the aircraft is becoming more established in the market. Boeing 767-300ER passenger new leases are down. There were no new leases for the Boeing 787-8 in 2017; the market is clearly favoring the Boeing 787-9

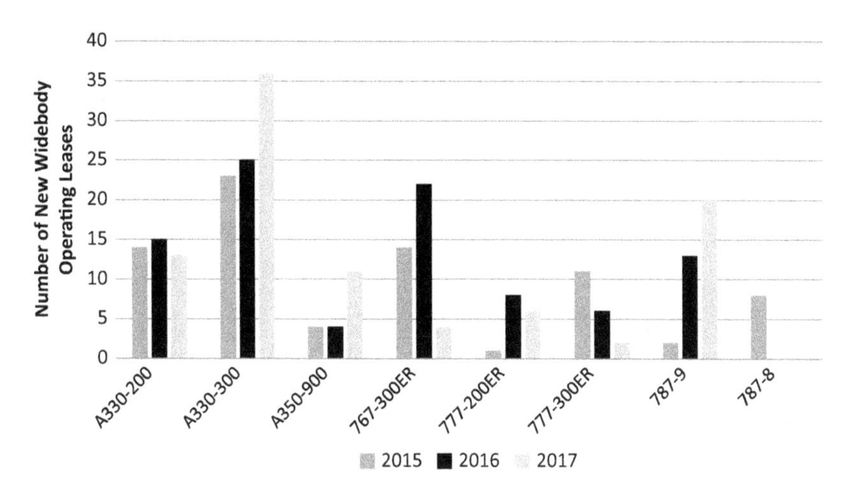

Fig. 3.29 New passenger widebody operating leases by type 2015–2017. (Source: IBA Group 2017)

and operating lease demand continues to grow for this type which is attractive for widebody lessors. There were also no new operating leases for the A380. All leased A380s so far have been completed through sale-leaseback transactions (Fig. 3.29).

Secondary trading volumes include aircraft sales with or without leases attached and sale-leaseback transactions. Overall, secondary trading volumes for narrowbodies have increased slightly in 2017 compared to 2016 as shown in Fig. 3.30.

Airbus A319-100 and Boeing 737-700 types have shown increased trading levels since 2015 as investors diversify from the more popular, and often highly priced, A320-200 and B737-800 models. Trades involving Airbus A320-200 and A321-200 aircraft declined in 2017 compared to 2016. The 2015 A320 trade total was bolstered by AWAS' sale of a subset of its portfolio which included 50 A320 aircraft. Boeing 737 classic aircraft show the vast majority of trades (88%) involving off-lease aircraft as shown in Fig. 3.31.

Some will be converted to freighters, however many face retirement. All of these observations exclude the large Avolon/CIT and DAE/AWAS M&A deals which distorts the market data.

Secondary trading volumes for widebodies have remained flat over the past two years as shown in Fig. 3.32. Although there are some differences

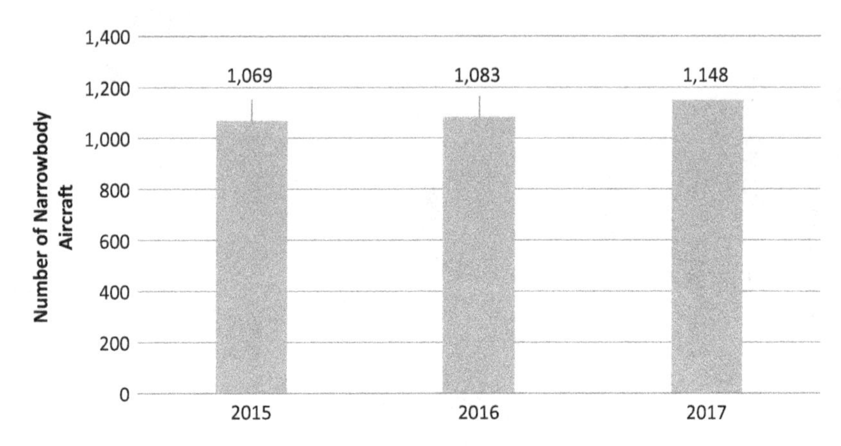

Fig. 3.30 Passenger narrowbody secondary trading volume: 2015–2017. (Source: IBA Group 2017)

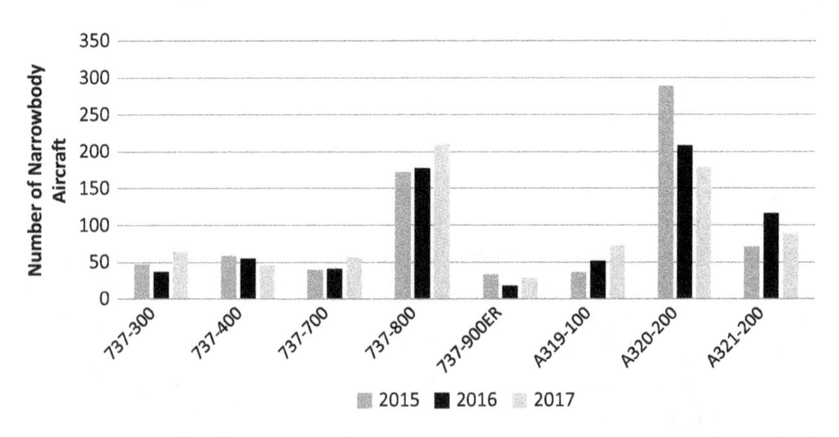

Fig. 3.31 Passenger narrowbody secondary trading volumes by type 2015–2017. (Source: IBA Group 2017)

based on the various aircraft models with the most popular type to trade is the 767-300ER as shown in Fig. 3.33.

Boeing 787-9 aircraft is growing in popularity with lessors and investors with sale-leaseback transactions driving increased secondary trades. The 787-9 now accounts for the majority of current 787 production. Secondary trades of the smaller 787-8 have declined as a result.

Fig. 3.32 Passenger widebody secondary trading volumes 2015–2017. (Source: IBA Group 2017)

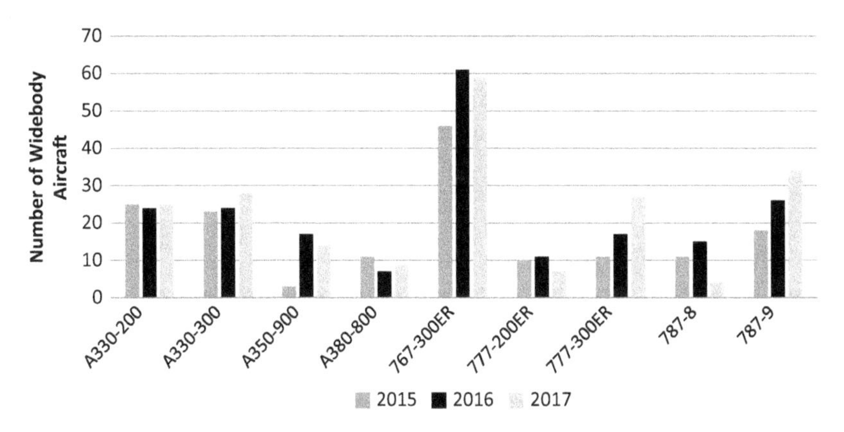

Fig. 3.33 Passenger widebody secondary trading volumes by type: 2015–2017. (Source: IBA Group 2017)

Sale-leaseback demand for the A380 remains strong for new and nearly new aircraft. Boeing 767-300ER aircraft continue to trade readily in the secondary market along with strong demand for freighter conversions. Boeing 777-200ER transactions have declined since 2016 with increased storage levels in 2017 compared to 2016 (Fig. 3.33). All of these observations exclude the large Avolon/CIT and DAE/AWAS M&A deals which distorts the market data.

6 Sidecars and Joint Ventures

There has been large growth recently in the number and the sizes of "sidecars" or joint ventures (JVs) in the aviation investment space. While these structures are not novel and have been in existence for a long time, it has now become a structure that is in vogue and has industry insiders and observers taking note. A sidecar or JV is an agreement to create a separate entity where the parties can coinvest with contributions that include cash, assets, other knowhow, or a combination of these. These JV entities generally run in parallel with the existing business scope of one of the partners, hence the sidecar analogy. This is very different from JVs with airlines on aircraft that are operated by the airline partner which has been around for some time.

Currently, there is a large supply of new capital with little or no experience entering the aircraft leasing industry looking for experienced personnel or partners to do deals with. This has created an environment where lease rate factors and returns have steadily been compressed over the last few years. Given this dynamic, established lessors rightly as risk managers have been happy to raise additional capital through sidecars to pursue more opportunities on a hedged basis. Sidecars enable established lessors to monetize their experience by putting down little equity while retaining benefits such as upside on the sale of the leased aircraft, having more stable cash flows, and operating leverage of their staff through servicing arrangements.

6.1 Parties Involved in Sidecars and Joint Ventures

In the case of the aircraft leasing industry, sidecars are generally formed with an established lessor and a financial party. While the terms of each agreement differ for each JV, for the most part, the operating lessor provides a combination of asset management, technical advice, and their network of established relationships to source and finance the deals while the financial party contributes equity, debt, or other means to fund the JV and the transactions. The financial party generally provides the majority of the equity capital while the operating party is a minority investor (generally 10–40% but more skewed to the smaller side).

The financial party tends to be investors such as investment funds, companies including pension, credit, hedge funds, private equity and family offices. Other types include large conglomerates with aviation interest, other financial institutions, and trading houses. By and large, these investors are risk averse and focus predominantly on more modern types and younger aged aircraft.

Usually, the financial party forms one JV while others take a more diversified approach with many sidecars with different established lessors. The latter includes Chow Tai Fook Enterprises Limited and NWS Holdings, which invested along with Investec in Goshawk Aviation and more recently both investing with Aviation Capital Group through Bauhinia Aviation Capital Limited. In the first case, Investec sold its entire 20% stake in Goshawk to its two partners. Sometimes the financial party is also a lessor in its own right such as Tokyo Century Leasing (TCL). They had a JV in aircraft leasing, TC-CIT, formed with CIT (TCL subsequently bought out CIT along with the sale of CIT to Avolon) and another JV focused on engine leasing with GA Telesis. Financial investors sometimes also have minority ownership stakes in the lessor partner such as the case with TCL's 20% equity ownership in GA Telesis and IBJ Leasing's association with Aircastle. IBJ Leasing and Marubeni are both members of the Mizuho keiretsu and through Marubeni own 15.25% stake in Aircastle.

6.2 Benefits Of Sidecars and Joint Ventures

In theory, sidecars sound simple and have benefits for both parties. They enable the established lessor to monetize their know-how and buy and manage assets that they might not have done themselves. It also provides the lessor a platform to become more of a servicer than an asset owner which is more stable. Owners, as the equity holders of assets, inherently have more risk and volatility on their returns due to unforeseeable events that might happen during the course of ownership and the lease. For example, at its most severe, this unforeseeable event can be an airline bankruptcy. During these periods, negative expected cash flow occurs due to additional investment capital required due to the stoppage of lease revenues from the operator, or other scenarios. Instead of these more volatile cash flows, in sidecars structures, established lessors have more certainty because cash flows are derived from servicing agreements with the financial investor. These sidecars also enable the established lessor to become more asset-light vehicles. This is one of the main considerations for

GECAS who have formed a sidecar, Einn Volant Aircraft Leasing, with Caisse de dépôt et placement du Québec. The new senior management of GE is more focused on an asset-light model and instead prefer more stable servicing model.

In addition, established lessors retain certain upside for transactions usually based on some hurdle rates. These hurdles might include a preferred initial return to the financial investor partner and then some subsequent split of profits afterward. These profit percentages are generally not the same as the equity ownership percentages. The cash flow waterfall sometimes includes other variations such as catchups where after the preferred return, the lessor gets all the cash flow until they arrive at the predetermined profit split percentages.

For the financial player, this structure provides for know-how from the operating partner that they cannot replicate themselves. It can take advantage of the operating partner's geographical presence and more seasoned personnel. It is arguable whether it is faster in time to form a sidecar versus starting a new lessor company from scratch with a sole party. While most sidecars takes less than a year to establish, a new lessor can be established, when funding is available, within three to six months. While there are many benefits, there are many pitfalls that need to be considered.

6.3 *Controls and Restrictions*

There are restrictions on one or both parties which protect the JV from competition with the parent companies. In some instances, the operating party provides other types of comfort such as guarantees including debt or hurdle rates for the JV entity and partner. As with any partnership, finding these mechanisms for the alignment of interests is very important. There will be natural conflicts such as the desire for a continued stable management fee versus finding the optimal time to divest an asset which creates an upside for the manager. Sometimes, this is not addressed fully just through the construction of the investment committee or board of directors of the JV. In most cases, the financial investors control the JV entity as they are the majority investor.

6.4 *Issues and Conflict Management*

In addition, thought needs to be put into on how to properly value the contribution of new or older aircraft. Is it contributed at cost or some sort of current market value especially as the initial batch of aircraft in a JV is generally bought from the established lessor partner's portfolio? Third party appraisers are often brought in to solve this question. As with any venture, the mechanisms of the management are important. Will the JV be staffed with full-time seconded management from the lessor or will it have its time split with the continued responsibilities of the parent lessor? Other conflicts that warrant consideration such as cherry-picking the best assets as well as conflicts over lease initiation, renewal, or sales, especially if the established lessor parent has existing aircraft at the same carrier. Outsourcing is very much part of the aircraft leasing landscape as there are opportunities for every single facet of the business. While most parties tend to retain only the most important services or where they are best suited along with outsourcing, others tend to do all the services in house. The logical follow-up question comes up as to whether the operating partner is the most suited in terms of costs, delivery timing, and quality versus outsourced options.

6.5 *Exits and Trends*

As with any joint venture, the management of the dynamics between two parties generally lends itself to the eventual exit between the two parties. Over time, partners change strategic directions. Some exits include selling the entire company to one of the partners, an initial public offering, through an asset-backed securities structure, or trade sale to another party either individual aircraft or wholesale. A prime example is Waha Capital's sidecar with AerCap in AerVenture. This stake along with all of Waha's aviation interests and $105 million cash were sold for a 20% stake in AerCap. There are numerous examples of these sidecar structures with almost every major aircraft leasing player including AerCap, Airlease Corp, Avolon, and others. The continued growth for more sidecars is analogous to that of other asset-heavy industries such as hotels, real estate, shipping among others which have gone towards a more asset-light service model with the distinct separation of the owner and the manager.

7 Airline Business Model Changes and Foreign Exchange

7.1 *Airline Business Model Changes*

Airline business models and the changes to it are another driver of the industry. Historically, most airlines were full service operators. Recently there has been a major shift away from full service to low cost in the business model. A low cost or LCC model is where the price for the customer is partitioned so that the airline can offer low cost for the consumer. Because of the desire of low cost, airlines have historically been based in non-main stream airports but this has slowly been changing. The consequences of these changes have many effects on the aircraft itself. Many low-cost carriers operate a single type of aircraft, originally a narrowbody such as a 737 or A320 and operate aircraft insignificant in numbers globally. These aircraft are generally higher density configuration and have low amenities, such as in-flight entertainment systems if any. Higher utilization and wear is another consequence as the goal is for low turnaround time and more flying time of the aircraft per day to maximize revenues. Low-cost carriers have slowly started to both diversify and expand their product offerings. Some have focused on long haul, while others have started to add premium seats with many of the same benefits as full service operators.

In many ways, the industry has gone through multiple models of operation over history. Originally the main business model is where the owner of the aircraft and the operator is the same, who commonly acquires aircraft by paying cash, raise debt, or a combination of the two. The business model has slowly evolved such that the roles of an owner and leasing company and the operator company, the airline, are split. This is very much like the evolution of the hotel business where one firm is the investor and owner that has a core focus on maximizing its investments while another company, the operator, specializes in maximizing the profits of operating such assets. The major difference as compared with hotels and other real estate is that the asset is movable and it is easier to repossess aircraft than hotels especially under distressed scenarios. Through this current model, both types of businesses can totally focus on what they know best in their core business and competencies.

While there are a lot of complexities surrounding this including business cycles, tax, structures, financing elements, at the very root of the fundamentals of the industry, aircraft leasing can be boiled down to how good the company is at managing the value and residual value of the aircraft. There is a wide range of business models which have different targeted end customers producing different returns. These business models can utilize aircraft differently which allows the more astute lessor players to select lessees more wisely.

7.2 *Foreign Exchange*

Another driver is the foreign exchange exposure. This is both a demand and supply side driver. On the demand side, the consumer-derived decisions and revenues are in the currency of the international country which can be different than the home currency. For the supply side, the cost of many inputs such as the cost of the aircraft (purchase or lease costs), fuel, maintenance, as well as some financings (export financing) is in US dollars while other costs are in home or other currencies. These changes in foreign exchange can drive corporate decision making for the current and long term which all flows back to the financial accounts. These are currency differences that are generally expressed in the annual reports through currency change gains or losses and can be quite significant and would depend on the scope of the airline's operations and corporate strategy. Since 2000, against the USD, Russian ruble and Brazil real have changed the most while the EUR and other currencies have fluctuated but not as much as shown in Fig. 3.34.

8 Current and Future Market Outlook

Boeing notes as shown in Fig. 3.35 from the 2016 Current Market Outlook, the fleet size at the end of 2015 consists of 22,510 passenger and freighter aircrafts in service globally. Fig. 3.35 also shows that at the end of 2035, there are 45,240 projected aircrafts of which 39,620 are new deliveries during the 20-year span. Fig. 3.36 shows that 71% of these 39,620 new deliveries are narrowbodies and 51% of the total list price of the total delivered aircraft. Fig. 3.37 shows in Asia, Boeing predicts that there are 15,130 narrowbody passenger aircrafts (38%) delivered by end of the 20-year projection in 2035 and there will be 40% by list value of all new deliveries (Boeing Corporation 2016).

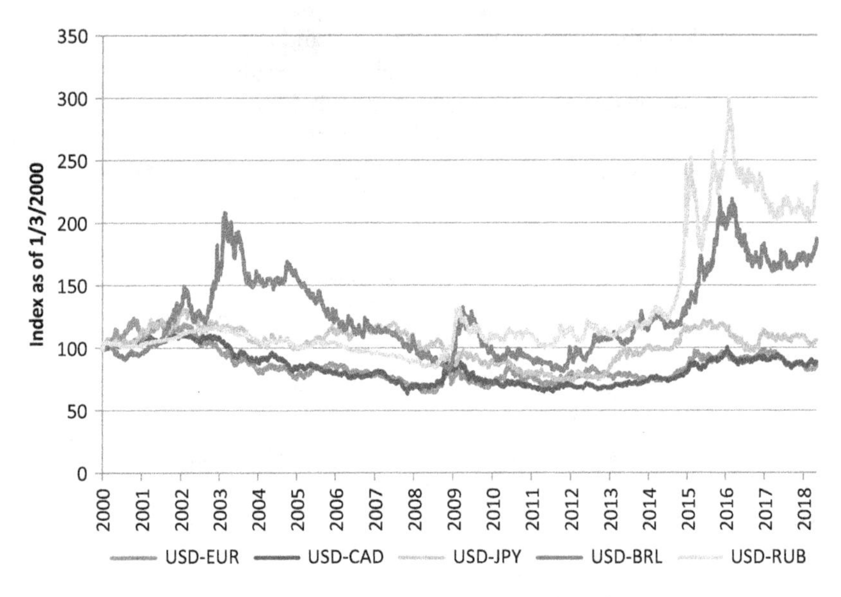

Fig. 3.34 USD versus various currencies. (Source: Bloomberg 2017)

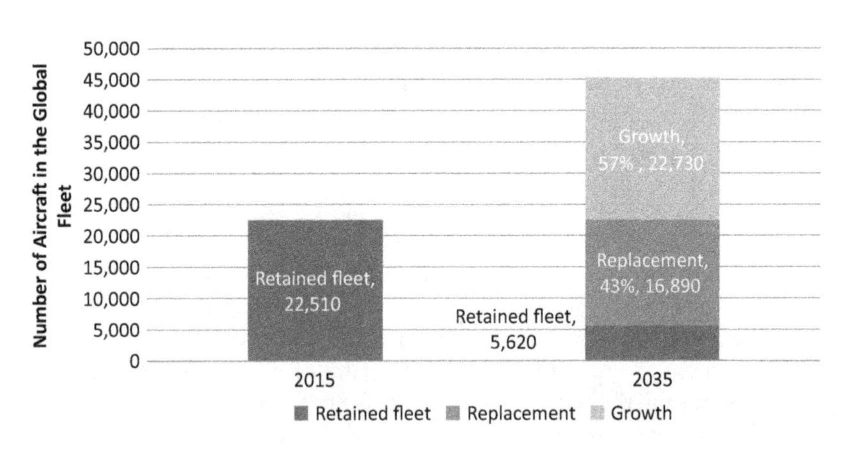

Fig. 3.35 Boeing global fleet: 2015–2035. (Source: Boeing Corporation 2016)

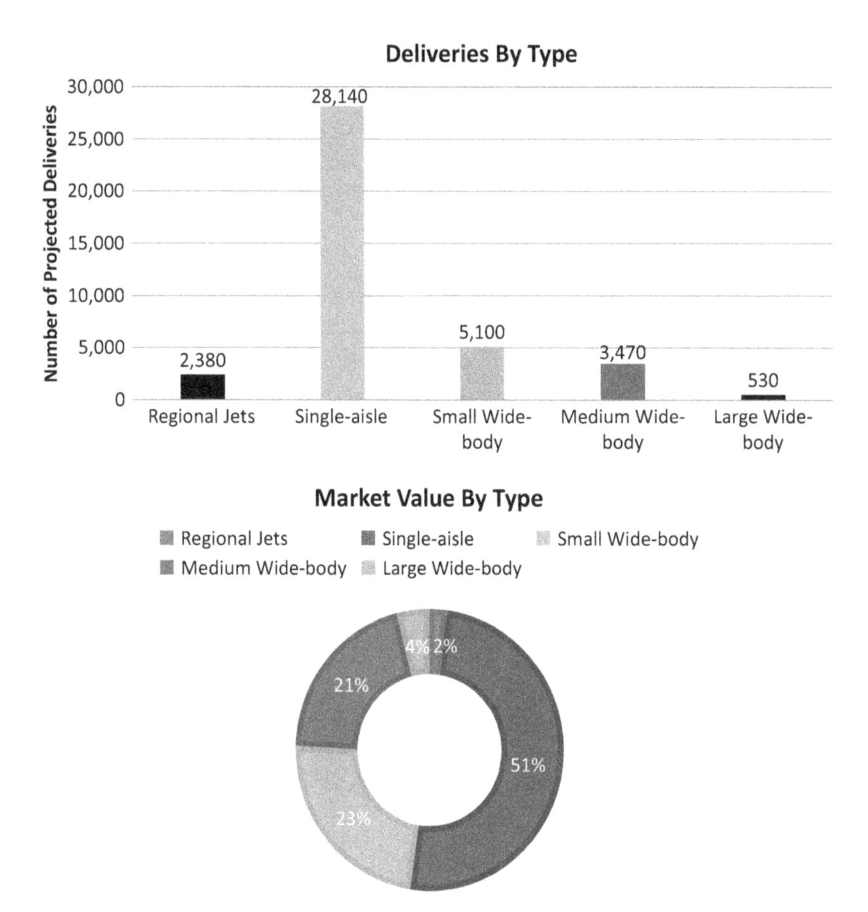

Fig. 3.36 Boeing projected deliveries and market value by type: 2016–2035. (Source: Boeing Corporation 2016)

Airbus notes as of the 2016 Global Market Outlook, the fleet size in 2015 is 18,019 passenger aircrafts and 1564 freighter aircrafts in service globally for a total of 19,580 aircrafts.

Figure 3.38 shows that Airbus predicts that there would be 39,819 aircrafts by the end of their 20-year projection in 2035 of which 33,074 would be new deliveries. In Asia, as of the end of 2015, there are 5659 passenger aircrafts (31%) and 302 freighter aircrafts (19%) resulting in a combined total of 5961 aircrafts in total (30%). Airbus forecasts that there

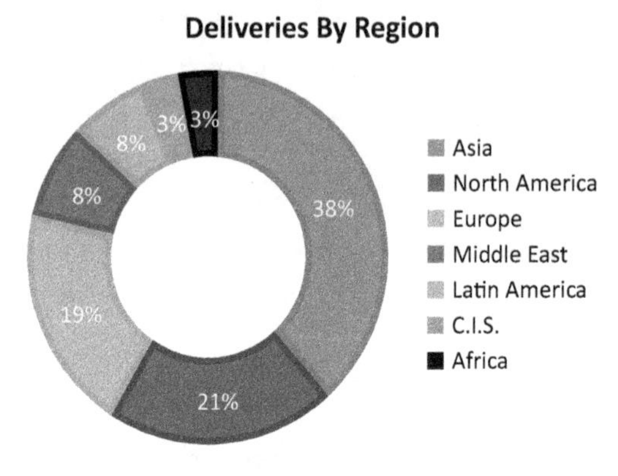

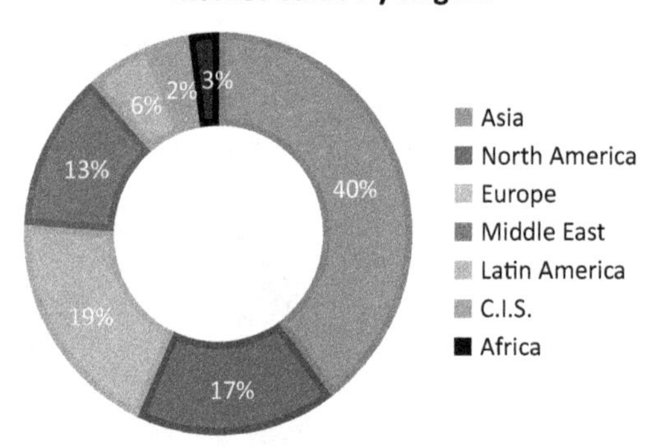

Fig. 3.37 Boeing projected deliveries and market value by region: 2016–2035. (Source: Boeing Corporation 2016)

will be 14,685 passenger aircrafts (39%) by end of the 20-year projection in 2035 during which there are 13,239 (41% total new deliveries) new aircraft deliveries. For freighter aircraft, Airbus predicts that there are 778 freighter aircrafts (37%) by end of their 20-year projection in 2035 during which there are 219 (34% total new deliveries) new deliveries. In total, Airbus forecasts that there are 15,463 aircrafts (39%) by end of their 20-year projection in 2035 during which there are 13,458 (41% total new deliveries) new deliveries (Airbus 2016).

In both Airbus and Boeing points of view, the current aircraft in use and traffic both globally and in Asia is quite significant as shown in Figs. 3.39 and 3.40. Both sources have projections within China and Asia as the top growth areas for traffic which are reflective of the demographic changes in the world. The differences between the two major

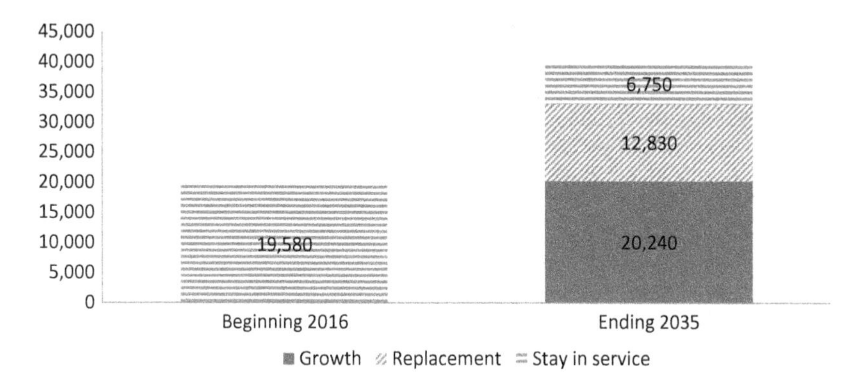

Fig. 3.38 Airbus global fleet projections: 2016–2035. (Source: Airbus 2016)

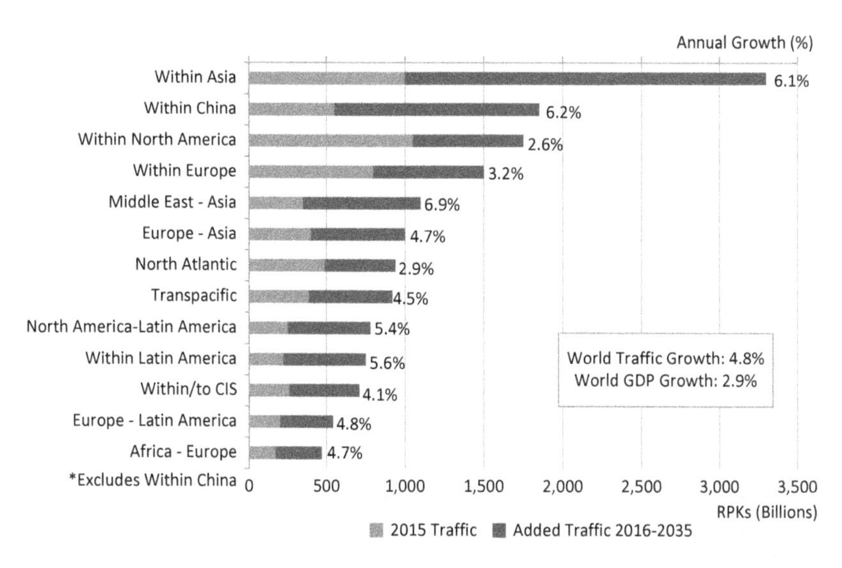

Fig. 3.39 Boeing projected traffic by region: 2015–2035. (Source: Boeing Corporation 2016)

manufacturers reflect each OEM's view on the future of airline traffic with Boeing more focused on point-to-point traffic growth while Airbus is more focused on the more classical hub and spoke model. These thoughts reflect the continued development and sales of new aircraft currently in production.

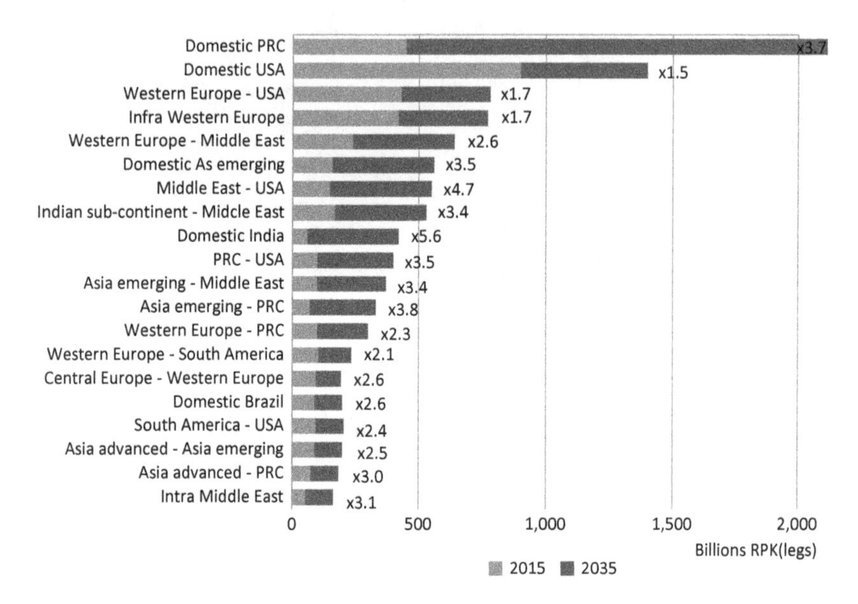

Fig. 3.40 Airbus projected annual traffic per leg flow by region: 2015–2035 (billion RPK). (Source: Airbus 2016)

9 Participants Characteristics

The international aircraft industry is made up of different participants including investors, lessors, airlines, traders, or a mix of multiple categories. The backgrounds and investors, promoters of the companies include high-net-worth individuals and families, public and private corporate, financial institutions, insurance, institutional investment companies such as private equity and hedge fund firms. The diverse variety of participants and geographies represented in the industry is driven by the multiple options that are available.

Not surprisingly, these are evidenced by the top 50 leasing companies which comprise of aggregate $260 billion of aircraft value including both owned and managed aircraft with the top ten leasing companies globally comprising of $158 billion or 61% of the top aircraft investors. Of the top ten, four are controlled by China-based investors representing $42 billion or 26% of the top ten and 16% of the overall respectively. Of the top 50 aircraft lessors, 11 Chinese lessors have $61 billion, representing 23% of the global top 50 lessors and 23% of the top 50 lessors, respectively.

Including Japanese and other Asian investors, five are in the top ten Asian investors which increased to $57 billion, 36% of top ten and 22% of top 50. Overall in top 50, 20 Asian investors represent $97 billion or 37% of the top 50 lessors' total value. Interestingly, public companies represent five of the top ten investors representing $105 billion or 66% of the top ten and 40% of the top 50 respectively. When looking at top 50, 11 are public companies representing $120 billion representing 46% of top 50 lessors' aircraft assets. Interestingly the other big regions represented are seven Western Europe and three Middle Eastern investors representing $51 billion, 20% of top 50 and $7 billion 3% of top 50 respectively. These current diverse representations shows the nature of the global aviation market along with the global economy as the historical American and European dominance starting to fade as entrants from new emerging economies such as Russia, Middle East, China, and other parts of Asia have started to have a greater representation in the top 50 list (Airfinance Journal 2016).

10 Characteristics and Review of M&A and Cross Border Trends of the Aircraft Leasing Industry

In recent years, the number and volume of cross-border investments has increased significantly across all industries and especially in aviation. While there have always been cross-border investments between countries, such deals are now more prominent in the news given their increasing size and frequency, as well as the profile of the targets.

One of the major themes has been the explosive growth of Chinese outbound deals over the past five years which saw outbound mergers and acquisitions (M&A) volume rise 33% per annum from $49 billion to $227 billion in 2016 according to McKinsey. While this growth has been more publicized, some other important facts might not be as acute. While Chinese companies were involved in ten of the largest deals worldwide in 2016, most deals were in the middle market with the median being approximately $30 million deal size. While the absolute volume has increased, there is arguably further room for growth as this volume as a percent of GDP is smaller for Chinese companies (0.9%) than its counterparts in Europe (>2.0%) and USA (1.3) in 2015.

While this trend has been the case up until this year, the latest figures as year-to-date third quarter 2017 show that the number of Chinese company M&A deals has decreased 14.8% to 572 compared the previous year while the overall value has decreased 38.9% to $97.7 billion according to PWC. These figures represent a decrease in the share of overall investment in Europe and USA from 77% for calendar year 2016 to 50% for the latest figures. It is noted that this decrease follows an almost 300% increase in the value of transactions to $214.9 billion for 2016.

Overall, according to law firm Baker and McKenzie, there have been 1320 cross-border deals in the second quarter of 2016, worth $214 billion. The breakdown is 798 deals were concluded, worth $137 billion across two different geographic regions versus 522 deals worth $77 billion within one geographic region. During this latest Q2 2016, a sizeable portion included Chinese acquirers, who completed 97 transactions worth $40.7 billion, compared to the previous year of $17.5 billion or 132% growth.

In a review of the recent activity in the aviation industry, most of the deals in 1H 2016 were in the aerospace sector and there have been a significant number of aircraft and airline acquisitions in terms of volume (~$3.5 billion) and number (~15) according to MergerMarket/ICF as shown in Fig. 3.41. In terms of cross-border investments in the aircraft space not originating from Asia, Nordic Aviation Capital's announced acquisition of Aldus Aviation and Jetscape, along with its purchase of 25 E-Jets from Air Lease, confirms the company's cross-border roll up acquisition[1] strategy in 40he regional aircraft space.

In the airline sector, there have been multiple deals by Middle Eastern carriers such as Qatar Airways, which announced the acquisition of a 49% stake in Italy's Meridiana, completed purchase of a 10% stake in LATAM, and purchased additional shares in IAG, bringing its holding to 20%. This trend is similar to Etihad's "equity alliance" strategy of investing in stakes in foreign operated airlines. Etihad is expected to increase its shareholdings in Jet Airways from 24% to 49% (the maximum amount of foreign investment allowed in Indian airlines) due to the relaxation of regulations. During the first half of 2016, CEFC China Energy Company exercised its option for additional 39.92% shares of Czech Travel Service Airlines to bring its stake to 49.92% after the initial investment in 2015. In addition, Nanshan Group and HNA Airlines separately invested

[1] Roll up is a specific M&A strategy that means companies acquire competing businesses with the goal growing bigger and also decrease competition and in this particular case comes with a cross-border point of view.

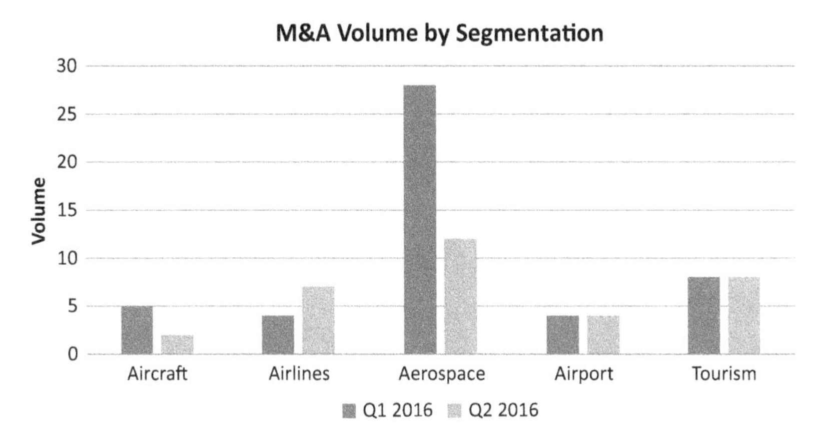

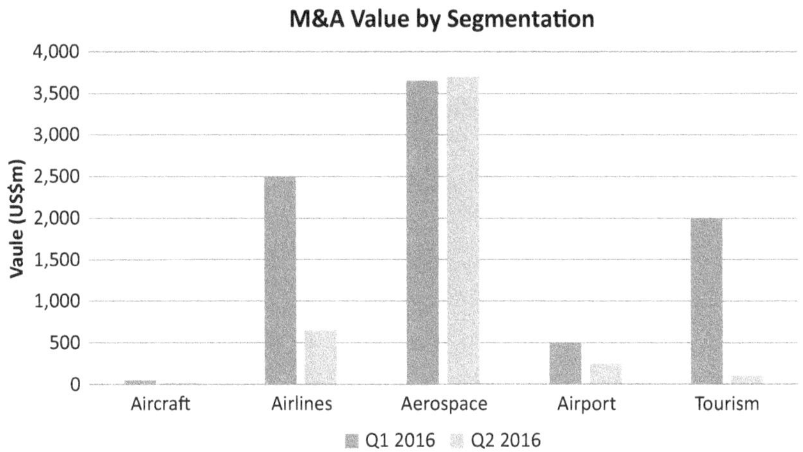

Fig. 3.41 M&A volume and value by segmentation. (Source: MergerMarket and ICF 2016)

$198 million for a 20% stake and $114 million for a 13% stake, respectively, in Virgin Australia. This continues the acquisition expansion of both groups in the aviation sector domestically and abroad. In Southeast Asia, Thailand's King Power Group bought 39.83% of Asia Aviation for $225 million.

In light of all the cross-border investment activity, there has been a growing and significant number of global outbound investments from Chinese investors. In Figs. 3.42 and 3.43, this can be partially attributed

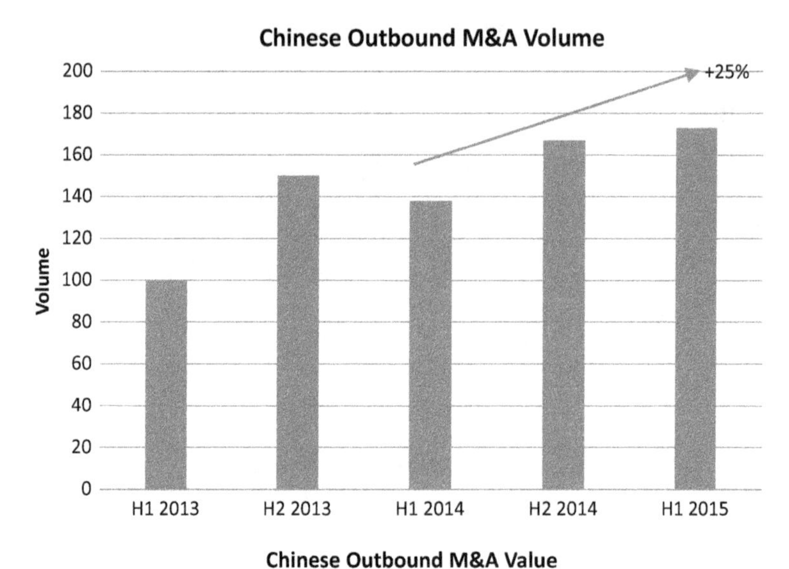

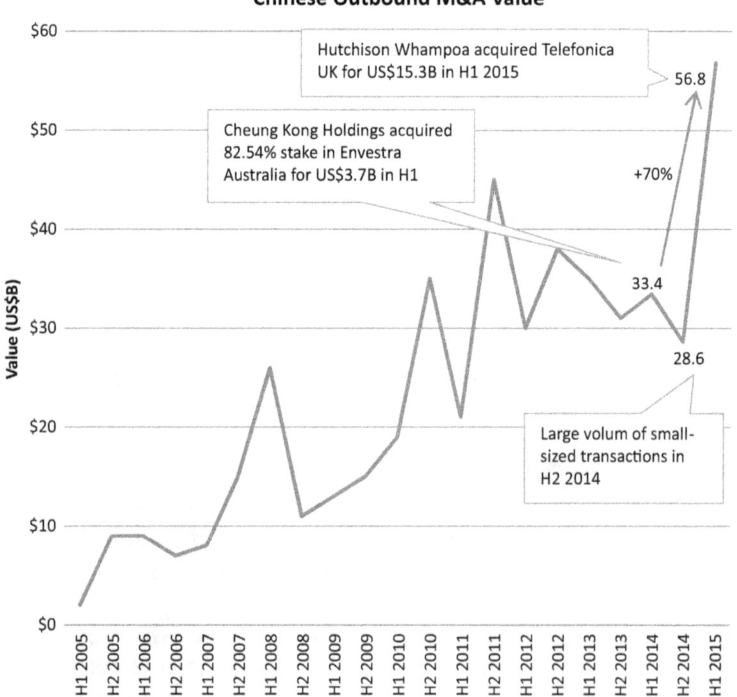

Fig. 3.42 China outbound M&A volume and value trends. (Source: Deloitte 2015)

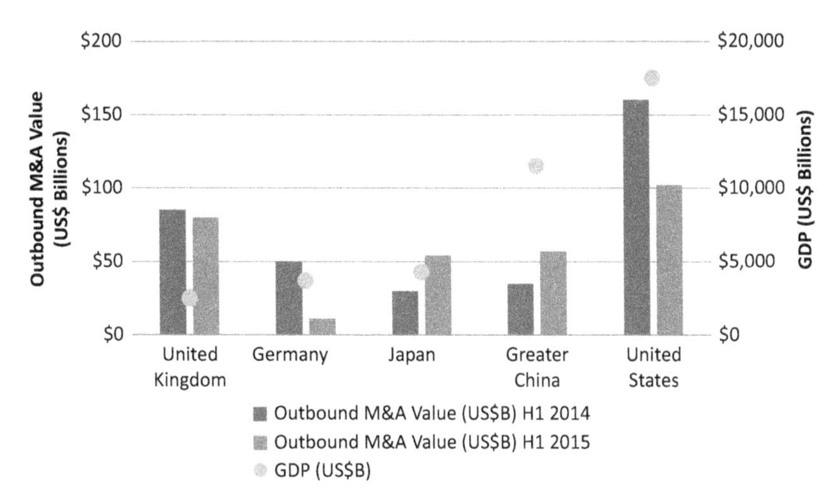

Fig. 3.43 China outbound M&A volume and value by region 1H 2014 and 2015. (Source: Deloitte 2015)

to the global economy and is highlighted by the fact this is a reversal of inbound investment flows that have occurred in the past few decades.

What are some of the drivers of the increased outbound investments from China? Some of the volume is reflective of the slowdown of the domestic market. While China's GDP has slowed from strong double digits for the past decade and more to a current target of 6.5–7% per annum growth, this slowdown has contributed to further desire for higher growth outside of China. Some points to note are that while some traditional industries are under pressure, the tourism, aviation, and transportation sectors are still attractive as the overall demographics of the population continue to improve and people have become wealthier with more disposable income to spend on discretionary travel. This has, in turn, increased interest in travel and related services especially abroad which has precipitated the increase for international acquisitions due to the investment strategy of "following the customer." In addition, there is a trend for increased vertical integration among Chinese companies, many of whom are conglomerates.

In addition, with low yields in traditional banks, much of the capital is moving towards higher yielding products. This demand is not being fully met by the slower domestic economy and has also driven investments internationally for higher growth and yielding deals as there are less and less attractive investment opportunities domestically. The continuing controlled depreciation of the RMB by the PBOC over the past few years compared to other major economies has reversed the one-way trade of appreciation since the relaxing of the official peg with the USD. Aircraft and other real assets denominated in USD or other major currencies are prime examples of this flow. This has also affected airline companies as costs and revenues are mismatched and no hedging is allowed by Chinese airlines. While there is a continued push by the authorities to restrain this outflow effect by restricting foreign exchange conversion per person per trip, shutting down grey channels, and so on, this trend continues. Another ramification of depreciating currency is more deals are being denominated in local RMB versus the US dollar.

Also, the slowing growth and the push for increased efficiencies have also driven the encouragement of consolidation and the creation of "national champion" companies to look for more expertise both domestically and internationally. One recent example is the state-owned enterprise (SOE), Aviation Industry Corporation of China (AVIC), which finalized the consolidation merger of all of its aircraft engine businesses worth RMB 129 billion ($19.71 billion) as part of the overhaul of China's SOEs. Other national champion businesses are also being formed in other strategic industries such as $21.9 billion shipping merger of COSCO and China Shipping Group. One such recent SOE news is that Air China is rumored to be behind an investment interest for 49% of LOT Polish Airlines. More activity from this segment should come in the future.

It is also important to note that the Chinese state can act as an investor, financier, or both in certain situations. Development banks, such as China Development Bank, China Export-Import Bank, and Agricultural Development Bank of China, have been the main drivers of policy-related lending. Also, SOE banks such as the big four—ICBC, China Construction Bank (CCB), Agricultural Bank of China (ABC), and BOC—have also contributed to this expansion overseas especially through their worldwide branch system in addition to provincial and quasi non-governmental organizations. Funding also comes from various sovereign funds including China Investment Corporation (CIC) and many other entities such as China's SAFE which manages the state foreign exchange reserves.

Another driver is government and regulation as this is always an important factor in the Chinese business. There are many points in the current five-year plan by the central government that encourages the transportation and tourism sector domestically, especially in the western growth regions. There have been multiple signals for the encouragement of companies to grow by going abroad for technology and resources. This has been an important part of the growth driver for cross-border activity. One prominent example of this policy is President Xi Jinping's One Belt, One Road policy and twenty-first century Maritime Silk Road which aims to rebuild the ancient Silk Road trade route between China and Europe and all the intermediate countries and sea routes along with South and Southeast Asia. With the purpose of driving economic development, this will inevitably be a driver for more overseas deals by Chinese companies and investors. One example of this resultant investment and development in airlines is in Georgia by the Hualing Group.

Chinese outbound investment and M&A have been generally focused on the pursuit of technology, expertise, and natural resources. This is overlaid upon normal expansion and acquisition growth strategies such as vertical or horizontal integration as well as conglomerate building activities. Acquirers come from a broad spectrum of backgrounds from private- and state-backed companies, both having existing industry experience and others diversifying their existing holdings. Cross-border investments come in all shapes and forms whether wholly, majority, or minority investments. While there are many factors that affect the size of the cross-border investment, the main drivers include the acquirer's strategy, existing shareholder desires, as well as other external regulatory factors such as specific country regulation. The rationale for minority interests is both strategic and forced upon by local regulation such as the 49% maximum foreign interest in European airlines.

While cross-border appetite, foreign direct investment, and minority purchases are generally welcomed worldwide, there are instances where this has created tensions among stakeholders including existing shareholders, labor unions, governments, or local populations. Tensions from these stakeholders have arisen by the acquisition of minority shareholdings in a few instances in the past few years including Etihad's 24% purchase of Jet Airways by the Indian regulators and its 33.3% stake in the Swiss carrier, Darwin Airline, both of which was finally approved. There are also instances of resistance from strong labor unions of state companies. An example of sustained large acquisition growth strategy is HNA Group, a quasi-private entity, which has had a strong continued appetite for acquiring companies both domestically and internationally in recent years in the aircraft leasing, airline, and other tourism-related industries. These include airline investments of $114 million for 13% of Virgin Australia, $450 million for 23.7% of Azul, 100% acquisition of Avolon for $6.4 billion, along with related investments in airline catering (Gategroup), ground/cargo handling (Swissport), and hotels (Carlson Rezidor Hotel Group and NH Hotel Group). In addition, there are rumors and announcements in the pipeline including Air France's Servair (airline catering), CWT (logistics), among others. In the aircraft space, HNA has also completed the acquisition of Allco Aviation, later renamed to Hong Kong Aviation Capital, in 2010.

There are also many lessons that can be learned from these recent cross-border deals. One recent board drama saw all of the HNA's board representatives at NH Hotels voted out due to shareholder-perceived conflicts of interests of its significant minority holdings given the announced acquisition of Carlson Rezidor Hotel Group. This holds further lessons in minority interest investments where companies are active compared with passive minority investments especially in the same competing industry and overlapping geographic areas. It is important to understand and appreciate the specific regulatory approvals for foreign exchange and approval of large overseas deals. This has in some instances taken a lengthy period and has been onerous on the investment process as it comes particularly in the way of auction-style sale processes where there are specific timeline targets for each round of bidding and deposits. This has been a particularly sensitive issue to be overcome as it has affected some of the recent industry sales. Another lesson is to know as much as possible about the counterparty. Not only does one want to know how the investment will be funded, other background information on their existing businesses and reputation is key in any potential transaction. While there is never 100% information available, some leap of faith is required. This is not only important for the transaction but also a telling indicator of the post investment integration and marriage afterward, if any. An example is the recent failed sale of Frankfurt-Hahn Airport to a mystery Chinese buyer by selling German federal state owners, Rhineland Palatinate and Hesse. After the failed sale, the airport owners ultimately sold the property to the HNA Group.

Like all acquisitions, prudent due diligence, post acquisition integration, and business planning is necessary whether done in-house or by experienced third-party advisors—one such aviation advisory group is IBA Group. There have been multiple examples of cross-border investments that quickly went bust as the expected turnaround situations or synergies were not realized as a lack of understanding of local culture and regulations proved to be too difficult of a barrier for the acquirers to overcome. Even with some of these sensitivities, it is en vogue to mention prospective Chinese buyers as this may be to drum up interest and valuation expectations in company sales. There have been several instances of this occurring recently. One thing is certain: there is more activity in cross-border investment activity, especially from China. Transactions will have a much more successful path if past lessons are learned and are the most prepared. All in all, signs are pointed to what seems to be a new “Aviation Silk Road” plan for the aviation industry similar to the back of the One Belt One Road policy.

11 Conclusion

The global aircraft leasing and aviation industries continue to grow as the overall global demographics and the drivers are positive for the industry. In addition, there is room for further uplift. Aircraft and the airline industry are intertwined. Globally, the demand drivers like general economics contribute to the increase to general wealth and fueled by the growth of the middle class, urbanization, liberalization of visas, and trade have contributed to the increase in leisure and business travel. There has been evidence of business cycles that are evident in the aviation industry along with large and small exogenous shocks that periodically affected the growth of the industry. These along with industry profitability which are particularly sensitive to interest rates, and fuel pricing have contributed to the traffic flows dynamics. Supply drivers are also significant drivers of the industry as the availability of various forms of aviation finance go hand in hand with the few global aircraft manufacturers who hold a duopoly in the ability to make and deliver new aircraft.

There are many types of aircraft financing available including equity or cash, commercial debt, capital markets, and manufacturer’s support. In addition to the financing environment, the lease rates and the residual values of these aircraft have significant effects on the use of aircraft by airlines and factors in the number of parked and stored aircraft. Intertwined throughout is the effect of foreign exchange rates specifically with the US dollar being used in conjunction with local currencies which has effects on the demand and supply side. There have also been changes to the existing industry structures and traffic dynamics due to the evolution and change of the airline business models to more LCC players instead of the traditional full service airlines.

Bibliography

High-flying Irishman who became global entrepreneur in aviation sector. (2007, October 6). *The Irish Times.*

A4A. (2017). *Passenger Airline Cost Index.* Retrieved from: http://airlines.org/dataset/a4a-quarterly-passenger-airline-cost-index-u-s-passenger-airlines/

Airbus. (2016). *Global Market Forecast 2016–2035.* Retrieved from www.airbus.com/aircraft/market/global-market-forecast.html

Airbus. (2017). *Orders and Deliveries.* Retrieved from https://www.airbus.com/aircraft/market/orders-deliveries.html

Airfinance Journal. (2016). Leasing Top 50 2016. *Airfinance Journal.*

Airline Monitor. (2016). *Airline and Aircraft Data.* Retrieved from: www.airline-monitor.com

Ascend. (2017). *Yearly Operating Lease Percentage as of September 14.* Retrieved from: www.flightglobal.com

Bjelicic, B. (2012). Financing airlines in the wake of the financial markets crisis. *Journal of Air Transport Management, 21*, 10–16.

Bloomberg. (2017). *Economic, FX, Fuel, and Funding Data.* Retrieved from: www.bloomberg.com

Boeing Capital Corporation. (2010). *Current Market Finance Outlook 2010.* Retrieved from Seattle, WA: www.boeing.com/resources/boeingdotcom/company/capital/pdf/2010_BCC_market_report.pdf

Boeing Capital Corporation. (2017). *Current Market Finance Outlook 2017.* Retrieved from Seattle, WA: www.boeing.com/resources/boeingdotcom/company/capital/pdf/2017_BCC_market_report.pdf

Boeing Corporation. (2016). *Current Market Outlook 2016–2035.* Retrieved from Seattle, WA: www.boeing.com/resources/boeingdotcom/commercial/about-our-market/assets/downloads/cmo_print_2016_final_updated.pdf

Boeing Corporation. (2017). *Orders and Deliveries.* Retrieved from Seattle, WA: http://www.boeing.com/commercial/#/orders-deliveries

Bowers, W. C. (1998). Aircraft Lease Securitization: ALPS to EETCs. Retrieved from http://pages.stern.nyu.edu/~igiddy/ABS/bowers2.html

CAPA. (2016). *Airline and Aircraft Data.* Retrieved from: https://centreforaviation.com

Dallos, R. E. (1986, February 9). Leasing Jets Takes Off, but May Be Throttled Back: Tax Reform a Threat to Airline Suppliers Such as GE, Xerox. *Los Angeles Times.*

Deegan, G. (2015, October 22). Aircraft firms pay just €23m in tax. *Irish Examiner.*

Deloitte. (2015). *China Outbound Momentum Unabated Despite Economic Uncertainty.* Retrieved from www2.deloitte.com/cn/en/pages/about-deloitte/articles/pr-china-outbound-investment-momentum-unabated-despite-economic-uncertainty.html

DIIO. (2017). *Airline Data*. Retrieved from: www.diio.net

Doganis, R. (1985). *Flying off course: the economics of international airlines*: Allen & Unwin.

EDB Singapore. (2012). Aircraft Leasing Scheme (ALS) 2012 Circular. Retrieved from www.edb.gov.sg/content/dam/edb/ja/resources/pdfs/financing-and-incentives/Aircraft%20Leasing%20Scheme%20(ALS)%202012%20Circular.pdf

Eisfeldt, A. L., & Rampini, A. A. (2008). Leasing, ability to repossess, and debt capacity. *The Review of Financial Studies, 22*(4), 1621–1657.

Franke, M., & John, F. (2011). What comes next after recession?–Airline industry scenarios and potential end games. *Journal of Air Transport Management, 17*(1), 19–26.

Goolsbee, A. (1998). The business cycle, financial performance, and the retirement of capital goods. *Review of Economic Dynamics, 1*(2), 474-496.

Haver Analytics. (2012). *Economic Data*. Retrieved from: www.haver.com

Holloway, S. (1997). *Straight and level: practical airline economics* (2nd ed.): Avebury.

Hong Kong Inland Revenue Department. (2017). Comprehensive Double Taxation Agreements Concluded. Retrieved from www.ird.gov.hk/eng/tax/dta_inc.htm

Hong Kong Legislative Council. (2016). Key drivers for developing an aircraft leasing centre. Retrieved from www.legco.gov.hk/research-publications/english/essentials-1516ise17-key-drivers-for-developing-an-aircraft-leasing-centre.htm

Hong Kong Legislative Council. (2017). Proposed Dedicated Tax Regime to Develop Aircraft Leasing Business in Hong Kong. Retrieved from www.legco.gov.hk/yr16-17/english/panels/edev/papers/edev20170123cb4-410-8-e.pdf

Humphries, C., & Hepher, T. (2014). Insight: Revenge of the Irish as mega-merger restores air finance crown. *Reuters*.

IBA Group. (2017). *Traffic and Aircraft Data*. Retrieved from: www.iba.aero

IDA Financial Services. (2017). International Financial Services. Retrieved from www.idaireland.com/business-in-ireland/industry-sectors/financial-services

IHS Economics. (2016). *Economics Data*. Retrieved from: www.ihsmarkit.com

International Air Transport Association. (2012). *Financial Monitor for Jan/Feb-2012 released on 01-Mar-2012*. Retrieved from www.iata.org

International Air Transport Association. (2016a). *Air Passenger Forecasts, October 2016*. Retrieved from www.iata.org

International Air Transport Association. (2016b). *Airline Industry Profitability*. Retrieved from www.iata.org

International Civil Aviation Organization. (2012a). *Airline Industry Profitability*. Retrieved from www.icao.int

International Civil Aviation Organization. (2012b). *WASA database*. Retrieved from: www.icao.intl

International Civil Aviation Organization. (2016). *Airline Demand*. Retrieved from: www.icao.intl

International Civil Aviation Organization. (2017). *Airline Operating Costs and Productivity*. Retrieved from: www.icao.intl

International Monetary Fund. (2016). *Country Economic Data*. Retrieved from: www.imf.org

Kasper, D. M. (1988). *Deregulation and globalization: Liberalizing international trade in air services*. Washington, DC: American Enterprise Institute

Laprade, D. B. (1981). *The Basic Economics of Air Carrier Operations*. Ottawa: Canada Transport Commission.

Lenoir, N. (1998). *Cycles in the air transportation industry*. Paper presented at the WCTR 1998, 8th World Conference on Transportation Research.

Liehr, M., Größler, A., Klein, M., & Milling, P. M. (2001). Cycles in the sky: understanding and managing business cycles in the airline market. *System Dynamics Review, 17*(4), 311–332.

Lyneis, J. M. (2000). System dynamics for market forecasting and structural analysis. *System Dynamics Review: The Journal of the System Dynamics Society, 16*(1), 3–25.

MergerMarket, & ICF. (2016). *Aviation and Aerospace M&A Quarterly Q2 2016*. Retrieved from www.mergermarket.com/assets/ICF_Q2_2016_Newsletter_Final_LR.pdf

Oxford Economics. (2016). *World Population*. Retrieved from www.oxfordeconomics.com

Pearce, B. (2012). The state of air transport markets and the airline industry after the great recession. *Journal of Air Transport Management, 21*, 3–9.

Pearson, R. J. (1976). Airline managerial efficiency. *The Aeronautical Journal, 80*(791), 475–482.

Pryke, R. (1987). *Competition among international airlines*.

Reuters. (1990). Company News; Guinness Peat In Joint Venture. Retrieved from www.nytimes.com/1990/06/15/business/company-news-guinness-peat-in-joint-venture.html

Robertson, D. H. (1915). *A study of industrial fluctuation: an enquiry into the character and causes of the so-called cyclical movements of trade*: London: PS King.

Shearman, P. (1992). *Air Transport: Strategic Issues in Planning and Development*.

Smith, A. (1776). *The wealth of nations*.

Smith, M. A. (1968, Dec. 19). New Douglas Finance Company. *Flight International, 81*, 1020.

Straszheim, M. R. (1969). *The international airline industry*: Brookings Institution, Transport Research Program.

Thomson Reuters Datastream. (2017). *Economic Data*. Retrieved from: www.thomsonreuters.com

UN Population Division. (2016). *Population Databases.* Retrieved from: www.un.org/en/development/desa/population/publications/database/index.shtml

Wayne, L. (2007, May 10). The Real Owner of All Those Planes. *New York Times.*

World Bank. (2017). *Country Economic Data.* Retrieved from: www.worldbank.org

CHAPTER 4

China Aircraft Leasing Industry Characteristics

1 Background and History of Aircraft Leasing in China

Historically, aircraft leasing in China has been by foreign lessors leasing aircraft for the Chinese market. In the beginning, airlines owned the aircraft outright with financing from local banking institutions. Local airlines also leased a small number of aircraft, mainly the aircraft of ILFC and GPA, the founders of the industry, but now all airlines lease aircraft and these aircraft are leased by all major global aircraft leasing firms. These developments have propelled the government to enact new legislation to encourage the development of the local financial leasing industry and later on more specifically aircraft leasing.

In July 1979, China passed the Law on Sino-Foreign Equity Joint Ventures, which had wide-ranging effects, including allowing a legal context for foreign investment into China, and spurred the formation of China International Trust and Investment Corporation (CITIC) in August 1979. They very soon established a leasing department. The first dedicated leasing company, China Orient Leasing Company Ltd., was set up in February 1981 by CITIC (20% equity), Beijing Machinery and Equipment Corporation representing the Beijing government (30% equity) and Orient Leasing Company (50% equity) from Japan later renamed ORIX jointly as a Sino-Japanese foreign joint venture and

D. Yu, *Aircraft Valuation*,
https://doi.org/10.1007/978-981-15-6743-8_4

approved by the People's Bank of China (PBOC) (Russell 2014). Formal ordinary leasing definitions were established in July 1982 in Article 23 of the Economic Contract Law of the People's Republic of China. As the country evolved, the government provided further guidelines and regulations that were helpful for more practical regulations and guidelines through the Contract Laws of the People's Republic of China.

From this time on, many leasing companies were established through joint ventures licensed either by the PBOC or by the Ministry of Trade and Economic Cooperation, especially with Japanese and European groups. In the 1980s, leasing capital represented about 10–20% of all foreign capital in China (Shi 2005). Most of the projects during the 1980–1990s were with state-owned companies or local governments. These were guaranteed by state-owned financial institutions such as the Bank of China or local governments. In September 23, 1980, the Civil Aviation Administration of China (CAAC), which was at that time both the regulator and the airline, signed the first Chinese aircraft lease agreement with the Manufacturers Hanover Leasing Corporation subsidiary of Manufactures Hannover (one of the largest banks in the US then and now a part of JP Morgan) for two Boeing 747 aircraft for a 15-year term. This was a leveraged lease (Li 2010; Ding et al. 2014). Tax exemptions and reductions were granted to CAAC for leased aircraft from 1980 and 1984 (Eaton 2016).

From 1980 to 2001, foreign investment represented about $30.3 billion in the aviation industry according to Mingchi YAN, Director of Policy Division of the CAAC, as disclosed in 2002. Breaking this down further, $1.879 billion was foreign government or preferred loans, 606 million foreign direct investment, and $1.77 billion capital from public listings; 437 aircraft were from leasing representing $26 billion of asset value ("Chinese Civil Aviation Set to Take Off" 2002). In addition, during this period, the CAAC also imported 117 aircraft through operating leasing (Ding et al. 2014). After strong growth in the early 1990s, many lessees defaulted due to the poor economic and political conditions and the guarantees by local governments were hard to enforce, which led to many going bankrupt including the first, China Orient Leasing. By the mid-1990s American and European multinationals entered the market but these overall negative events led the government to suspend the issuance of new licenses for more than ten years.

The growth in leveraged leases in the 1980s and 1990s was also driven by the tax incentives structures. These tax-driven and optimized leveraged lease structures were very popular from different jurisdictions in Japan, the US, Europe, and among others, which fluctuated with the changes in the respective country tax codes. Given the Japanese background of the joint venture partners, the strong growth of Japanese-optimized tax and Hong Kong-leveraged leases were the most preferred as a result of the changes. By November 1990, Hong Kong amended its tax laws to prohibit non-Hong Kong leases the tax benefits, which caused the structure to be unattractive. The 1990s saw a very poor Japanese economy, because of which there was a lack of investors for the Japanese optimized leverage lease structure, which resulted in the shift to other forms of leasing in China.

This suspension of new licenses did not change until preparations were under way for China to enter the World Trade Organization (WTO). These included government-provided pledges to open the local financial markets to foreign players. One of the goals of the government was to diversify financing sources for local companies, especially for small and medium-sized enterprises organizations. These considerations are all drivers and factors that initiated changes in the local regime to better promote the financial leasing industry. The year 2004 saw the start of leasing law formulations by the National People Congress through the formation of a legislation steering committee, which established a working group to solicit opinions from all stakeholders to draft the outlines of the leasing laws. MOFCOM also made it clear in 2004 that foreign capital can establish wholly owned foreign enterprises in leasing, not just joint ventures as previously required.

The Leasing Law official draft was published in 2005, which was then submitted to the Finance and Economic Commission for further evaluation, resulting in the final law in the Measures for Administration of Financial Leasing Companies, which became effective as law in March 2007. This is a big step forward as the legislation allowed commercial banks, both local and overseas, to set up financial leasing companies which were at first regulated by the China Banking Regulatory Commission

(CBRC) and this officially saw banks reenter the leasing industry after the ban post the Asian Financial Crisis. The legislation required that banks, big leasing companies, established manufacturers, and other authorized business entities be eligible for being the "major shareholders" or shareholders holding more than 50% of shares. Table 4.1 is a summary of the regulatory requirements for financial leasing companies.

The first batch of financial leasing licenses were governed by the CBRC and are considered the real restart of the local leasing regime. Like the regulatory name suggests, the CBRC is also the regulator for banks and this first batch of licenses were subsidiaries or affiliates of large banking groups. Groups are more likely to be government- or state-owned enterprises. The other batch of leasing companies that have grown is regulated by the Ministry of Commerce (MOFCOM). MOFCOM-regulated leasing companies have both SOE company groups and private enterprises. The barriers of entry for a license regulated by MOFCOM are less onerous and easier to obtain than for CBRC-licensed leasing companies.

The major differences are that leasing companies regulated by the CBRC have high barriers of entry and generally higher capitalization requirements. They are generally to be under more rigorous management and oversight by regulators. These companies are also more likely to be state-owned and banking groups. Because of this, there are a large number

Table 4.1 China leasing company structures

	Local banks	*Local leasing companies*	*Leasing companies associated with manufacturers*
Capital adequacy ratio	Not less than 8%		
Total assets	Not less than RMB80 billion	Not less than RMB10 billion	Not less than RMB5 billion
Profitability	Achieved profit successively for 2 years		
% of leased assets/ Total assets			Not less than 30%
% of leased income/ Total income			Above 80%
Minimum registered capital	RMB 100 million		

Source: National Development and Reform Commission of China 2016; Ministry of Commerce of the People's Republic of China 2016; China Banking Regulatory Commission 2016

of companies incorporated under MOFCOM rather than CBRC but the total capitalization is roughly equal.

While leasing companies could be formed under the MOFCOM, bank-owned financial leasing proved even more popular as new Administrative Measures for Financial Leasing Companies (Order of China Banking Regulatory Commission No. 3) were issued in 2014 that lowered the barriers even more for entry into the bank-owned financial leasing industry (China Banking Regulatory Commission 2016; Ministry of Commerce of the People's Republic of China 2016). See Fig. 4.14 for more data about asset growth in the two different types of leasing companies.

In addition, due to the regulation, most of the leasing companies are financial leasing companies and the leases for multiple industries are not by stand-alone aircraft leasing companies as compared with counterparts globally. There are many reasons for this as leasing has been viewed mostly akin to debt but this view is slowly evolving with more operating leases. In addition, the capital requirements for aircraft are still quite high compared to other assets such as medical equipment, which have asset sizes of under $1 million. These leasing companies view aircraft as one segment within the overall company but constitute a large percentage of overall activity.

2 China Jurisdiction: Tax and Regulations

Since the start of leasing in China in the late 1970s, the vast majority of aircraft leasing transactions have been using the international leasing structure. This was especially true in the first decades of existence. In 2000, there was the emergence of domestic private capital using domestic leasing companies in aviation with Xinjiang-based Delong International Group's two leasing companies, New Century Financial Leasing Co., Ltd., and Xinjiang Financial Leasing Co., Ltd., launching their aircraft leasing businesses in May ("Private Sector's Flying Dream" 2002).

Then in November 2000, Shenzhen Financial Leasing Co., Ltd., signed an order contract of 60 MA-60 regional passenger turboprops worth Chinese renminbi (RMB) 4 billion with AVIC subsidiary, Xian Aircraft, which were subsequently leased to domestic airlines including Sichuan Airlines, marking the official start of the domestic aircraft leasing business

("Xian Aircraft Wins" 2000). This marked the start of the domestic aircraft leasing business.

As a further enhancement of the promotion of the leasing industry in China from the restart in 2007, MOFCOM issued the Guidance on Promoting the Development of Financial Leasing Industry during the 12th five-year plan at the end of 2011. This encouraged the development and formation of the leasing industry centers and it also enhanced the functions of financial leasing in free trade zones (FTZs), namely, Qianhai in Shenzhen, Shanghai and Tianjin and Hong Kong were mentioned, as shown in Fig. 4.1. It also opened up the factoring business in Shanghai Pudong, Qianhai, Guangzhou, Tianjin Binhai, and Beijing. Another new jurisdiction that is attracting aircraft leasing special purpose vehicles (SPVs) is the Henan Zhengzhou Airport Economic Zone. This zone is like the other special zones set up to attract aircraft assets. This jurisdiction landed its first aircraft leasing SPVs in January 2018 and looks to attract other aviation activity to the Zhengzhou area.

With respect to the jurisdiction, China as of 2017 had 105 signed double tax treaties (Deloitte 2017). Shanghai has risen to become a finance hub in Asia. One of the major issues is that any foreign exchange transactions require State Administration of Foreign Exchange (SAFE) approvals such as all transactions involving aviation (there have recently been a few transactions in RMB). This additional hurdle has raised some concerns for some international investors. In addition, the country lacks specialized personnel and service firms to support the industry. This is especially acute as personnel and service providers with both domestic and international experience are lacking. Currently, tax depreciation allowance is ten years. The tax on profits is the statutory 25% corporate tax rate. There are current efforts to increase the desirability of the jurisdiction on a competitive basis for aircraft leasing (Deloitte 2017).

The FTZ zones each separately provide enhanced benefits that are negotiated on a bilateral basis with the leasing company. Some factors that affect the benefits include the size and scope of the leasing company. As these are negotiated terms, while the agreements all differ, some included aspects can be lowered or outright absent in terms of things such as duties, rebates on local or total taxation, personnel

Fig. 4.1 Free trade zones and jurisdictions in China. (Source: PWC 2012)

income tax abatement, ease of foreign exchange and other regulatory matters, among others. There are also significant reforms to taxation regulations so this is something to watch out for. In addition, some aspects to be aware of is that sales within each FTZ zone can have beneficial tax effects and this has been established but sales to other FTZ zones can create tax liabilities that are not defined. As such, there have been very few instances of buying and selling and there has been trading intra-FTZ zone but not cross-zone; nor has there been cross-border trading activity with other international jurisdictions. Given that the taxation regime is still constantly changing and there is continued dominance by Ireland and Singapore, this has yet to be factored in significantly for the global aviation leasing community. This will change as investors and leasing companies better understand the local regulatory and tax regime.

3 Drivers

There are many drivers for the tremendous growth and diversification in the aircraft and aircraft leasing business in China. While all of the global drivers for aviation as described previously apply to the Chinese market, there are several main drivers specific for the Chinese market for aviation leasing.

3.1 Economics (Urbanization, Growth of the Middle Class, Globalization)

3.1.1 Economics

The state of the economy and the growth dynamics driving the economy are important considerations for any jurisdiction.

Chinese GDP was $11.2 trillion as of 2016, which has grown from $92.6 billion in 1970, representing a 120× increase. In GDP per capita PPP constant USD terms, in 2016 China's GDP was $15,397.39 while it was $309.96 in 1980, representing a nearly 71× increase, as shown in Fig. 4.2 (World Bank 2017; IMF 2017; Bloomberg 2017). This represents a large growth in the wealth measure of the nation as a whole.

Figure 4.3 shows that real GDP measures are increasing while one component, private consumption, has kept relatively steady and projected forward percentage, thereby creating a larger amount of capital for private consumption including more opportunities and demand for air travel.

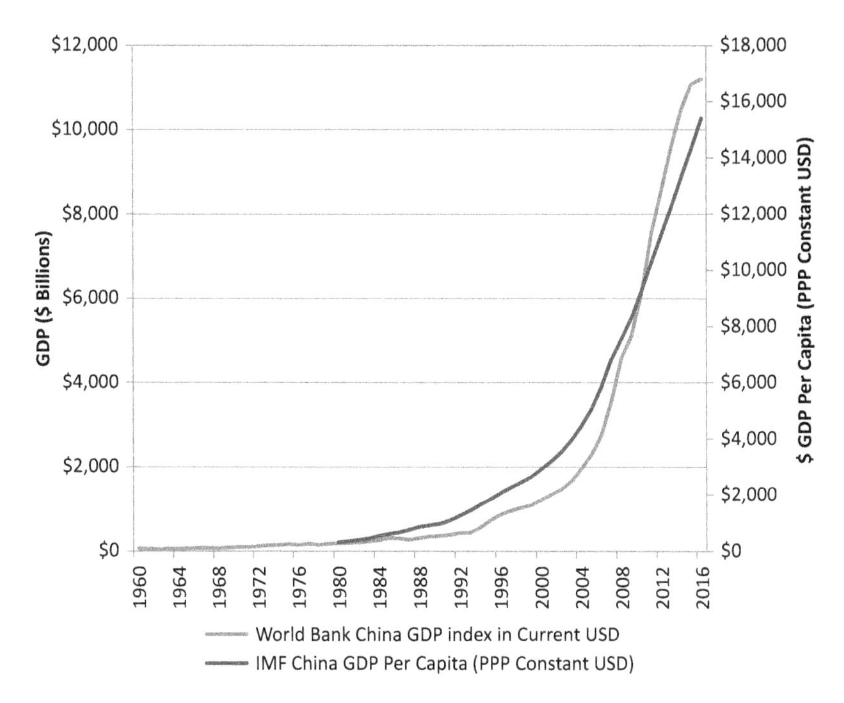

Fig. 4.2 China GDP and GDP per capita. (Source: World Bank 2017; International Monetary Fund 2017; Bloomberg 2017)

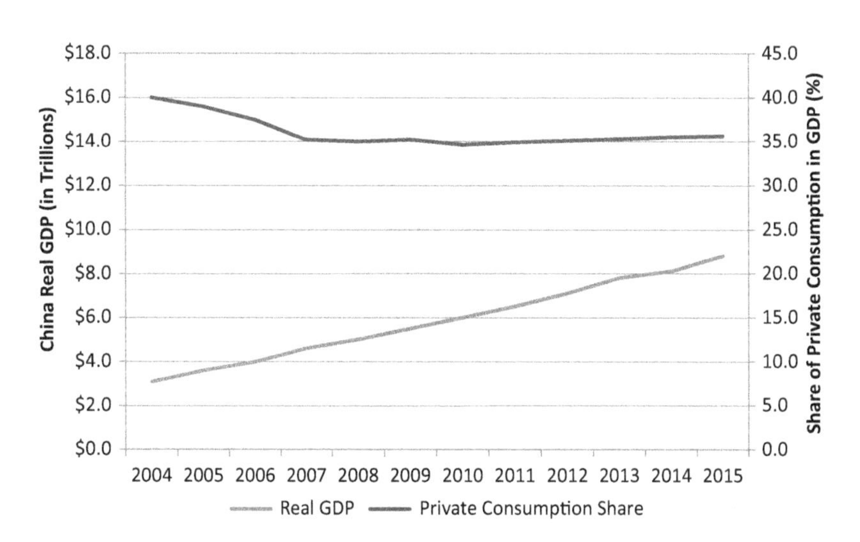

Fig. 4.3 China private consumption growth. (Source: IHS Economics 2016; Airbus 2016)

3.1.2 *Urbanization and Globalization*

Wealth creation and the increased demand for air travel have been brought on by certain dynamics. First, urbanization has created many wealthy citizens. In China, the eastern coastal cities and provinces are the powerhouse economic drivers. They include Beijing, Shanghai, Tianjin, Jiangsu, Zhejiang, Hebei, Shandong, Guangdong, Fujian, and Hainan. The eastern zone represents roughly 1/3 of the population (525 million out of 1.375 billion) but represent 51% of the economy and wealth (National Bureau of Statistics of China 2016). The eastern region's GDP per capita basis was $10,903 with a foreign exchange rate of 6.5138 CNY/USD (Exchange-rates.org 2015; National Bureau of Statistics of China 2016).

From 13.3% in 1953 to 56.1% in the 2015 urbanization rate, the overall urban populations have also grown at a much faster CAGR of 3.7% compared to the overall population CAGR of 1.4% from 1953 (National Bureau of Statistics of China 2011). Table 4.2 contains additional urbanization and population figures. The target is 60% urbanization rate by 2020. In comparison, in the US roughly 80% of the population lives in cities.

Urbanization is very much linked to the hukou system, which is a local household registration. This conveys local benefits and rights such as education (important for families), healthcare, housing, and other social benefits. This hukou system and conversion of rural to urban cities has a direct impact on urbanization rates. There are currently 260 million migrant workers who are urban residents. To help achieve the urbanization rates, reforms have continued and the 2020 target is to convert 100 million rural workers to an urban *hukou* (An 2013). Progress is slow as large cities need to invest in more infrastructure and the funding to accommodate these new citizens is in an already stretched budget environment.

This trend has been continuing as there has been a strong increase in the population in the coastal mega cities that are dominated by migrant workers. Historically, migrant workers come from more rural counties for factory jobs and they have propelled the growth of the economy since its opening in 1989 by Deng Xiaoping. These trends will also increase the demand for aviation.

3.1.3 *Growth of the Middle Class*

In addition, the leisure market has also grown heavily due to the growth of the middle class in China. According to McKinsey, this middle class is represented by an annual income of RMB60,000 to RMB229,000 ($9000 to $34,000) and will number more than 400 million or more than 75% of

Table 4.2 China urbanization and population figures

	1953	*1964*	*1982*	*1990*	*2000*	*2010*	*2015*	*2020E*[a]
Urbanization rate	13.26%	18.30%	20.91%	26.44%	36.22%	49.68%	56.10%	60%
Urban population (10,000 people)	7726	12,710	21,082	29,971	45,844	66,557	77,116	88,292
Total population	58,260	69,458	100,818	113,368	126,583	133,972	137,462	147,154
CAGR from 1953 urban		4.2%	3.4%	3.6%	3.8%	3.8%	3.7%	
CAGR from 1953 pop		1.5%	1.8%	1.8%	1.6%	1.4%	1.4%	

Source: National Bureau of Statistics of China 2011, 2016

[a]2020 target of 60% urbanization rate based on the same total population growth rate

urban consumers by 2020. This middle class has grown from 4% of Chinese households in 2000 to 68% in 2012 (Barton et al. 2013). These factors along with the growth of working-age populations have driven private consumption growth further. These statistics and growth have propelled aviation in terms of both leisure activity and business travel, which in turn have driven up demand for both domestic and international travel.

3.1.4 Visas

One trend is the relaxation and liberalization of visa free or light measures toward Chinese citizens. Traditionally there have been strict policies due to the country's developing and low-income status compared with other more established countries such as those of the OECD. These updated policies have included the lengthening of terms, relaxation of return air tickets, employment and financial abilities, and have even provided visa waivers or free status.

3.1.5 Passports

As of November 2016, there were 120 million passports issued out of China's population of 1.411 billion people, which represents only 8.5% of the population (Ministry of Public Security of the People's Republic of China 2016). This growth in the issuance of passports is quite fast given that there were only 38 million passports issued in 2012 ("38 Million Chinese Citizens" 2012). However, this is low compared with other OECD countries; for example, the US has 136.1 million passports in circulation based on a population of 327.0 million, which represents 41.6% of the population (US Census Bureau 2018; US State Department 2017). In the UK, 83% had passports as of 2011 (Office of National Statistics 2013). Some caveats about this figure are that not everyone is eligible for a passport. This number is distorted as certain government bureaucrats and employees of SOEs need to get permission before applying for a passport. The trend is toward upward growth and higher penetration.

3.2 CAAC, CAAC Airlines, and the "Big 3" State-Owned Airlines

3.2.1 CAAC

In China, airlines are both intertwined and strongly influenced by the CAAC. The CAAC's history has been very different compared to other regulators globally such as the US's FAA or Europe's Aviation Safety

Agency (EASA). The frameworks which became the CAAC were established in 1949 with the founding of China and, initially, it was set up not only as a regulator but as a regulator of all non-military aviation as well as a provider of general and commercial services as an operator. Only about 20% of the current airspace in China is available for civil aviation while the rest is still controlled by PLAAF (Wang 2010). It is not surprising that the CAAC was controlled by the PLAAF from its founding to 1980.

3.2.2 *CAAC Airlines and the "Big 3" Airlines*

CAAC Airlines, the airline component, started domestic service from the onset and started international services in 1962. It was not until 1980 that the management and control of the airline component was transferred to China's State Council. Then in 1988, the airline operations were split into six separate independent airlines (Air China, China Eastern Airlines, China Southern Airlines, China Northern Airlines, China Northwest Airlines, and China Southwest Airlines) with different geographies. The CAAC has acted as the sole aviation regulator since 1988. In 2008, the Ministry of Transport was created and the CAAC became its subsidiary and the CAAC's status was downgraded from a previously held ministry-level agency.

The six original airlines are the predecessors of the dominant SOE airline players (Air China, China Southern, and China Eastern) in existence today, which resulted from a CAAC administrative action in July 2000 to combine ten airlines into three major groups. The state-owned airlines are still the main drivers and have a dominant market share (greater than 80% of traffic) of the aviation market in China (Ballantyne 2017).

3.2.3 *China Airspace and Delays*

Only about 20% of the current airspace in China is available for civil aviation (Wang 2010). Due to the low amount of airspace, there are congestion issues especially high-traffic areas corresponding to large metropolitan areas. The amount of airspace available to civil aviation in the US is 80% (Hsu 2014). During peak times such as the Chinese New Year, the PLAAF might open some additional airspace temporarily (M. Yu 2013). There also times when the PLAAF temporarily closes civilian airspace on short notice for its air exercises. "Other" or military exercises contributed to more than a quarter of all delays in 2016 according to the CAAC (Civil Aviation Administration of China 2016b). There are policies in place for further liberalization, especially for low-altitude general aviation.

China's airlines and airports have been in the bottom ten for punctuality in international surveys. This is compared with US airports and airlines, which are in the top 20 of punctuality. The total passenger traffic in China is only about 30% of the US. The small airspace results in delays in air travel when congestion is an issue. Overall, in 2016, the CAAC notes that on-time performance was 76.8%, which was higher by 8.4% compared with 2015. Other factors both locally and in other regions that could affect delays include congestion of airspace, at or near capacity airports, weather, smog, and efficiencies of the air traffic control. These factors are in addition to the normal airline and airport operational issues that occur such as lack of staff, appropriate aircraft, and so on. Increasing air traffic efficiencies by allowing for additional aircraft that can safely take off and land is important coupled with the expansion of existing airports, and construction of new ones is imperative. Airlines have also increased the estimated time to take into account more delays compared with the actual flight time.

3.3 Traffic Flows, Business Models, High-Speed Rail, and Foreign Exchange

The growth in the number of airlines and airline traffic both intra and inter China has contributed to the growth of aviation in China. Of the top 50 airlines in the world by fleet size, 7 Chinese carriers now represent 13.1% and 12 airlines based in Asia ex-China represent 19.7% of the top 50 airlines and together Asia carriers represent 33% of the top 50 world fleet. It was not until the mid-1980s that non-CAAC airlines were established. In the following decades, many types of new airlines were introduced including joint venture airlines with CAAC, airlines owned by various regional governments, and independent airlines, some of which had collaborations with regional governments.

This demand for air travel has created many new airline entrants and they have, on average, grown 27% since 2011 according to Innovata. There were 46 airlines operating in China as of 2016; of these 20 are new airlines that were approved from 2011 to 2015. In the previous periods such as the one in 2000, there was a consolidation of the regional airlines into what is now known as the "Big 3" Airline Group. In this new way, many regional governments have been new players both in establishing their own airlines and in partnering with existing airline groups. The growth in domestic travel driven by new airlines resulted in increased competition among existing players.

In addition to higher domestic travel demand, there has been a significant increase in the number (over 150%) and frequency (over 140%) of international destinations on offer by the Chinese airlines since 2011 (Table 4.3).

Also, international passenger growth, as well as the ratio compared with domestic passengers, has grown steadily. These statistical trends are aligned with the share of new international routes versus domestic routes. Figure 4.4 shows the growth of passengers for domestic and international routes. While domestic passengers have grown significantly in absolute terms, international passengers have grown but not in the same absolute

Table 4.3 China's airlines long haul network since 2011

Year	*Nonstop markets*	*Weekly frequencies*	*Marketshare*
2011	46	350	36%
2016	114	830	48%
Growth	*+150%*	*+140%*	*+12%*

Source: Innovata 2016

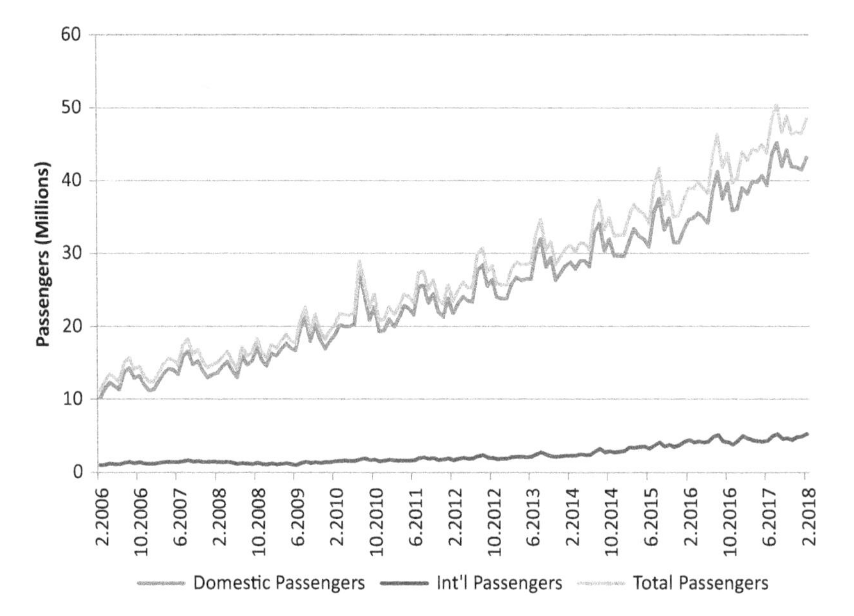

Fig. 4.4 Monthly China air domestic and international passenger numbers. (Source: Civil Aviation Administration of China 2018)

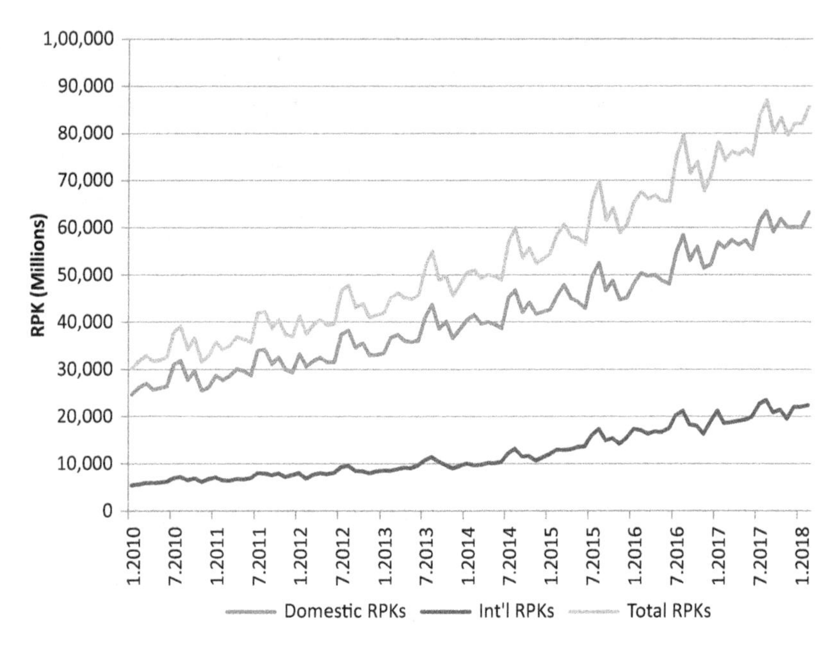

Fig. 4.5 Monthly China air domestic and international RPKs. (Source: Civil Aviation Administration of China 2018)

terms. Figure 4.5 shows the same route breakdown in terms of RPKs and air traffic terms. It shows a similar breakdown as the passenger demographics relationships in Fig. 4.4.

Like the changes in the global business model mix, there has been the addition of the Chinese version of LCCs in the market with the formation of Spring Airlines in 2004, and airlines now include Lucky Air, Colorful Guizhou, West Air, 9 Air, and the reformulation of existing airlines such as China United Airlines. As the CAAC has the authority to manage the creation of new airlines and the management of existing ones such as the routes and number of airlines that can be inducted into the airline, there has now been a control of the growth in new airlines and routes. This has been due to an increased emphasis on safety and tightening of management of the airlines after such a period of explosive growth.

Airline traffic is expected to continue growing very rapidly. Airbus projects a 6.8% per annum domestic growth and 6.2% per annum inter-China traffic growth over 20 years, as shown in Fig. 4.6. Overall, this equates to a 6.9% per annum increase in overall traffic to, from, and within China. In

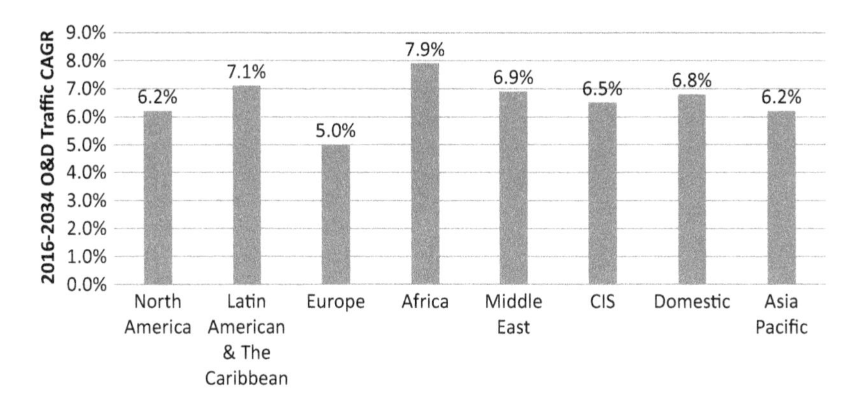

Fig. 4.6 Airbus' China regional projected traffic growth over 20 years (2016–2035 CAGR). (Source: Airbus 2016)

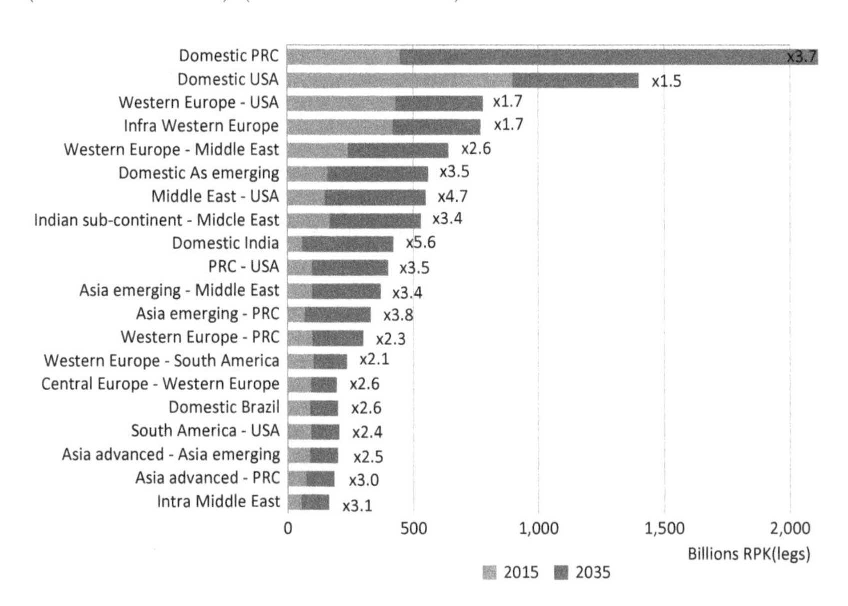

Fig. 4.7 Airbus-projected top traffic flows in 2035. (Source: Airbus 2016)

addition to this projected high growth, domestic China traffic is projected to increase 3.7× from slightly over 400 billion RPKs to over 1500 billion RPKs and become the most trafficked area in the world in 2037, as shown in Fig. 4.7. Other international routes out of China rank lower in terms of

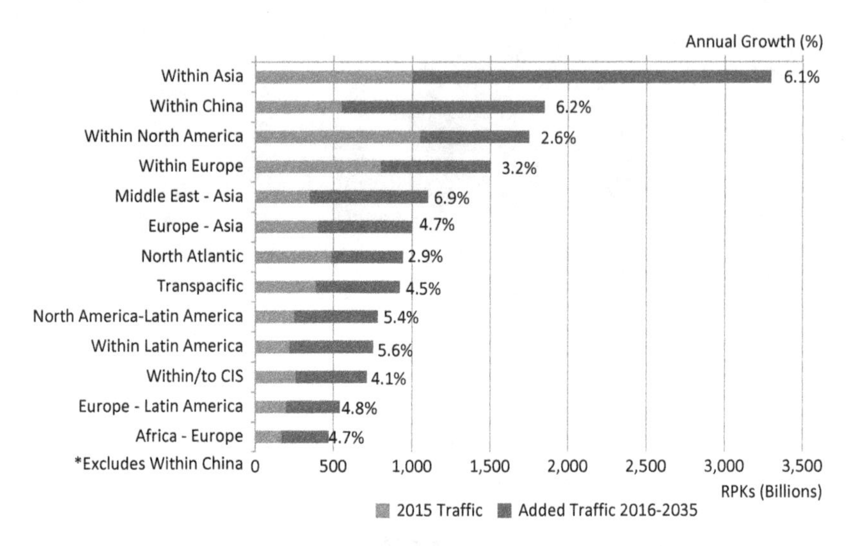

Fig. 4.8 Boeing-projected top traffic flows in 2035. (Source: Boeing Corporation 2016)

growth (Airbus 2016). Boeing also forecasts 6.2% growth per annum within China over 20 years and it is lower than that for the larger Asian continent in absolute terms, as shown in Fig. 4.8. Likewise, the industry association IATA concurs that China is in the top growth traffic area with similar projections of 5.2% growth over the 20-year timeframe, as shown in Fig. 4.9.

3.3.1 China Aircraft Leasing Industry Characterist

HSR is a passenger rail service with a travel rate greater than 250 km per hour. The service was introduced in April 2007. By the end of 2016, the country had more than 22,000 km of HSR with the overall rail network covering 124,000 km. The HSR network will increase to 30,000 km by 2020, which will connect more than 80% of the big cities ("China to Start Construction" 2017).

This service can be viewed as a competitive threat as well as a complement to air travel. This can be a threat since travelers can consider this for fast and reliable service when compared to domestic air traffic. Some of the factors that are considered are the convenient location of the station (usually in the city for rail versus outside for air), security delays, and the

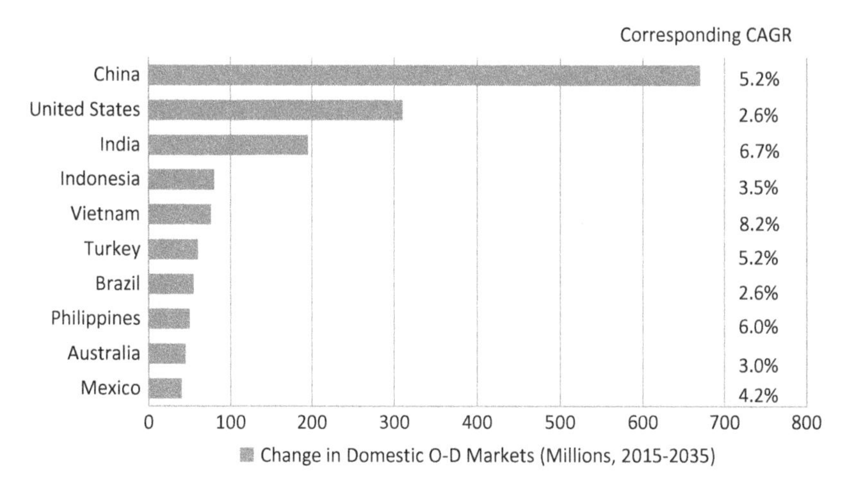

Fig. 4.9 Projected traffic growth by region (2015–2035). (Source: International Air Transport Association 2016)

potential travel delays which are more prevalent with air travel. Of course, price is another factor where the HSR has static pricing and with lower number of comparative cabins while the pricing for air travel is more dynamic and higher. The total time is a factor especially for long distance travel where air travel still has the advantage. As a complement, there are more rail and air combination offers especially from smaller cities or locations that do not have as many routes.

Even with the rise of HSR, there has been no evidence of rail eroding air travel's share of overall travelers. The portion of air travel has gained share versus rail. The proportion of rail passengers to airplane passengers was 88:12 in 2007 while in 2016 it was 85:15 as the rail passengers' percentage has decreased in this period, as shown in Fig. 4.10. If distance travel is compared, the same trend as in the case of passenger numbers is evident where the share of air travel has gradually increased from 28% to 40% from 2007 to 2016, as shown in Fig. 4.11.

3.3.2 Foreign Exchange in China

The RMB is not a fully convertible currency such as other global currencies like the US dollar or euro. The conversion is regulated by SAFE and this has significant effects on airlines as a lot of the aviation industry is denominated by US dollars.

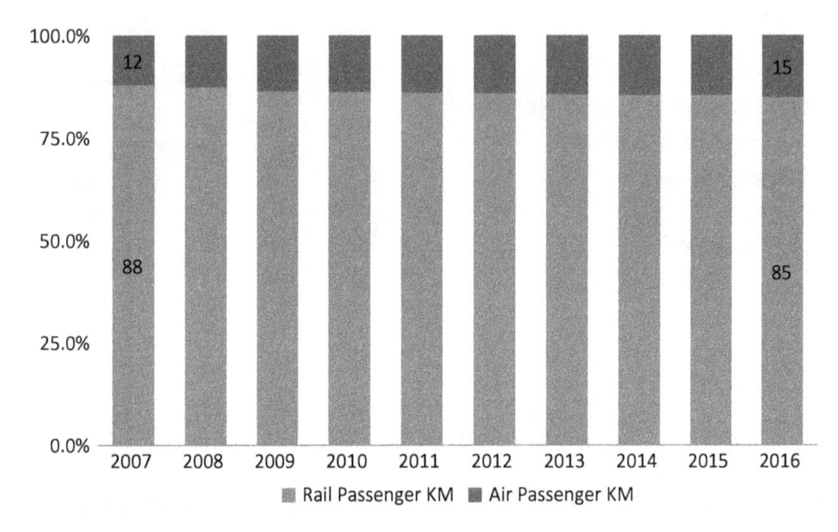

Fig. 4.10 China air and rail passengers market share. (Source: Haver Analytics 2017)

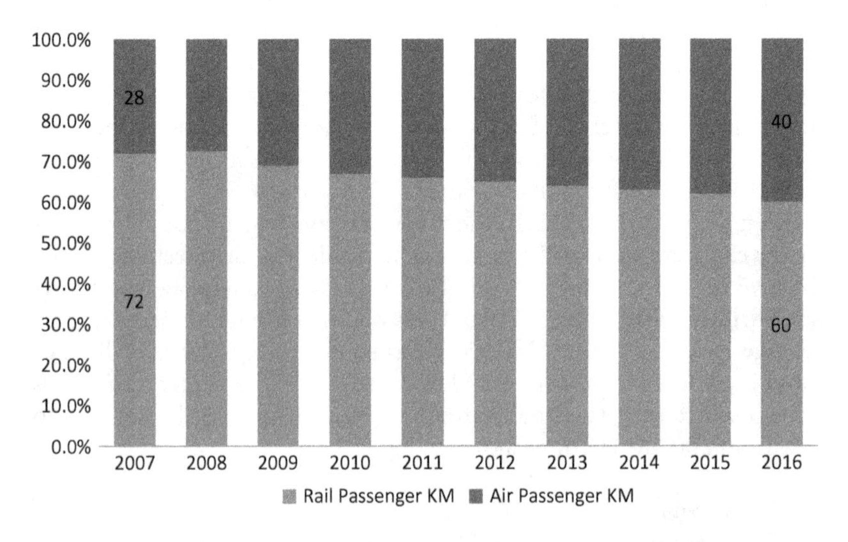

Fig. 4.11 China air and rail passengers traveled distance market share (Source: Haver Analytics 2017)

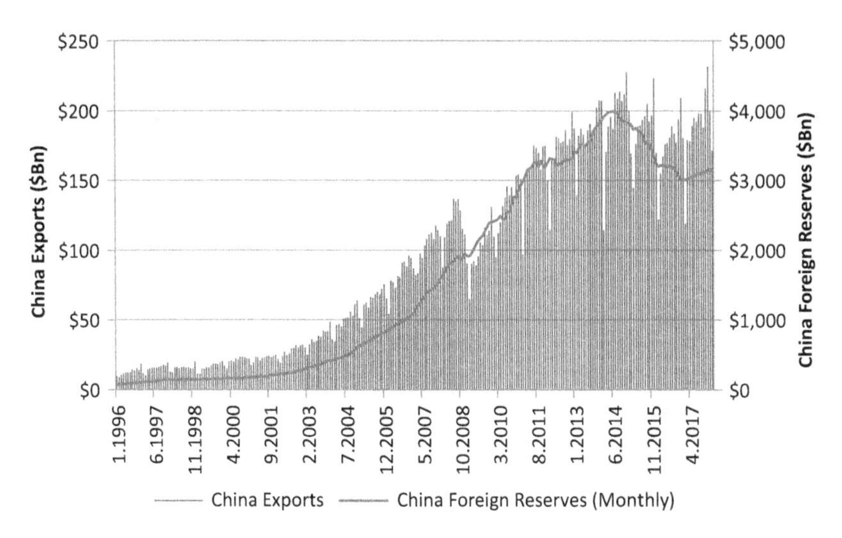

Fig. 4.12 China's economic development. (Source: People's Bank of China 2017; State Administration of Foreign Reserves of China 2017; Bloomberg 2017)

As shown in Fig. 4.12, China's foreign reserves have expanded alongside China's export volumes since 1996 and have risen considerably. Chinese airlines also have a significant percentage of their cost structure in US dollars such as aircraft purchasing, rental, debt, fuel, and other services. While there is income in dollars and other foreign currency through ticket sales, this is small in comparison to the costs along with the mismatch in other costs such as labor paid in local RMB. While the RMB currency became an official IMF reserve currency in 2016, and there is an increase in the use of the RMB in aviation contracts that was previously in US dollars, this is a slow adoption trend and is still very small compared to the use of US dollar in aviation contracts. This foreign exchange rate can create significant swings in the profitability of the airlines as demonstrated by SOE Chinese airlines that have gone public in Hong Kong.

These drivers have created additional demand for leasing companies by supplying the growth of domestic aviation but also the growth of international airlines globally. On top of these drivers are the increase in China's foreign reserves and the growth of the banking sector.

3.4 *Local Government Policies Affecting Aviation*

In this ever-changing world, there are constantly changing policies and other drivers that have an impact on aviation. While some changes have a more direct impact than others, the main drivers analyzed below are the ones to watch out for going forward.

3.4.1 *One Belt One Road (OBOR)*

There are many broad themes and subthemes behind these impressive figures. One of the major Chinese policy drivers is President XI Jinping's One Belt One Road economic investment initiative in infrastructure and increased trade, which aims to bridge the countries that comprise the route, land and sea, along the old Silk Road. While this concept was first announced in 2013, the real traction kicked off in earnest in 2015 and was developed alongside the founding and backing of the Asian Infrastructure Investment Bank in 2013 and the Silk Road Fund in 2014. This focus on trade has driven a lot of cross-border and outbound investment to countries along the routes. This has amounted to $33 billion in 68 countries this year as of August compared with $31 billion in 2016 according to Thomson Reuters with $124 billion pledged at the May 2017 summit. In the aviation space, there has been a surge of investment in tourism activities, airlines, airports, and aircraft leasing.

3.4.2 *Supply Side Structural Reforms*

Another major theme and driver are the series of supply side reforms that have been implemented since 2015. All of these different components are the overarching policy directions which have large effects on aviation and cross-border investments. The main subcategories include cutting excess industrial capacity, deflating real estate inventory and bubble, corporate deleveraging, lowering corporate costs (taxes, fees, etc.), and Made in China 2025.

Some of these subthemes are aimed at realigning the domestic economy and transforming it from old to new while making the sources of growth more sustainable. This includes cutting excess older industrial capacity and encouraging more clean energy projects. This goes hand in hand with overall deleveraging and control of the growth of the credit exposure in China as well as deflating the real estate bubble and excess inventory especially outside of top tier cities. This desire includes slowing down the credit growth due to normal and shadow banking activities and thus far has focused on larger private companies and smaller banks and has now moved on to local municipal and provincial SOEs. Credit growth has intensified since the 2008 global financial crisis and infrastructure spending and borrowing has been a key method to drive economic growth by local governments but lately initiatives such as for new subways in farther out regions, and so on, have wound up due to the concern for the amount of additional debt required.

3.4.3 Airports Infrastructure and Construction

These policies, however, do not dampen the growth of airports and related aviation infrastructure especially in more Western and more underdeveloped regions in China which have continued to see steady growth. Airports are another domain regulated by the CAAC. There were 218 airports as of 2016. Historically, China planned to build about six airports per year from 2011 to 2016. In the 12th 5 Year Plan unveiled in 2011, the stated target was 230 constructed airports while only 207 were built. In the updated 13th 5 Year Plan unveiled in 2016 the target was construction of 260 new airports by 2020 and 136 by 2025 with the largest in Beijing and Chengdu, currently under construction, each handling over 90 million passengers annually as shown in Fig. 4.13 (Ge 2017). These figures don't include the general aviation airports, which currently stand at 310 with

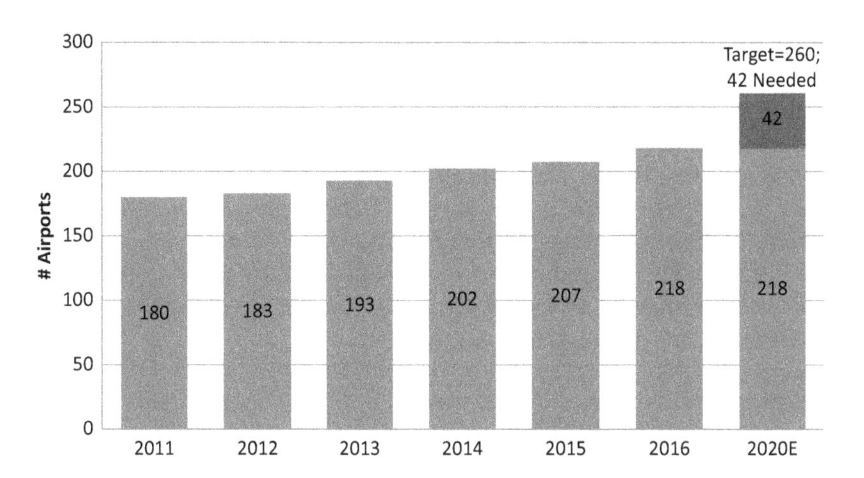

Fig. 4.13 China airport construction—actual and targeted. (Source: Civil Aviation Administration of China 2016a, b)

the goal of reaching 500 by 2020 or the targeted 139 expansion projects at existing airports (Civil Aviation Administration of China 2016a).

The lowering of corporate costs has not been a big point of emphasis. The focus so far has been on reducing fees and bureaucracy rather than major tax cuts while trying to stimulate the economy through encouragement of "mass entrepreneurship" by Premier LI Keqiang. This has especially been true of the free trade zones which are considered the policy test regions including the Tianjin Dongjiang Free Trade Port zone (DFTP) and Shanghai FTZ. These two jurisdictions are home to the most active jurisdictions for aircraft leasing in China.

The Made in China 2025 initiative is also part of the overall supply side reforms that have their roots in 2013, which focuses on the upgrade of the Chinese industry similar to the "Industry 4.0" initiative by Germany. This initiative increases the competitiveness of industry and encourages investments in high-technology sectors and has recently been given a renewed push. In the aviation world, this focuses on new biofuels, clean technologies implemented for airlines, aircraft, and airports. This theme is sometimes combined with the OBOR initiative. This is the case of domestic-made COMAC C919 and ARJ21 and aircraft leasing, both of which are being rolled out under both initiatives.

As a result of these initiatives and the desire for more offshore assets due to the exchange rate and the perceived difference in regional investment returns, these have driven up cross-border investments in aviation and tourism with the Wanda and HNA Group, which among others have been among the most acquisitive groups. These high-profiled deals have caused the CBRC to review the credit exposures for "systematic risk" to these four outbound groups including the aforementioned and Anbang and Fosun. In addition, the government has stepped up scrutiny of both the convertibility of RMB to other liquid currencies and the transfer of funds to offshore locations. The aim has been to slow down and dampen the more frivolous offshore investments that are outside the scope of the main business competencies and refocus on more core policy-driven investments that still have government backing. These actions also help support the RMB exchange rate and total foreign reserves which have steadily grown from the 1980s to their all-time high of $4.0 trillion, which was reached in June 2014 but has since receded to its recent low of $3.0 trillion in January 2017 to the current levels of $3.1 trillion in December 2017 (People's Bank of China 2017).

3.4.4 Opening Up of the Domestic Economy

Recently, there have been more policies relating to the opening up of the domestic economy to foreign capital, which includes publications of a national unified negative list of industries for inbound investments that were previously under more locally administered policies as well as a pledge for the opening up of investments in the financial institution space along with asset management companies. This provides more stability for foreign investors especially in aircraft leasing and airlines and also prevents retroactive unraveling of deals due to unanticipated policies. This has seen more foreign lessors establish local onshore subsidiaries in China to attract more local customers. These newer policies go hand in hand with the opening of the Hong Kong jurisdiction as a base for foreign and Chinese capital investing in aircraft leasing.

While some of these policies and drivers can seem a bit contradictory at times, what is certain is the encouragement of technologies and growth platforms that enhance the sources of GDP growth and composition, especially as China continues to emerge as a leading driving force in the global integrated economy. Aviation, including airports, leasing, technology, and tourism, is still a much favored industry that is promoted not only because of the policy considerations but also because of the strong underlying stability and economics. There will be continued challenges as these policies and drivers evolve to changing the global economy and industry, which bodes well for more opportunities for nimble and creative players.

4 Financing Environment and Current Trends

Similar to the global mix, locally in China, there are a variety of financing sources for aircraft with manufacturers' support, cash or equity, export credit agencies, bank debt by commercial banks, institutional players such as private equity and hedge funds, tax equity, and capital markets. The market is mostly funded by cash and commercial debt, while export credit agencies and manufacturers' support having some small market size. There are also very few capital markets transaction as these types of transaction are just starting to pick up in the market.

4.1 *Export Credit Financing*

In comparison to a long history in the West, China's history of a formal export credit bank is far shorter as the Exit-Import Bank of China (CEXIM), one of the three policy banks under the State Council, was founded more recently in 1994 with the mission to implement state policies to promote the export of Chinese products and services. Unlike US EXIM, CEXIM funds projects directly instead of through guarantees or insurance, which is akin to what US EXIM did in its early years.

After the gradual reduction of tax benefits in the 1990s, the preference for tax-driven leveraged leases changed with an increased interest for export credit-backed lease structures. The year 1996 saw the first US EXIM-guaranteed deal with China on McDonnell Douglas MD-11 s valued at nearly $900 million (Export-Import Bank of the US 2018). European ECAs also started to see increased activity after the first A300-600R was completed for China Eastern Airlines on October 22, 1993. European ECAs provided support for Airbus-manufactured products.

As these structures evolved, creative structures such as a package of Japanese-optimized leveraged lease with an ECA guarantee emerged. This, of course, had both advantages and some disadvantages but the main point was to lower the overall borrowing cost compared with using a single structure alone.

Another structure was the securitization of US EXIM-guaranteed leases, which provides a credit enhancement for a normal securitization product. This form of financing first appeared in Asia in early 1994, but Chinese airlines keenly seized this form. For example, China Southern Airlines acquired seven B737-300 aircraft in April 1994 using this form of business securitization, with financing amounting to 201 million US dollars.

While direct aggregate funding numbers are not published, the author has seen a significant rise in new financings by CEXIM especially over the past few years. This increase in funding is in line with the continued increased share of China's new deliveries in the global aviation market. CEXIM will continue to have a larger impact with respect to the global ECAs and the overall global aircraft financing market. This in addition to the hopeful resolution of the quorum issue at US EXIM will have global ECAs play a larger role in the overall aircraft finance market to come.

While direct aggregate funding numbers are not published, the author has seen a significant rise in new financings by CEXIM especially over the past few years. This increase in funding is in line with the continued increased share of China's new deliveries in the global aviation market. CEXIM will continue to have a larger impact with respect to the global ECAs and the overall global aircraft financing market. This, in addition to the hopeful resolution of the quorum issue at US EXIM, will have global ECAs play a larger role in the overall aircraft finance market.

4.2 Capital Markets, Sidecar, and Joint Ventures

Another change is the sources of the funding in aviation and aircraft leasing. Newer entrants providing both equity and debt capital have emerged such as large private, regional, and SOE conglomerates, insurance companies, and more creative onshore ABS like financing structures. These innovations will continue to drive the changing dynamics of the industry.

In addition, the trend toward the creation of new dedicated equity funds and the resurgence of sidecars and joint ventures has been a significant portion of growth especially involving Chinese companies. Sidecars are a form of joint venture which involves a capital source in a joint venture along with a more established aircraft leasing company, which creates benefits for both parties. These new funds include committed and announced from JVs constitute more than $26 billion in the aviation space since 2008 (D. Yu 2017). The latest example is Standard Chartered with their SDH Wings International Leasing Limited JV with Sichuan Development Holding and separately their establishment of an onshore China leasing platform in Tianjin DFTP.

4.3 Differences Between Domestic Onshore and International Structures

There are many differences between domestic onshore China structures and more international offshore structures as described previously. With each option, there are multiple jurisdictions that one can choose from. These develop further observations and reasons why each developed.

One of the biggest challenges is the short period in which the domestic jurisdiction has been in existence compared to its international counterparts. China's market is roughly 20 years behind the US market. The Chinese market has not caught up with older markets. In addition, historically there have only been few, smaller players who have lower penetration rates, resulting in this immature market.

Government policies and regulations, especially in terms of taxation, play an important role in the development of the leasing business in any country. The passing of the Investment Tax Credit Law in 1962 was instrumental in the US that enabled a tax credit for interest and dividend payments. With the enacted advancements in tax policies and regulations, the US, Europe, and Japan all enable leasing transactions to recognize interest deductions, depreciation of the asset (including accelerated depreciation), and other tax credits and allowances. China's policies and regulations advancements with regard to leasing, not counting one-off company-specific benefits, however, have fallen far behind especially as a result of slow depreciation, non-deductibility and other allowances for trading, and capital gains as compared to other jurisdictions. In addition, the taxation is quite high as well; also, other fees associated with employees have resulted in high taxation on leasing transactions. These result in lower returns on investment compared with other jurisdictions, making the jurisdiction not as competitive on a global basis.

In addition, the domestic financial markets have historically not been well developed and where the financing channels require increased cost of funds or interest as compared to other markets. In addition, the market is less flexible in terms of financing higher-cost assets as well as long-term liquidity. There is still the preference for shorter-term financings. As with all foreign-denominated assets, SAFE strict controls and regulations also exist. While the domestic circumstances are changing quite rapidly, international financing markets, especially the capital markets, are more mature and flexible to handle these transactions.

China's aircraft leasing market consists mainly of finance leases rather than operating leases. In part, this is due to history as this was first introduced and it accounted for the majority of leasing transactions for the first few decades. Especially in the first ten years of leasing aircraft, all leasing methods used by China's civil aviation were financial leasing. From the 1980s to 2001, China introduced 437 aircraft using leases representing $26 billion of asset value. Of these, 320 aircraft used financial leasing or 73% and 117 aircraft used operating leasing or 27 aircraft. China's operating leasing did not begin in earnest until 1990 with China Southern Airline's first package of ten aircraft consisting of five Boeing 737-300 and five Boeing 737-700 aircraft with the GPA Group, then the largest aircraft leasing company in the world. In addition, the package included the sale of ten older Boeing 737-200 aircraft (Unsworth 1990). This set a precedent for not only the largest operating leasing in China but also a familiar structure still used today for packing the lease of new aircraft along with the sale of older aircraft from the airline's inventory.

In addition to finance or operating leases, there are a variety of structures that utilize local tax benefits that have become popular. These include various forms of tax-optimized leveraged lease structures in the US, Germany, Netherlands, Sweden, Hong Kong, France. China has adopted many of these innovative leasing structures.

5 Current and Future Market Overview

Since 2007 when legislation allowed banks to invest in leasing, leasing has boomed. Turnover has increased from 8 billion RMB in leasing in 2006 to 4.4 trillion RMB in 2015 and all aspects of financial leasing, domestic financial leasing, and foreign financial leasing have shown similar growth. In addition, the number of leasing companies now totals 44,400 nationally with almost an equal number distributed between the three different types mentioned, with an astonishing growth rate of 12,400% (Yang 2016).

See Table 4.4 for a more detailed Chinese leasing segment revenue breakdown. In addition, the total assets under management in both types of leasing companies have grown quite rapidly as well, with 68% CAGR from 2008 to 2014 under both of their respective regulators. See Table 4.5

Table 4.4 Chinese finance leasing segment's revenue, 2006–2015

Chinese finance leasing segment's revenues 2006 to 2015 (in 100 million RMB)				
Year	*National turnover*	*Financial leasing*	*Domestic financial leasing*	*Foreign financial leasing*
2006	80	10	60	10
2007	240	90	100	50
2008	1550	420	630	500
2009	3700	1700	1300	700
2010	7000	3500	2200	1300
2011	9300	3900	3200	2200
2012	15,500	6600	5400	3500
2013	21,000	8600	6900	5500
2014	32,000	13,000	10,000	9000
2015	44,400	17,300	13,000	14,100

Source: Yang 2016

Table 4.5 Chinese finance leasing company statistics by type in 2015

	Chinese finance leasing company statistics by type in 2015			
	Number of enterprises	*Business volume (billions RMB)*	*Growth rate (%)*	*Ratio*
Finance lease	17,300	13,000	4300	33.08
Domestic finance lease	13,000	10,000	3000	30
Foreign finance lease	14,100	9000	5100	56.7
Total	*44,400*	*32,000*	*12,400*	*38.8*

Source: Yang 2016

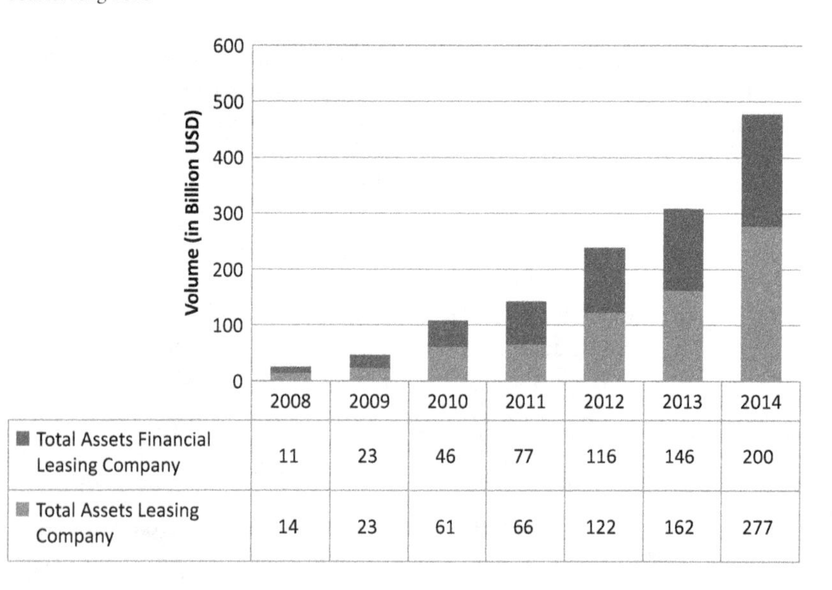

Fig. 4.14 China leasing company assets. (Source: China Banking Regulatory Commission 2016; Ministry of Commerce of the People's Republic of China 2016)

for statistics on types of Chinese leasing companies. Figure 4.14 shows the substantial increase of both leasing companies and financial leasing companies in terms of asset growth over the 2008 to 2014 timeframe.

As of 2015, there were 2880 large commercial aircraft in China, with narrowbody aircraft representing more than 80% of the fleet mix. According to Airbus, the current breakdown of the 2711 aircraft in China is 49% and 51% market share by Airbus and Boeing, respectively. This figure comes to 2377 or 88% overall narrowbody aircraft and the market share is broken down to 48% and 52% for Airbus and Boeing, respectively. Widebodies represent 334 aircraft or 12% overall mix and the market share breakdown is 55% and 45% for Airbus and Boeing, respectively. Ascend's data on China's current fleet shows similar breakdowns with 2800 aircraft as at the end of 2016. These figures are similar to that of the global mix; both manufacturers have an almost equal share of the aircraft market in China. Given the aircraft on order and under LOI, we can conclude that the majority of the aircraft are narrowbodies and a large number are next generation new technology aircraft.

While there are many thoughts about the growth of the market by either party, what is agreed is that there will be tremendous growth in the Chinese market in the next 20 years. Boeing projects in its 20-year forecast that there will be 7720 large commercial aircraft in China as of 2035 and narrowbody aircraft will represent 75% of the aircraft mix, as shown in Fig. 4.15.

There will be 6810 deliveries in the 20 years representing $1.025 billion of market value with an average of $150 million per aircraft as shown in Fig. 4.16 (Boeing Corporation 2016).

While a bit more conservative than Boeing, Airbus still forecasts a very large market size of 5970 new aircraft being delivered, representing $945 million of market value in their 20-year projections, as shown in Fig. 4.17. Airbus differs slightly from Boeing in the market projections and shows that 71% of new deliveries over 20 years are narrowbody aircraft.

6 Industry Participants' Characteristics

The Chinese leasing landscape is changing quite rapidly given the encouragement of the industry by the government. There are many new trends being witnessed by industry participants. In China, almost all of the top 15 banks by assets are active as investors through their owned leasing companies, except for Postal Savings Bank of China and Agricultural Bank of China, a policy bank (Relbanks.com 2016a). With the exception of Bank of China through its acquisition of SALE in 2006, all the other banks' activities are newly formed financial leasing entities created after the 2007 creation of financial leasing entities owned by banks.

Top Chinese lessors are among the top 20 globally, according to Airfinance Journal's Top Lessors 2016 by aircraft: BOC Aviation is at

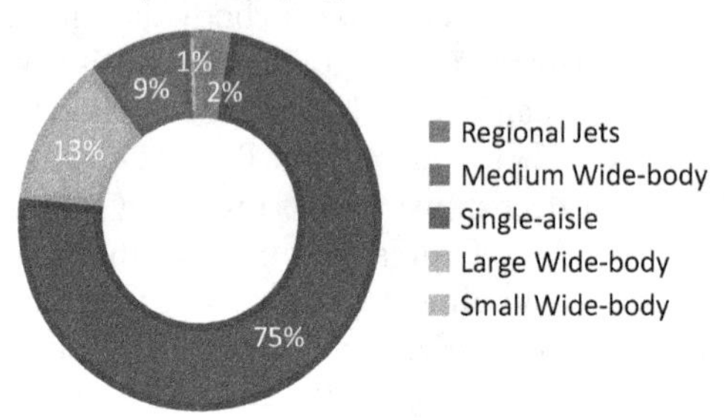

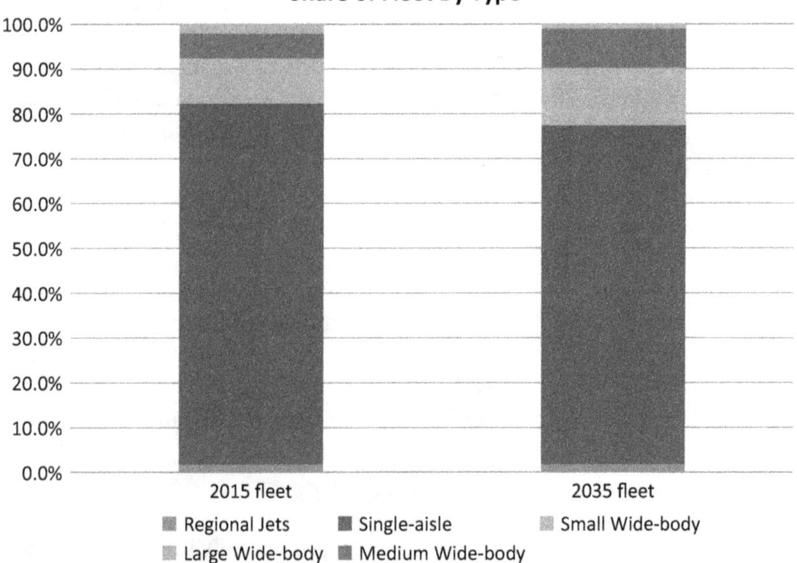

Fig. 4.15 Current China fleet and future projections by type. (Source: Boeing Corporation 2016)

eighth place with 267 aircraft; ICBC Leasing at 12th place with 218 aircraft; and CDB Leasing at 17th place with 148 aircraft. Meanwhile, Minsheng Financial Leasing is at 46th place with 39 aircraft. Looking at the top 50 lessors by value (2016), BOC Aviation is at 6th place ($11.4 billion), ICBC Leasing is at 8th place ($10.2 billion), CDB Leasing is at 13th place ($6 billion), Bocom Leasing is at 19th place ($4.2 billion), and

Growth Measures (%)			New airplanes	Share by size (%)
Economy(GDP)	5.1	Large widebody	60	1
Traffic(RPK)	6.4	Medium widebody	630	9
Airplane fleet	5.1	Small widebody	870	13
		Single aisle	5,110	75
		Regional jet	140	2
		Total	6,810	

Market size			2015 fleet	2035 fleet
		Large widebody	60	80
Deliveries	6,810	Medium widebody	160	680
Market value	$1,025B	Small widebody	290	990
Average value	$150M	Single aisle	2,320	5,830
		Regional jet	50	140
		Total	2,880	7,720

Fig. 4.16 China fleet and Boeing future projections segmentation. (Source: Boeing Corporation 2016)

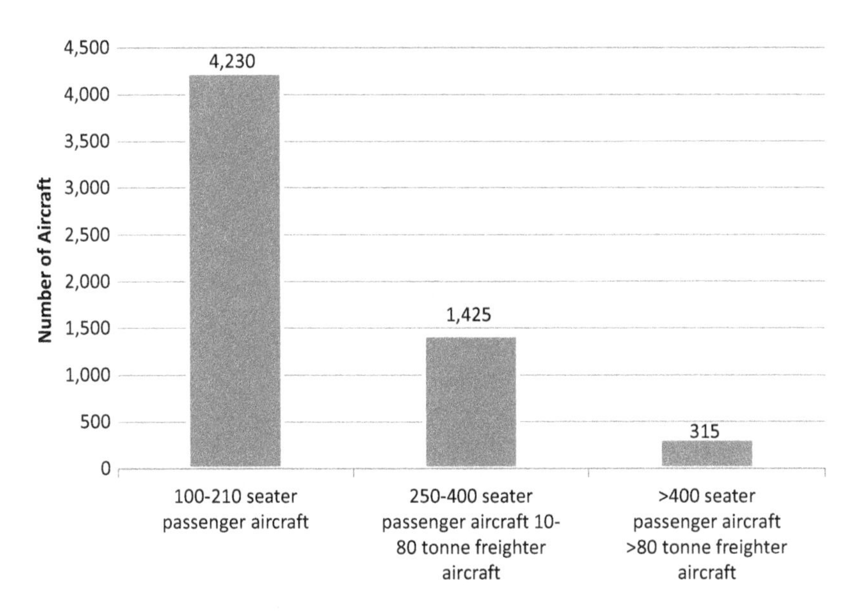

Fig. 4.17 Current Airbus China 20-year projections. (Source: Airbus 2016)

Minsheng Financial Leasing is at 41st place ($900 million) (Airfinance Journal 2016). This trend for increasing number of Chinese leasing companies in the global top 50 lessors list and higher assets under management by these companies will continue to develop as the demand for aircraft in the region increases.

In addition, there is a new trend that is emerging in aircraft leasing of global commercial banks, which is that of downsizing their investments. RBS Bank sold its aviation division to SMBC of Japan. Investec, for example, recently sold its 20% share of Goshawk Aviation to Hong Kong-based co-shareholders Chow Tai Fook Enterprises (CTFE) and NWS, both of which now have 50% shareholding. Goshawk is Investec's third aircraft leasing platform, along with Global Aircraft Fund and Aircraft Syndicate Limited. It is also interesting to note that Goshawk was originally set up with backing from Investec, CTFE, and Cheung Kong (CK), but CK subsequently sold its stake in the lessor to NWS. CK has since established Accipiter and several other joint ventures with global aircraft lessors. This trend also shows the interest of Hong Kong-based aircraft lessors in CTFE, CK NWS, along with China Aircraft Leasing Group, which listed in Hong Kong along with newer entrants such as Asia Pacific Aviation Leasing Group backed by Hong Kong-based private equity interests.

Increased interest in the sector from large insurance companies in China has also emerged. Like its compatriot banks, most of the major insurance firms in China have created financial leasing companies under China's Ministry of Finance regulations and have primarily focused on finance leases, although some have started specific aviation divisions and others have diversified into operating leases. Insurance companies are also a growing force in terms of financing and investments as they are being allowed to diversify their holdings by the various regulators, mostly China Insurance Regulatory Commission (CIRC), CSRC, and PBOC. Traditionally, insurance groups have invested, and still have a bulk of their assets, in conservative fixed income such as local bonds and have expanded to equity as well as more alternative investments such as real estate and financial leasing. Examples include the formation of aviation leasing arms at Ping An, China Life, and Taiping insurance groups (the latter in a JV with Sinopec, an SOE oil major).

Some reasons insurance companies globally and regionally are attracted to aircraft leasing assets include the benefit of depreciation of the aircraft assets, which offsets other earnings from a tax perspective; the fact the investment is backed by physical assets with long life and stability of cash flows is attractive. Insurance companies try to increase their investment returns as well as match the duration of their liabilities. In addition, one can deploy sizable amounts which can be significant for the insurance groups. All of these rationales apply to Chinese insurance groups as they are also backing the expansion of their financial leasing activities in China and abroad to find higher-yielding and lower-risk returns. Aircraft leasing's structural characteristics are similar to real estate with large transaction sizes backed by physical assets, and debt, which is optional, and involves rental returns and the potential of asset appreciation for operating leases.

These new Chinese insurance players are big including Ping An Insurance ($753 billion in assets in 2015 and ranked the number 5 top global insurance company), China Taiping Insurance ($63 billion in assets in the financial year 2015) through its joint venture with Sinopec (by itself a large company ranking 4th in the Global Fortune 500 rankings in 2016), and China Life ($378 billion in assets in 2015 and the number 20 top global insurance company) through its joint venture entity (Relbanks.com 2016b). It will not be a surprise if over the next few years most of the top insurance players will have leasing arms and will be invested in aviation.

This is not a surprising trend given that most financial groups and large conglomerates have joined the bandwagon in investing directly in aircraft leasing assets. As the industry continues to grow in 2017 and beyond, even more players from insurance companies and other sectors will be entering the industry. These new capital sources will continue to change the composition of finance capital globally and increase weight toward insurance companies and Asian-based companies along with the continued growth of the global aircraft leasing industry.

7 Conclusion

The Chinese aircraft leasing and aviation industries continue to grow at a rapid pace compared with the overall global perspective. The demographics and the drivers are positive for the industry in China and there is significant room for further uplift. Aircraft and the airline industry are intertwined. The rapid changes and growth in China are affecting the global economic state. Specifically, the rapid growth of China-based airlines and aircraft leasing companies has had significant effects on the global aviation and leasing space. These drivers in China have also contributed to the increased significant volume and activity in cross-border M&A and investment trends in the aviation industry.

Bibliography

Xian Aircraft wins order for 60 MA60 twin-turboprops. (2000, November 14). *Flight International.*

Chinese Civil Aviation Set to Take Off. (2002, September 10). *Xinhua News Agency.* Retrieved from http://china.org.cn/english/BAT/42435.htm

Private Sector's Flying Dream. (2002, November 28). *China Daily.*

China to start construction on 35 railway projects: report. (2017, February 19). *Xinhua News Agency.*

Airbus. (2016). *Global Market Forecast 2016–2035.* Retrieved from www.airbus.com/aircraft/market/global-market-forecast.html

Airfinance Journal. (2016). Leasing Top 50 2016. *Airfinance Journal.*

An, B. (2013, December 18). Hukou reforms target 2020: official. *China Daily.*

Ballantyne, T. (2017, May 1). The long and winding road. *Orient Aviation.*

Barton, D., Chen, Y., & Jin, A. (2013, June). Mapping China's middle class. *McKinsey Quarterly.*

Bloomberg. (2017). *Economic, FX, Fuel, and Funding Data.* Retrieved from: www.bloomberg.com

Boeing Corporation. (2016). *Current Market Outlook 2016–2035.* Retrieved from Seattle, WA: www.boeing.com/resources/boeingdotcom/commercial/about-our-market/assets/downloads/cmo_print_2016_final_updated.pdf

China Banking Regulatory Commission. (2016). *Leasing Companies.* Retrieved from www.cbrc.gov.cn

Civil Aviation Administration of China. (2016a). *China Civil Aviation Development 13th Five-Year Plan (2016–2020).* Retrieved from www.caac.gov.cn/XXGK/XXGK/ZCFBJD/201702/P020170215595539950195.pdf

Civil Aviation Administration of China. (2016b). *Statistical Bulletin of Civil Aviation Industry Development in 2016.* Retrieved from www.caac.gov.cn/en/HYYJ/NDBG/201706/W020170602336938087866.pdf

Civil Aviation Administration of China. (2018). *Monthly Statistics of Civil Aviation Industry.* Retrieved from www.caac.gov.cn/

Deloitte. (2017). Taxation and Investment in China 2017. Retrieved from www2.deloitte.com/content/dam/Deloitte/global/Documents/Tax/dttl-tax-chinaguide-2017.pdf

Ding, Y., Ge, X., & Lv, F. (2014). Requirement Analysis and Prediction of Aviation Finance in China. *Information Technology Journal, 13*(1), 140–146.

Eaton, S. (2016). *The advance of the state in contemporary China : State-market relations in the reform era.* Cambridge: Cambridge University Press.

Exchange-rates.org. (2015). USD-CNY Exchange Rate 12.31.15. Retrieved from www.exchange-rates.org/Rate/USD/CNY/12-31-2015

38 million Chinese citizens hold ordinary passports. (2012, May 15, 2012). CCTV. Retrieved from http://news.cntv.cn/20120515/118225.shtml

Export–Import Bank of the US. (2018). Historical Timeline. Retrieved from www.exim.gov/about/history-exim/historical-timeline/full-historical-timeline

Ge, L. (2017, February 21). China to Build 74 New Civil Airports by 2020: CAAC. *China Aviation Daily*. Retrieved from www.chinaaviationdaily.com/news/60/60893.html

Haver Analytics. (2017). *China Transportation Data*. Retrieved from: www.haver.com

Hsu, K. (2014). *China's Airspace Management Challenge*. Retrieved from www.uscc.gov/sites/default/files/Research/China%27s%20Airspace%20Management%20Challenge.pdf

IHS Economics. (2016). *Economics Data*. Retrieved from: www.ihsmarkit.com

Innovata. (2016). *LCC Data*. Retrieved from: www.innovata-llc.com

International Air Transport Association. (2016). *Air Passenger Forecasts, October 2016*. Retrieved from www.iata.org

International Monetary Fund. (2017). *Country Economic Data*. Retrieved from: www.imf.org

Li, X. (2010, December 13). China aircraft leasing makes new steps. *China Economic News*.

Ministry of Commerce of the People's Republic of China. (2016). *Leasing Companies*. Retrieved from www.mofcom.gov.cn

Ministry of Public Security of the People's Republic of China. (2016). *Twenty-fourth Meeting of the Standing Committee of the Twelfth National People's Congress*. Retrieved from www.mps.gov.cn/n2253534/n2253535/n2253536/c5538068/content.html

National Bureau of Statistics of China. (2011). *Basic Statistics on National Population Census in 1953, 1964, 1982, 1990, 2000 and 2010*. Retrieved from: www.stats.gov.cn/tjsj/Ndsj/2011/html/D0305e.htm

National Bureau of Statistics of China. (2016). *China Statistical Yearbook 2016*. Retrieved from: www.stats.gov.cn/tjsj/ndsj/2016/indexeh.htm

National Development and Reform Commission of China. (2016). *Leasing Companies*. Retrieved from www.ndrc.gov.cn

Office of National Statistics. (2013). *Detailed country of birth and nationality analysis from the 2011 Census of England and Wales*. Retrieved from http://www.ons.gov.uk/ons/dcp171776_310441.pdf

People's Bank of China. (2017). Foreign Exchange Reserve Figures. Retrieved from www.pbc.gov.cn

PWC. (2012). *Overview of China's Aircraft Leasing Industry Promoting Industry Development*. Retrieved from www.pwccn.com

Russell, D. W. (2014). *Good risks : discovering the secrets to ORIX's 50 years of success*. Singapore: John Wiley.

Shi, Y. (2005). *Foundations and practices of leasing*, Beijing, China: University of International Business and Economics Press.

State Administration of Foreign Reserves of China. (2017). *Foreign Reserves.* Retrieved from www.safe.gov.cn

Unsworth, E. (1990, June 20). GPA Leases 1st Aircraft To China. *JOC.com.* Retrieved from www.joc.com/gpa-leases-1st-aircraft-china_19900620.html

US Census Bureau. (2018). Population Clock. Retrieved from www.census.gov/popclock

US State Department. (2017). Passport Statistics. Retrieved from https://travel.state.gov/content/travel/en/passports/after/passport-statistics.html

Wang, W. (2010, March 10). Airline boss in plea for more airspace. *Shanghai Daily.*

World Bank. (2017). *Country Economic Data.* Retrieved from: www.worldbank.org

Yang, H. (2016). *China financial leasing industry development report for 2015.* Tianjin: Nankai University Press.

Yu, D. (2017). *Aviation Joint Ventures, Sidecars & M&A.* Unpublished manuscript, Johns Hopkins University.

Yu, M. (2013, January 31). Inside China: No airspace for holiday travel. *Washington Times.* Retrieved from www.washingtontimes.com/news/2013/jan/31/inside-china-no-airspace-for-holiday-travel/.

CHAPTER 5

Empirical Aircraft Asset Pricing

1 Introduction and Research Questions

It has been a long-established thought and rumor of the industry that aircraft asset investment returns are moderate with lower volatility compared with other asset-backed classes. No systemic, rigorous comprehensive empirical study with real historical data has been done to substantiate this mere hypothesis and this research intends to fill that gap through the hand-collected data set. Previous thoughts have been incomplete based on sample sets too small, short, biased, or based on non-publically available proprietary data sets that have not been conducted with the scientific rigor to make convincing augments and conclusions. There are gaps in the academic literature that can be filled by examining empirical analysis especially in regards to aviation through analysis of 21-year historical representation of the aircraft asset class and construction of various indexes to test the current theories on shocks and examine the effects and characteristics before and after the 2008 global financial crisis.

To put things in perspective in terms of the relative size of the aviation industry, there are currently 6350 and 22,510 aircrafts in Asia and globally, respectively, as of the end of 2015. Deliveries of new aircraft over 30 years as of 2035 is projected to be 15,130 and 39,620 aircrafts in Asia and globally, respectively, representing market values of $2.35 trillion and $5.93 trillion in Asia and globally, respectively, based on list prices (Boeing Corporation 2016). This is a very big market and also has general

D. Yu, *Aircraft Valuation*,
https://doi.org/10.1007/978-981-15-6743-8_5

significance to the overall economy and it does not to have any significant robust academic research relating to asset pricing and its dynamics.

The basis of the empirical study is the unique data set that has been hand collected from a significant subset of the world's foremost aircraft appraisal firms. This subset group of aircraft appraisers has specifically agreed to provide the aircraft historical appraised data and value projections for this research. While not every single source data is available for the entire time frame, the overall time period comprises aircraft pricing data by various types from June 1996 to June 2007. This empirical research is novel in that the data has never been systemically provided by so many of the world's aircraft appraisers, who represent the bulk of the aviation market and industry. While the actual historical transaction data is the ideal, this is not available due to its competitive, proprietary trade secret characteristics so this is the closest proxy to actual real market aircraft pricing. None of the appraisers providing data has previously contributed data towards any group-based research project nor have they contributed to such an encompassing collective study on the aircraft industry.

The main question addressed is determining the market characteristics of the aircraft asset class in terms of its returns, volatility and trends. In this empirical study, several indexes were created to group the various aircraft types, namely, Index: (1) all aircraft types, (2) all narrowbody aircraft, (3) all widebody aircraft, (4) all narrowbody classic aircraft, and (5) all narrowbody next generation (NG). A more thorough discussion is in Methodology Sect. 4.5. Aircraft pricing data is also segmented into four time segments. For each of these five aircraft type groups, Index 1–5, the four time periods analyzed are Time: (1) all available historical information 6.1996–6.2017, (2) pre-Global Financial Crisis (pre-GFC) period 6.1996–6.2007, (3) during the GFC period 6.2007–6.2010, and (4) post-GFC period 6.2010–6.2017. To clarify the data time series throughout, 6 denotes June throughout the book. The time segments were chosen to analyze the effects and trends by the GFC on aircraft asset pricing.

The main hypothesis is that the aircraft market, represented by all aircraft types (Index 1), will have higher value depreciation compared to the accounting depreciation and low standard deviation throughout the four time segments.

The second hypothesis is that narrowbody aircraft, represented by Index 2, will have lower value depreciation and standard deviation throughout the four time segments than widebody aircraft, represented by Index 3.

The third hypothesis is that NG narrowbody aircraft, represented by Index 5, will have lower value depreciation and standard deviation throughout the four time segments than classic narrowbody aircraft, represented by Index 4, as well as all narrowbody aircraft, represented by Index 2.

The fourth hypothesis is that the five aircraft market segments, represented by Index 1–5, will have lower value depreciation and standard deviation in the pre-GFC period, Time 2, than during and post the GFC in Time 3 and Time 4, respectively. Conversely, compared to other time segments, all the aircraft market segments, represented by Index 1–5, will have higher value depreciation and standard deviation during the GFC Time 3.

The fifth hypothesis is that there is a slight to moderate correlation between market value (MV) and base value (BV) for all the aircraft types, represented by Index 1–5, throughout the four time segments, while the highest relative correlation occurs in pre-GFC Time 2 and the lowest occurs during the GFC Time 3.

In addition to the five hypotheses, both cross aircraft type and cross time subset characteristics are analyzed. These empirical studies add on both the academic conversation about the subject but also show that there are many potential applications of the study results. One example is that investors can have an accurate view of the historical returns and volatilities of the aircraft asset space as well as public market pure-play aircraft leasing companies. This might enable the investors to decrease, hold, or increase their overall exposure to the asset space given their understanding of other compared asset classes. Another example of a possible implication of the results is that the investors can adjust their percentage allocation holdings of different types of aircraft given the findings, that is, single-aisle versus twin-aisle aircraft or commercial versus regional aircraft. Any, even a small, potential movement of attitudes based on the proposed research results might have significant dollar size ramifications and impact given the size of the market sizes.

2 Aircraft Asset Values and Valuation Methods Discussion

Aircraft is a major asset for an airline and/or the owners such as lessors, and managing this asset with the goal of maximizing its value is a complex process with many factors as discussed in Chap. 3.

2.1 *Types of Aircraft Values*

The main categories of values are theoretical, market, and relative values. In commercial aviation, the most widely recognized and accepted industry organization is the non-profit International Society of Transport Aircraft Trading (ISTAT) and the standards of valuation are governed by its ISTAT Appraisers' Program (IAP). Not only does it have the gold standard certification program, of which this author is one of the approximately 50 certified appraisers, the IAP has created common value definitions, which are generally accepted by most of the industry. The most common ISTAT definitions of values are Base Value (BV), Market Value (MV or CMV or FMV), Residual Value (RV), Distress Value (DV), Securitized Value or Lease-Encumbered Value (LEV), and those are found in Table 5.1.

The BV definition is a somewhat idealized aircraft and viewed under long-term market trends and conditions while MV has some similarities to BV but differs in the relaxation of the reasonable supply and demand equilibrium constraint and reflects the market at the specified time. DV is lower than BV and MV and assumes harsher counterparty conditions according to the definition. With LEV, this is the equivalent of a simplified discounted cash flow value where the lease cash flow streams along with a terminal future RV are present valued with an appropriate discount rate. This approach is the most suited analysis from an aircraft lessor or owner's approach to value. There has been much academic literature on discount rates to use (Beechy 1969; Johnson and Lewellen 1972; Bower 1973; Harris and Pringle 1985, etc.). The drawbacks of this approach is that the valuation appraiser may not have all the facts as to the credit risks with the lessee and parties involved nor all of the side provisions of the lease, some of which can have significant value impacts such as security deposits, return conditions, repossession rights, term extensions, sublease rights, reserve, and other payments.

Salvage Value is the part out value and is appropriate with older aircraft as well as during times when the secondary parts market is very robust as

Table 5.1 Value definitions

- *Base Value*, the appraiser's opinion of the underlying economic value of an aircraft in an open, unrestricted, stable market environment with a reasonable balance of supply and demand and assumes full consideration of its "highest and best use." An aircraft's Base Value is founded in the historical trend of values and in the projection of value trends and presumes an arm's-length, cash transaction between willing, able and knowledgeable parties, acting prudently, with an absence of duress and with a reasonable period of time available for marketing. In most cases, the Base Value of an aircraft assumes its physical condition is average for an aircraft of its type and age and its maintenance time status is at mid-life, mid-time (or benefiting from an above-average maintenance status if it is new or nearly new, as the case may be).
- *Market Value*, or *Fair Market Value* or *Current Market Value* if the value pertains to the time of the analysis, is the appraiser's opinion of the most likely trading price that may be generated for an aircraft under the market circumstances that are perceived to exist at the time in question. Market Value assumes that the aircraft is valued for its highest, best use, that the parties to the hypothetical sale transaction are willing, able, prudent and knowledgeable and under no unusual pressure for a prompt sale and that the transaction would be negotiated in an open and unrestricted market on an arm's-length basis, for cash or equivalent consideration and given an adequate amount of time for effective exposure to prospective buyers.
- *Residual Value* is the value of an aircraft, engine or other item at a future date, often used in connection with the conclusion of a lease term.
- *Distress Value*, *Forced Sale Value*, *Liquidation Value* are terms to describe the appraiser's opinion of the price at which an aircraft (or other assets such as an engine or spare parts) could be sold in a cash transaction under abnormal conditions—typically an artificially limited marketing time period, the perception of the seller being under duress to sell, an auction, a liquidation, commercial restrictions, legal complications, or other such factors that materially reduce the bargaining leverage of the seller and give prospective buyers a significant advantage that can translate into heavily discounted actual trading prices. Depending on the nature of the assignment, the appraiser may be asked to qualify his opinion in terms of disposition within a specified time period, for example 60 days, 90 days, or six months as the needs may be. Apart from the fact that the seller is uncommonly motivated, the parties to the transaction are otherwise assumed to be willing, able, prudent, and knowledgeable, and negotiating at arm's-length, normally under the market conditions that are perceived to exist at the time—not an idealized balanced market.
- *Securitized Value* or *Lease-Encumbered Value* is the appraiser's opinion of the value of an aircraft, under lease, given a specified lease payment stream (rents and term) and estimated future residual value at lease termination and an appropriate discount rate.
- *Salvage Value* is the actual or estimated selling price of an aircraft, engine, or major assembly based on the value of marketable parts and components that could be salvaged for reuse on other aircraft or engines. The value should be determined and stated in such a way to make it clear whether it includes adjustment for removal costs.
- *Scrap Value* is the actual or estimated MV of an aircraft, engine, or major assembly based solely on its metal or other recyclable material content with no saleable reusable parts or components remaining. The scrap value is usually expressed as net of removal and disposal costs. In some cases scrap value could be zero if the dismantling and disposal costs are high, as for example hazardous materials or composite assemblies that might be impossible to recycle.

Source: International Society of Transport Aircraft Trading 2013

the sum of the salvageable parts can be higher than the MV of the whole aircraft. Salvage Value differs from Scrap Value as there are no more reusable parts left and is based on the metal or recyclable contents. These valuation terms and approaches are used more towards the end of life as Salvage Value and Scrap Value are compared to the valuation of the aircraft as a rental producing asset.

Appraisal values are asset values that are generally provided by a third party and should be independent. These are the definitions of values followed by the large valuation consultancy firms and stated compliance these conditions are written in the reports. There are some points to note about appraisal values. The purpose or objective of the appraisal is important as there can be many reasons to undertake an aircraft appraisal. Some objectives and the various types of appraisals listed by ISTAT are found in Table 5.2.

The most public figures are the manufacturer's published list prices, which are published yearly for each of the models for current production aircraft available for sale. For a historical view of Boeing and Airbus' list price growth, see Figs. 5.1 and 5.2 which are hand collected data from Boeing and Airbus.

These figures are usually in the form of minimum, average, and high list prices as each aircraft can be customized based on airline, lessor, or investor, requirements. Some of these include the engine type, thrust, different certified maximum aircraft weights such as maximum take off weights (MTOW), seat manufacturer, number of seats, and so on. As part of these negotiations, the aircraft original equipment manufacturer (OEM) can provide various forms of incentives and concessions that are written in the form of various support letters when acquiring an aircraft. These support letters can describe discounts to the delivered list price, pre-delivery

Table 5.2 Various kinds of appraisals

1. To determine the value of the aircraft for sale or lease;
2. To determine present and future aircraft value for loan collateral;
3. To determine future value of aircraft for establishing residual insurance premiums;
4. To determine future value of aircraft as a basis for lease rates;
5. To determine airline fleet values to support equity stock offerings, values in mergers, bankruptcies, and so on;
6. To determine aircraft values as a basis for state and local property taxes;
7. To determine values acceptable to the IRS [Internal Revenue Service or tax authorities] of contributions to entities such as aviation schools and museums.

Source: International Society of Transport Aircraft Trading 2009

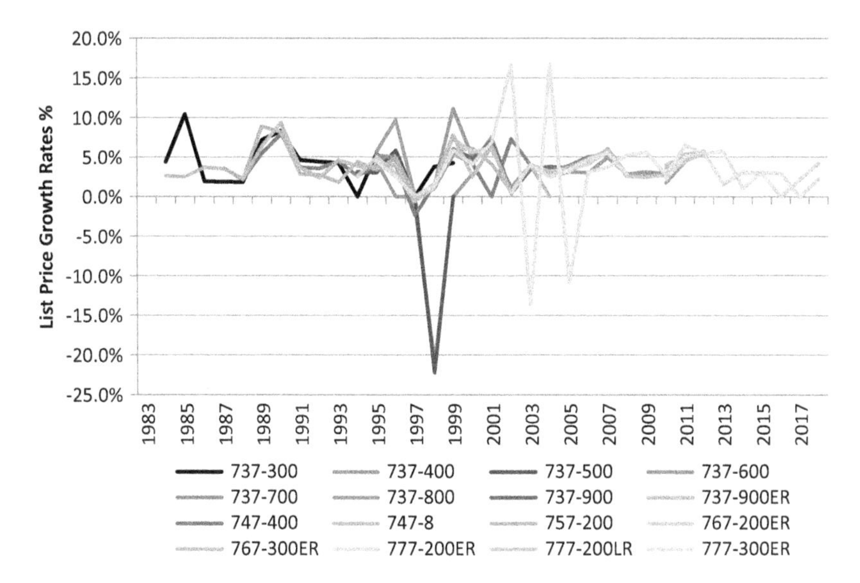

Fig. 5.1 Boeing mean list price growth rates. (Source: Boeing Corporation 2017)

payment terms and conditions, caps on escalation rates, special terms for training, spare parts, and other technical assistance.

The list price is just a starting point for the negotiation and the final delivery price depends on a variety of factors with some examples such as the customer profile, number and type of aircraft, OEM's order book, market conditions, competitive and strategic dynamic, and so on. These list values can be different from the realized and also appraisal values. The list prices of aircraft by both Boeing and Airbus are generally set every year with an upward trend. They have historically increased every year with a few exceptions as noted in the Figs. 5.1 and 5.2. The year-to-year increase percentage is independent of the set yearly escalation rate.

2.2 *Valuation Methods*

There are several ways to value an aircraft. This is a complex process, given that there are many different factors that can be included in the valuation model. The basic concept like all valuations is to determine the direct revenue net of the total operating costs that can be derived from the aircraft.

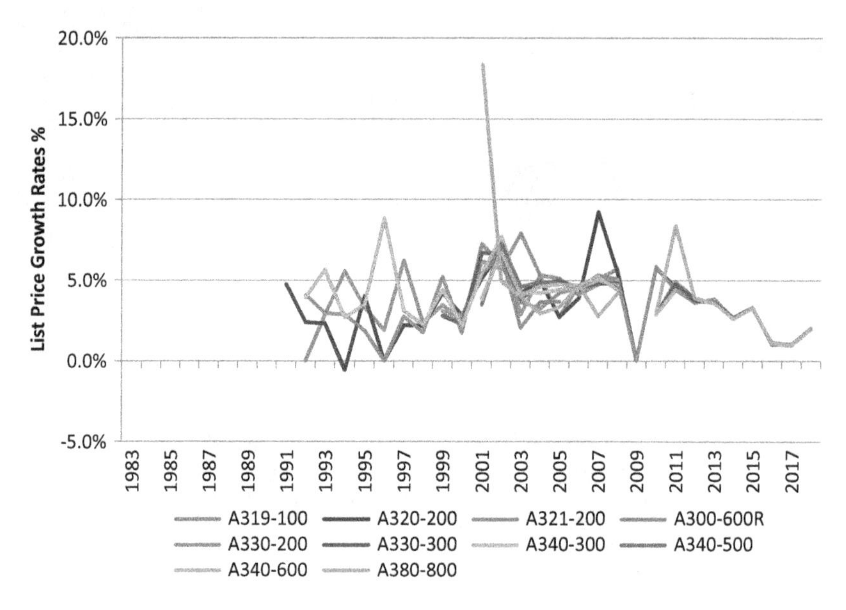

Fig. 5.2 Airbus mean list price growth rates. (Source: Airbus 2017)

From this basic form, many adjustments are made, both internal and external to the aircraft itself. Some of these internal factors include the manufacturer, aircraft type, specifications, age, the useful life of the asset, seating capacity, size, maintenance, and technical status. Some of these external factors include market supply and demand, number of airlines and their stage of development, fuel price, and aviation policy factors. Figure 5.3 is an example of a scatter plot of a dataset of about 4000 various aircraft resale values from 1978 to 1998 with a regression curve based on the percentage of cost to estimate the life span depreciation of aircraft assets.

There are many similarities and differences in aircraft asset valuation compared with traditional company-level corporate valuation techniques. These corporate valuation techniques, such as the discounted cash flow method (DCF), adjusted present valuation method (APV), the sum of the parts valuation, liquidation or bankruptcy valuation, and relative value approaches including comparable companies' multiples analysis, precedent transactions analysis, can be combined and adjusted for corporate-level

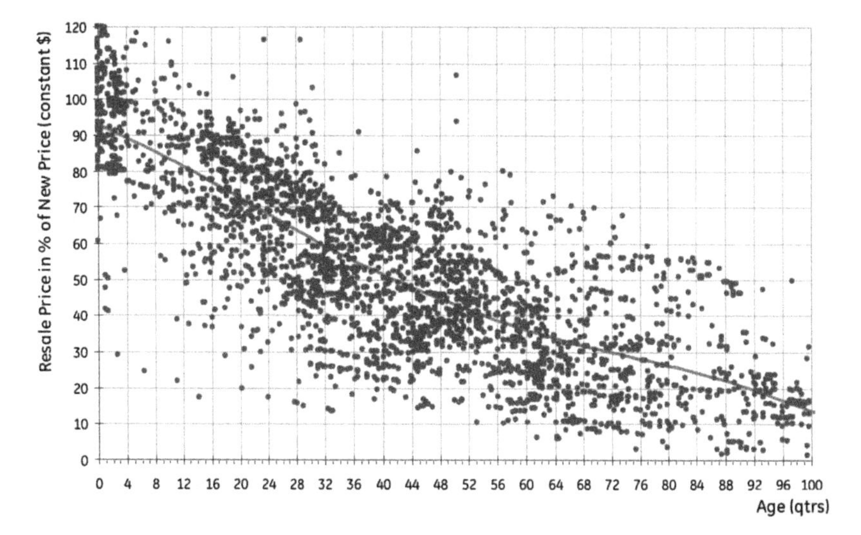

Fig. 5.3 Scatter plot of constant dollar resale prices as a percentage of new price for jet aircraft collected: 1974–1998. (Source: Hallerstrom 2013. Note: The dataset contains about 4000 data points. Some sales may have had leases attached to the aircraft.)

application and can also be adjusted for asset valuation (Myers 1974; Myers et al. 1976; Gibson and Morrell 2004, 2005).

The most traditional valuation method is the DCF approach (Gibson and Morrell 2004, 2005; etc.). In the company valuation method, one calculates the present value of the projected free cash flows from both operations as well as the accounting effects with an appropriate discount factor. In the aircraft asset valuation approach, one envisions a basic single-aircraft charter company. A cash flow approach can be derived with the effects of the net sources of revenue or yields including passenger, freight, and ancillary components and total costs including direct operating, indirect operating, and nonoperating costs. Some examples of direct operating costs are crew, fuel, maintenance, ownership, and landing fees while indirect operating costs include cargo and baggage handling, passenger service, sales, and reservation systems. Nonoperating costs mainly consist of items administrative in nature such as general management. Both revenue and costs can be fixed or variable costs. Some examples of fixed costs are salaries and insurance. Some examples of variable costs are fuel, maintenance, and landing fees. Some examples can either be direct or variable are crew and ownership costs (fixed or benchmarked leased or pay by amount utilized).

$$DCF = \frac{FCF_1}{(1+r)^1} + \frac{FCF_2}{(1+r)^2} + \ldots + \frac{FCF_n}{(1+r)^n} + \frac{RV_n}{(1+r)^n}$$

FCF = free cash flow	RV = residual or terminal value	r = discount rate (WACC)	n = number of periods

FCF for Company = FCF for aircraft = TR – TC.

TR = total revenue	TC = total cost

FCF for Aircraft Asset from Lessor's Point of View
=TR–TC=gross aircraft lease
payments – maintenance-crew-fuel-insurance
FCF for Aircraft Asset from Operator's Point of View = TR–TC
$TR = Revenue_{passenger} + Revenue_{cargo} + Revenue_{ancillary}$
$TC = Cost_{direct\ operating} + Cost_{indirect\ operating} + Cost_{non\text{-}operating}$
Company and Aircraft Asset Cost of Capital
$WACC = w_{debt1}k_{debt1}(1\text{-}t) + \ldots + w_{debtn}k_{debtn}(1\text{-}t) + w_{equity}k_{equity}$

k = cost of equity or $debt_n$	w = weight of equity or $debt_n$	n = type of debt number

These valuation techniques can also be combined with other techniques and incorporate adjustments to account for market dynamics, and so on. While the traditional company DCF approach takes into account debt, this asset approach does not and is considered the all equity or asset approach similar to the data. To incorporate the debt effects, the modification of the APV method for aircraft assets is more suitable from academic literature. The APV approach is the summation of the value of all equity or unlevered asset with the incremental value of the debt effects on the asset. The debt effects can be further separated for each tranche or type of debt such as senior debt, mezzanine debt, and so on.

Asset Price = Unlevered Asset Value + $(T_1 - C_1) + (T_2 - C_2) + \ldots + (T_n - C_n)$

T = tax benefit of debt type n	C = cost of debt type n	N = different types of debt

There are other valuation techniques that are similar between company valuation and asset value. Other than comparable companies' multiples analysis, precedent transactions method, sum of the parts valuation, and liquidation or bankruptcy valuation can all be adjusted to suit the asset. Distress Value is similar to liquidation or bankruptcy value while Salvage Value is analogous to the sum of the parts value for companies. A similar relative value-based method, precedent transactions approach, can also be used for asset valuation based on the adjusted maintenance status and price of the aircraft. As there is not a similar public comparables that are easily accessed, this is not a useful method for asset valuation. A reason for this is similar to that of academic research due to the lack of data available.

2.3 Aircraft Depreciation and Economic Useful Life

As the aircraft is used and age, there is real, book, and market depreciation. Real depreciation happens as physically the aircraft costs more man-hour repairs, downtime, and financial cost in maintenance as the aircraft is used and life limited parts are replaced and flight cycle and flight hour limits items are reached. This causes operating costs to increase over its lifetime. An aircraft is similar to other machines like an automobile whereas older cars are more expensive to maintain than newer ones.

Book depreciation is the accounting depreciation of the aircraft. An aircraft's accounting or economic useful life is 25 years with a 15% residual for value recapture. This accounting depreciation happens as the aircraft is depreciated on a straight line basis but actual market depreciation is much more volatile from the study. Figure 5.4 is an example of the actual market depreciation curve of two different vintages (one early and one late) of the MD-83 type aircraft which can be seen to have very different depreciation from a straight line accounting framework.

Market depreciation is based on many things such as real and accounting depreciation but it also takes into more market-based pricing inputs such as market dynamics. These market dynamics include the relative performance of the aircraft and the relative supply and demand of the aircraft. It also includes the technological competitiveness of the specific aircraft type relating to other types. As technology advances and the aircraft type

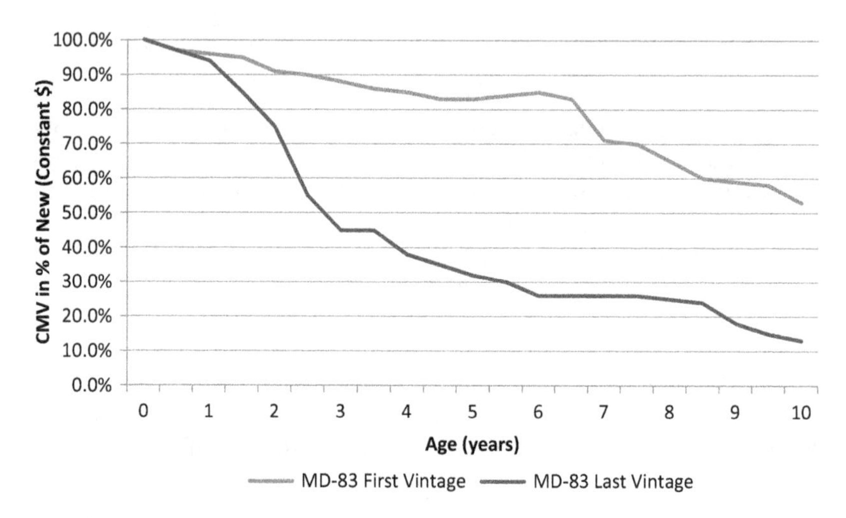

Fig. 5.4 Relative constant dollar ascend CMV for earliest and last vintage of MD-83. (Source: Hallerstrom 2013. Note: Relative constant dollar Ascend CMV for the earliest and last vintage of the MD-83. Note that the early vintage has substantially better value retention)

technologically depreciates and along with the competitive nature of markets creates the introduction of a new replacement type and technology. This is in relation to the life cycle of the product and technology. For example, the Boeing 737 classic series of aircraft are replaced by the newer Boeing 737 NG series, which are then replaced by the Boeing 737 MAX series. As a duopoly of the aircraft manufacturers exists, especially for large aircraft, this competitiveness in technology is alive and well (Oum et al. 2000).

Aircraft assets have a finite economic useful life. This is due to a couple of factors. One is the technology time and use constraints of its parts for safe operation. In theory, the useful life of an aircraft is as long as there is economic benefit and the technical requirements and regulatory certifications can be maintained. This time of useful life is reached when big components of the aircraft such as airframes have reached their limits either due to absolute time, flight hour, or cycle limits, or a combination. While some can be continuously repaired safely and restored such as engines, others such as the airframe would not be economically feasible which forms the basis of economic useful life. The other factor for finite life is the technological cycle and its competitiveness aspect where the future newer generations would accelerate the retirement of older technologies. With this economic useful life constraint, more visibility is needed in cash flows to account for both depreciation and an adequate return. There is a lot of debate in the industry as to whether this 25-year economic useful life is still a good metric or is the useful life less than 25 years, given the increasing speed of technology cycles and replacement, among other factors.

2.4 *Salvage and Scrap Value*

At all points of the aircraft life cycle, owners can elect to salvage the parts and scrap the remaining aircraft for its part out value. While this is historically at the later end of the life cycle when it is difficult to find lessees or consideration of other market forces, this option always exists in the entire life cycle of the asset. Salvage Value differs from Scrap Value as there are no more reusable parts left and is based on the metal or recyclable contents so there is a need to consider the removal costs for Salvage Value. The equation for considering continued use versus scrapping is below, whereas MP_a represents the expected value of the aircraft and its cash flow as described in Section 5.2.2.

MP_a = DCF (SUM (Revenues$_1$, Costs$_1$) + ...+ SUM (Revenues$_n$, Costs$_n$)+ Residual Value) $\geq$SV – RC + SCV

MPa = market price of aircraft a	SV = salvage value	SCV = scrap value	RC = removal costs

This consideration can also be viewed to the highest limiting large part which is the closest to needing repairs as the cost for major restorations such as the engines is significant and the investment is more than the expected value afterward. As aircraft type families have many interchangeable parts and components, this can result in increased demand where the summation of the part values can be more than the entire aircraft sold as is. Likewise, this case also produces instances where lessors acquire a functional aircraft for the purpose of parting out the aircraft where they view that the total value of the parts minus costs is more than the acquisition cost.

2.5 Inflation

Inflation has effects on valuation. First, the OEMs set yearly escalation rates for the pricing of delivered new aircraft. This escalation is an approximation of the inflation in the cost of the production of aircraft. While the exact formula is undisclosed, it is said to be based on a basket of metrics such as consumer price index inflation, labor, and other factors. For new aircraft with a long future anticipated delivery period, the escalation component of the price is a significant portion of the increased price from the base year period reflected in the contract. Secondly, there can be inflation in the components and subcomponents parts which all have inflation pricing reflecting the increases in production cost. These do not account for market increases due to higher margins requirements for OEMs or others. In turn, these inflation costs are reflected in the cost of the aircraft—both new and used. Given that aircraft and its components are priced in US Dollars, aircraft can be viewed as a movable US Dollar based real asset. Real assets like real estate can act as inflationary hedges (Froot 1995).

2.6 Maintenance Condition Value

The US Federal Aviation Authority's definition of an aircraft is a "device used or intended to be used for flight in the air" while an airplane is defined as an "engine-driven fixed-wing aircraft heavier than air, that is supported in flight by the dynamic reaction of the air against its wings" (Federal Aviation Administration 2018). The basics of a commercial airplane is an airframe that is supported by engines—normally two or four—and various critical parts such as landing gears and the auxiliary power unit (APU). Each component, assemblies, subassemblies, and parts have specific usage regulations in terms of absolute time, flight hour or cycle limits, or a combination. When an aircraft is delivered brand new from the OEM, all of the components are at full life. Each of the components has an absolute or relative time or utilization, in-flight hours or cycles, limitation, or both. As the aircraft is used and the parts are in use, each of these components, assemblies, subassemblies, and parts' technical life counts down. Some of these items are repairable and can be repaired and restored as if new, others are expendable, where no repair procedures exist and some items are life limited.

The main cost components of a whole aircraft are the airframe, engines, APU, and landing gears. Airlines maintain different types of approved maintenance schedules. The most commonly used maintenance checks are the letter checks—A Check, C Check, and D check or the heavy maintenance check. For each aircraft type and depending on the utilization, the intervals for the light and major checks are different. Airframe has time limitations for this heavy maintenance check. For engines, it is made up of many components and there are many life limited parts (LLP) that need to be replaced after certain limits are achieved. The engine also needs to be overhauled for the performance restoration. The time interval between every subsequent performance restoration after the first will be less as the overhaul will not be as effective. Theoretically, the overhaul repair is where the engine is removed when all of the aircraft LLPs reach zero life but in practice, as each of the parts has different limits, the need for repair is based on the lowest common denominator or the first LLP to reach zero life as the cost to remove and the downtime of the engine is costly. The total cost of each of the LLPs and the refurbishment can be significant amounts of capital. Aircraft maintenance evolution is more geared towards less removal of engines for heavy checks and towards more smaller, frequent checks. Landing gears have specific time and cycle limits to overhaul and similar to APUs. Figure 5.5 is an illustrative example of the negative value changes over time of an aircraft based only on its maintenance consumption and refurbishment from the lessor's point of view.

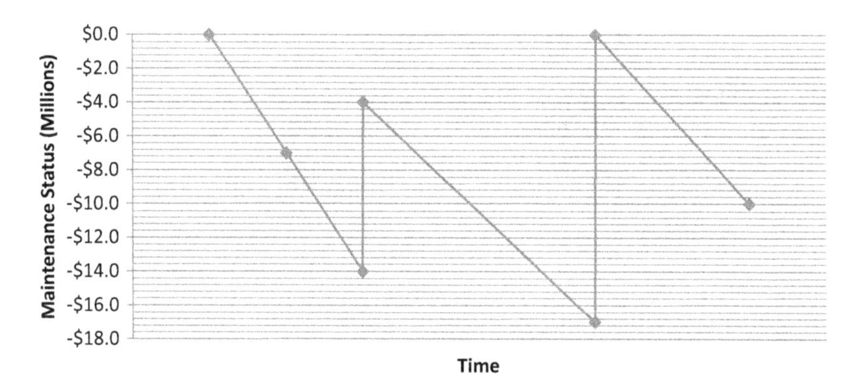

Fig. 5.5 Value change from maintenance consumption and refurbishment

All of the maintenance cost items have a significant impact on the value of the aircraft asset as each part costs significant sums as well as downtime for the replacement. Each of these items needs to be adjusted from the full life status in the valuation to reflect the current condition in the aircraft. Forecasting and estimating future maintenance costs are both a science and an art form. One of the main information required is the aircraft route and utilization assumptions. While some items are relatively easy to forecast and model such as the airframe heavy check, APU, landing gears, and engine LLPs, other things such as engine performance restoration and even the different maintenance programs can make projections challenging. In terms of the valuation, the industry standards for comparison purposes are full time/life, half time/life and zero time/life.

2.7 *Sensitivity of Inputs and Variables*

There are many inputs and variables that would have some effect on the valuation of the aircraft. These items include both demand and supply items which some of them include revenue yield, cost items such as fuel price, maintenance, and capital cost. These along with the business cycle, produce volatility for the value. Overall Demand Drivers Section 5.4 and Overall Supply Drivers Section 5.5 discuss many of these variable inputs in more detail as they have some effect on the valuation on the aircraft asset.

3 Data and Methodology

The author collected data from selected top appraisal firms around the world and other publically available sources. Most of the top appraisal firms for commercial aviation are physically located in USA or Europe with London being the location with the highest concentration. Each appraisal firm provides worldwide estimates and future projections of aircraft valuation globally on a published basis for some periodical time frame.

Each appraiser provides for each particular aircraft type, a current BV and MV for each of the different yearly vintages and a forward series of projected values in the future for each dated issue. Table 5.3 below is a representative snapshot of the type of datasheet available.

Table 5.3 Representative A330-300 historical valuation snapshot as of December 31, 1999

Manufacturer	*Engines*	*Hush*	*EFIS*	*FADAC*	*MTOW*	*Thrust*	*Other*
Airbus	*CF6-80E*				*478,400*		

Manufacturer	*Type*	*Model*	*Engine(s)*	*TOW (lbs.)*
Airbus	*A330*	*–300*	*CF6-80E1A1,PW4164/4168 or Trent768/772*	*478,400*

Year. of build	*Current market value*	*Distressed value*	*Base values*																			
	1999	*1999*	*1999*	*2000*	*2001*	*2002*	*2003*	*2004*	*2005*	*2006*	*2007*	*2008*	*2009*	*2010*	*2011*	*2012*	*2013*	*2014*	*2015*	*2016*	*2017*	*2018*
1992	xx.xx	xx.xx	xx.xx	xx.xx	xx.xx	xx.xx	xx.xx	xx.xx	xx.xx	xx.xx	xx.xx	xx.xx	xx.xx	xx.xx	xx.xx	xx.xx	xx.xx	xx.xx	xx.xx	xx.xx	xx.xx	xx.xx
1993	xx.xx	xx.xx	xx.xx	xx.xx	xx.xx	xx.xx	xx.xx	xx.xx	xx.xx	xx.xx	xx.xx	xx.xx	xx.xx	xx.xx	xx.xx	xx.xx	xx.xx	xx.xx	xx.xx	xx.xx	xx.xx	xx.xx
1994	xx.xx	xx.xx	xx.xx	xx.xx	xx.xx	xx.xx	xx.xx	xx.xx	xx.xx	xx.xx	xx.xx	xx.xx	xx.xx	xx.xx	xx.xx	xx.xx	xx.xx	xx.xx	xx.xx	xx.xx	xx.xx	xx.xx
1995	xx.xx	xx.xx	xx.xx	xx.xx	xx.xx	xx.xx	xx.xx	xx.xx	xx.xx	xx.xx	xx.xx	xx.xx	xx.xx	xx.xx	xx.xx	xx.xx	xx.xx	xx.xx	xx.xx	xx.xx	xx.xx	xx.xx
1996	xx.xx	xx.xx	xx.xx	xx.xx	xx.xx	xx.xx	xx.xx	xx.xx	xx.xx	xx.xx	xx.xx	xx.xx	xx.xx	xx.xx	xx.xx	xx.xx	xx.xx	xx.xx	xx.xx	xx.xx	xx.xx	xx.xx
1997	xx.xx	xx.xx	xx.xx	xx.xx	xx.xx	xx.xx	xx.xx	xx.xx	xx.xx	xx.xx	xx.xx	xx.xx	xx.xx	xx.xx	xx.xx	xx.xx	xx.xx	xx.xx	xx.xx	xx.xx	xx.xx	xx.xx
1998	xx.xx	xx.xx	xx.xx	xx.xx	xx.xx	xx.xx	xx.xx	xx.xx	xx.xx	xx.xx	xx.xx	xx.xx	xx.xx	xx.xx	xx.xx	xx.xx	xx.xx	xx.xx	xx.xx	xx.xx	xx.xx	xx.xx
1999	xx.xx	xx.xx	xx.xx	xx.xx	xx.xx	xx.xx	xx.xx	xx.xx	xx.xx	xx.xx	xx.xx	xx.xx	xx.xx	xx.xx	xx.xx	xx.xx	xx.xx	xx.xx	xx.xx	xx.xx	xx.xx	xx.xx

Note: where xx.xx represents $ in millions

This example presents an Airbus A330–300 type aircraft with its current and future pricing values set for each aircraft manufactured vintage year from 1992 to 1999. All of the current and future values represent the values as of the historical snapshot issued on December 31, 1999. For each of the data sources, there is a snapshot for each type of aircraft for the various years of built available. This data snapshot is available every year or half year, depending on the appraiser, for every particular type of aircraft that is included in the sample set. Each subsequent snapshot date also includes similar information incremental of the current production year so it would include all the manufactured vintage years available up to that snapshot date, that is, at the end of the year 2000, 2000 vintage data is included. For clarity, this is only an example data subset from one appraiser of one particular aircraft type at one point in time.

There are multiple appraisers providing data on multiple aircraft types with the information as the type found in Table 5.3. For the Table 5.3 example, this particular aircraft only started to be manufactured in 1992 hence the vintage data started in 1992 and not earlier. The total dataset timeframe analyzed is from June 1996 to June 2017, covering 21 years of data. During this time, not all aircraft types will have existed or still being manufactured for the entire timeframe with the replacement of older technology aircraft by newer aircraft happening in the later stages of the time series. This time series represents a robust data set both in terms of current snapshots and also their predictive thoughts in the future which can be analyzed. While the predictive nature of the future values is not the objective of this research, this can be considered for additional subsequent studies.

The data consists of figures from the following aviation appraisal firms—Avitas, Back, BK Associates, Inc., Collateral Verifications LLC, and IBA Group. BK data is partial type data that is provided and Avitas data is also partial as the sources are in the public domain. Together, the data from these firms, while not totally complete, represents the most complete set of data assembled of the top aircraft appraisal firms in the world and constitutes continuous data from 1996 to 2017 with data from aircraft manufactured back to the 1980s.

3.1 *Methodology*

There are several methodologies that have been incorporated in this overall study. Figure 5.6 shows the single isle classification by manufacturer and type while Fig. 5.7 shows the widebody classification by manufacturer

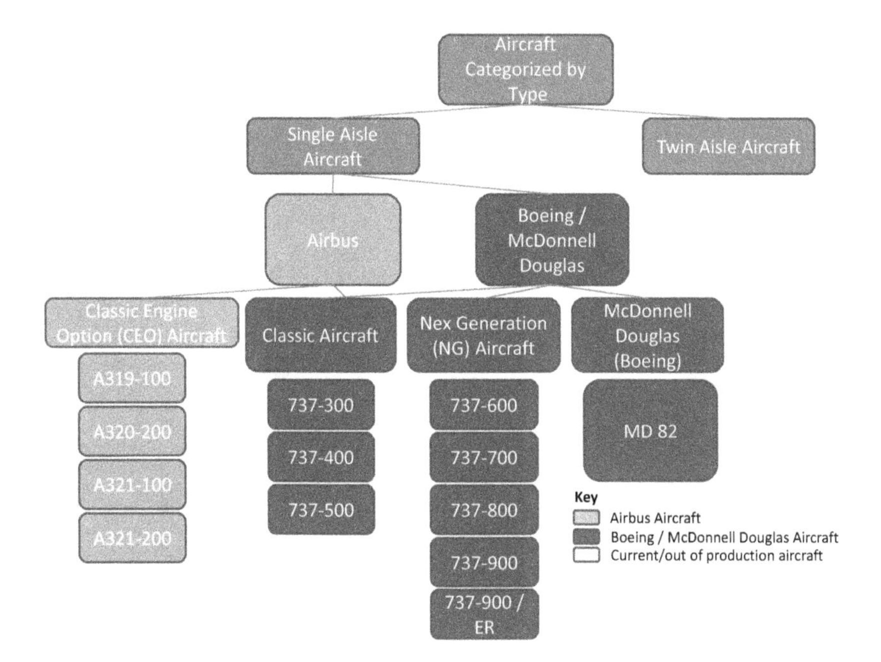

Fig. 5.6 Single aisle aircraft classifications by type

and type. In this case, the research will focus on both commercial narrow-body and widebody jet aircraft larger than 100 seats and manufactured by Boeing and Airbus. As there is little historical data available for new generation technologies such as Airbus NEO and Boeing MAX aircraft due to the low number of actual deliveries, this is excluded from the study. Also, regional jets and propeller aircraft are not a part of this study as they are smaller than the seat capacity threshold. These could be possible continuation and extension empirical studies conducted in the future.

For the first part of the analysis, taking into account all of the various types as shown in Figs. 5.6 and 5.7, a subset of aircraft types is selected to create the overall index which is shown by classification by manufacturer and type in Fig. 5.8. Then a subsample of these aircraft types would be used to create more specific indexes by characteristics.

These particular aircrafts were chosen as given the possible aircraft type data available, many aircrafts not selected have very limited production numbers and therefore low usage by airlines. The remaining selected aircrafts are

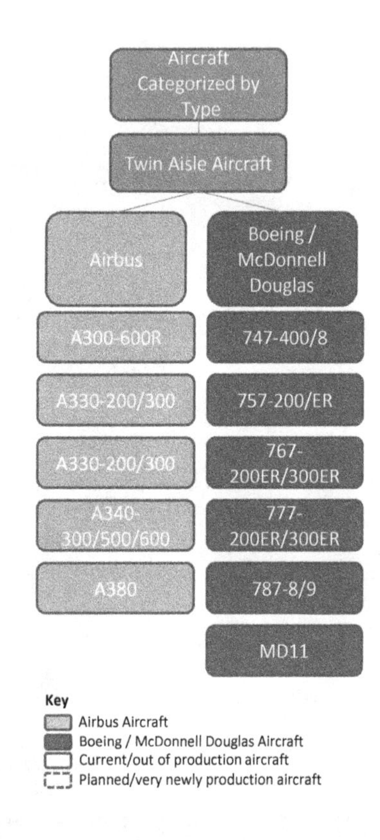

Fig. 5.7 Twin aisle aircraft classifications by type

more commonly prevalent and used aircraft types which become the focus in the study. This selected aircraft subset includes both narrowbody and wide-body aircrafts and comprise the following: Boeing aircraft 737–300, 737–400, 737–500, 737–600, 737–700, 737–800, 737–900, 737-900ER, 747–400, 747-8I, 757–200 (ETOPS and non-ETOPS), 767-200ER, 767-300ER, 777-200ER, 777-200LR, 777-300ER, 787–8, 787–9; McDonnell Douglas aircraft MD-82 and MD-11; and Airbus aircraft A300–600, A300–600R, A319–100, A320–200, A321–100, A321–200, A330–200, A330–300, A340–300, A340–500, A340–600, A380–800.

All of the data snapshots (Table 5.3 is a representative example) are sorted and collated into a standardized format. Each data point includes the Appraiser, Date of Observation, Manufacturer, Year of Build, Type, Sub-type, Narrowbody or Widebody, Current MV, Current BV, and any

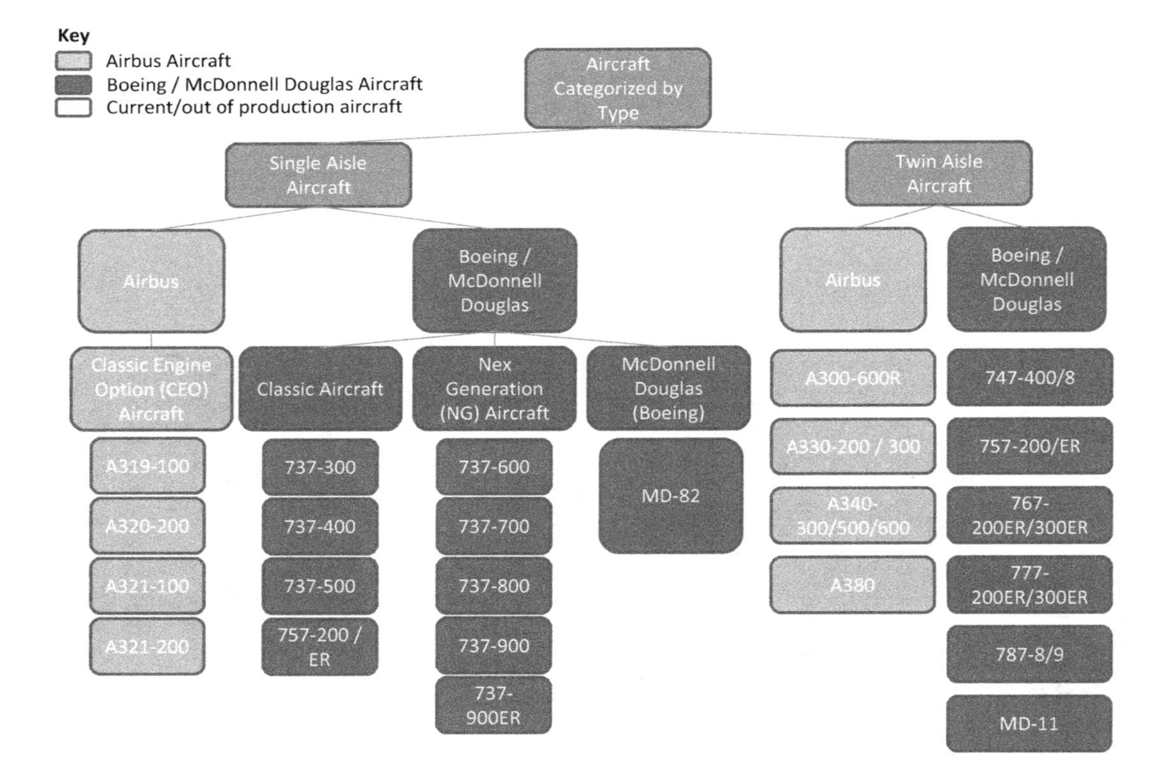

Fig. 5.8 Aircraft types included in the study

Future Base Values. Standardization of the data enables the creation of multiple indexes that will be used as the inputs for the statistical models to be used and analyzed. These statistical methods have been employed for analysis of other asset classes such as real estate (Quigley 1995).

Each data point in the sample set represents a specific source, historical date, specific aircraft type, manufactured vintage year, and historical asset BV and MV. Using different combinations and subsets of this data, multiple indexes are created to be used as the inputs for further multiple analyses. These indexes represent the whole of the aircraft asset market and particular descriptive subsets such as single-aisle or twin-aisle aircraft in order to make conclusions based on the analyses.

In this particular research, the five indexes are created which include: (1) all aircraft types, (2) all narrowbody aircraft, (3) all widebody aircraft, (4) all narrowbody classic aircraft, and (5) all narrowbody next-generation aircraft. The types of aircraft in the indexes are indicated with Indexes 1–3 being self explanatory while Index 4—all narrowbody classic aircraft—includes Boeing aircraft types 737–300, 737–400, 737–500, 757–200 (ETOPS and non-ETOPS); McDonnell Douglas aircraft types MD-82; and Airbus aircraft types A320–200 with CFM 5A/IAE V2500-A1 engines (1988 to 2003 year of built), A321–100 as these are considered the "classic" aircraft generation by technology. Index 5, all narrowbody NG aircraft, includes Boeing aircraft types 737–600, 737–700, 737–800, 737–900, 737-900ER and Airbus aircraft types A319–100 with CFM 5B/IAE V2500-A5 engines, A320–200 with CFM 5B/IAE V2500-A5 engines (1988 to 2003 year of built), A321–200. While A320–200 aircraft has been in existence since 1988, the upgrade of the engines differentiates the first and second generation of the aircraft. These are all considered aircraft types in the "next generation" based on its technology. Indexes 4 and 5 are created to gauge the differences between the older age and technology aircraft or the "classics" compared with the newer age and "next generation" technology aircraft. Some items to note about the data—Index 3 widebody data starts from 2.1997 and Index 5 NG narrowbody aircraft type data starts from 10.1999.

With each of the aircraft types, both MV and BV are observed and utilized in the study. For the index composition, a simple arithmetic average (SA) and a weighted average (WA) approach using both MV and BV values are utilized to calculate the respective indexes. This weighted average approach takes into account the smaller weighting of the smaller values, such as the older and smaller aircraft, versus the larger values with larger weighting for the newer or large aircraft. The simple arithmetic average approach treats all aircraft values equally.

Month Over Month (MoM) Simple Average Calculations

$$SimpleAverage_t = \left(\frac{1}{n}\right) * \sum_{n}^{i=0} AircraftReturn_{i,t}$$

where $AircraftReturn_{i,t} = \frac{MarketValue_{i,t}}{MarketValue_{i,t-1}} - 1$ OR $\frac{BaseValue_{i,t}}{BaseValue_{i,t-1}} - 1$.

i = specific aircraft type and vintage
t = time, month, and year of the valuation date (in months)
Month Over Month Weighted Average Calculations

$$WeightedAverage_t = \sum_{n}^{i=0} AircraftWeight_{i,t} X AircraftReturn_{i,t}$$

where $AircraftReturn_{i,t} = \frac{MarketValue_{i,t}}{MarketValue_{i,t-1}} - 1$ OR

$$= \frac{BaseValue_{i,t}}{BaseValue_{i,t-1}} - 1$$

$$AircraftWeight_{i,t} = \frac{MarketValue_{i,t}}{\sum_{i=0}^{n} MarketValue_{i,t}} \text{ OR } = \frac{BaseValue_{i,t}}{\sum_{i=0}^{n} BaseValue_{i,t}}$$

i = specific aircraft type
t = time, month, and year of the valuation date (in months)
Year Over Year (YoY) Calculations for Both Simple Average and Weighted Average

Use MoM Weighted Average Calculations and MoM Simple Average Calculations and replace with

$$AircraftReturn_{i,t} = \frac{MarketValue_{i,t}}{MarketValue_{i,t-12}} - 1 \text{ OR } = \frac{BaseValue_{i,t}}{BaseValue_{i,t-12}} - 1$$

As each data source is given as of the month and year of issuance and can be different, a straight line approach is used for the intermediate months in between the data points for each data source. In addition, two types of return calculations are utilized. The first technique utilized is a month-over-month return where the percentage difference month is calculated compared with the month prior. The second technique utilized is a year-over-year return where the percentage difference month is calculated compared with the same month the year prior. This YoY method reduces the seasonality of the data if it exists. All data is calculated and derived based on monthly figures.

For each of the different indexes comprising different weighting, return, and value types, different time scenarios are performed to see if there are differences due to major economic events such as the global financial crises. The time series scenarios (Times 1–4) selected are as follows: Time 1) 6.1996–6.2017; Time 2) 6.1996–6.2007; Time 3) 6.2007–6.2010; and 4) Time 6.2010–6.2017. For each index under the different conditions, statistical analyses are performed including mean, median, standard deviation, and correlation to understand the characteristics.

4 Results and Analyses

While all four scenarios (MoM, YoY, weighted, and simple averages) have been calculated, the most relevant case is the weighted average on a MoM basis. While YoY is useful to adjust for data that is seasonally driven which compares the same month year on year such as comparing January 2016 to January 2017, and so on. This is not the case with aircraft values as the value of the aircraft are not dependent on a particular month or season. Thus, MoM is a more appropriate form of calculating the differences directly from the previous month and is the focus of the analysis. For comparing weighted average versus simple average, the case of simple averages results in larger, newer, or higher valued aircraft with the same weighting as smaller, older, or lower valued aircraft. Utilizing the weighted average method, this decreases the natural bias towards smaller, older, and lower valued aircraft by weighting each case according to its value. As such, the weighted average case is the focal point of analysis.

In terms of time frame segments, the data is carved out into four time groups (Times 1–4). Time 1 encompasses all of the historical time data available (6.1996–6.2017), Time 2 represents the pre-GFC time period, Time 3 represents the period during the GFC, and Time 4 represents the time after the GFC. This is segmented to show explicitly whether or not the GFC had any effect on aircraft values. With each of the four time periods, the five indexes represented the five different types of aircraft, namely, (1) all aircraft types, (2) all narrowbody aircraft, (3) all widebody aircraft, (4) all narrowbody classic aircraft, and (5) all narrowbody next generation aircraft. The different indexes are chosen as they each represent distinct categories for aircraft types in terms of characteristics. This is used by industry as a general categorization of the aircraft type. These are also the categories that are used by both the asset investors and end users for aircraft. In each of these Indexes, both MV and BV are examined as having differing definitional context.

In addition to the Index 1 and Time 1 baselines, the accounting definition of aircraft depreciation is also used as a baseline for comparison. The definition states that accounting book depreciation provides for a useful life of 25 years with a 15% residual scrap value at the end terminal. That results in 3.4% depreciation per year or 0.283% deprecation per month on a straight line depreciation basis. The full MoM summary statistics for Index 1–5 are found in Table 5.4 while the full YoY summary statistics for Index 1–5 are found in Table 5.5. Figures 5.9–5.28 provide a graphical view of the weighted and simple average market and base values of each of the MoM Indexes 1–5 over the four different time scenarios.

4.1 Baseline Index 1 (All Aircraft Types) and Time 1 (6.1996–6.2017)

First, the Index 1 and Time 1 case is analyzed to establish the baseline for other comparisons. Looking at the all data encompassing scenario with the entire historical timeframe, Time 1, along with Index 1 (all aircraft types) on a MoM basis, the weighted and simple average MV means are respectively −0.6327% and −0.4861% and the MV median is −0.3724% and −0.0288%, respectively. The corresponding standard deviations are 2.6313% and 2.8609% respectively. For the BV, weighted and simple average mean are respectively −0.4689% and −0.3919% and the median is −0.1245% and −0.2930%, respectively. The corresponding standard deviations are 2.2450% and 2.7458% respectively. As values for aircraft are decreasing for the most part as an aircraft ages, a decreasing monthly return figure whether mean or median is not uncommon as a lower number means a higher decrease in value and vice versa from the previous month.

Table 5.4 Index 1–5, time scenarios 1–4 MoM summary statistics

Index 1—All aircraft MoM summary statistics

Time scenario	*1*		*2*		*3*		*4*	
Time	*6.1996–6.2017*		*6.1996–6.2007*		*6.2007–6.2010*		*6.2010–6.2017*	
	Weighted ave	*Simple ave*	*Weighted ave*	*Simple ave*	*Weighted ave*	*Simple ave*	*Weighted ave*	*Simple ave*
n	40,507	40,507	12,015	12,015	7243	7243	21,249	21,249
MV:								
Min	(10.2745%)	(9.8618%)	(10.2745%)	(9.8618%)	(5.2569%)	(4.9216%)	(8.0354%)	(7.9413%)
Mean	(0.6327%)	(0.4861%)	(0.4730%)	(0.3223%)	(0.7875%)	(0.7920%)	(0.8094%)	(0.5861%)
Standard deviation	2.6313%	2.8609%	2.6050%	2.0581%	1.5157%	1.2822%	2.9732%	4.1209%
Median	(0.3724%)	(0.0288%)	(0.0241%)	0.0077%	(0.4361%)	(0.1351%)	(0.8896%)	(1.3562%)
Max	11.8712%	11.0411%	11.8712%	9.8277%	3.1365%	0.4146%	9.1773%	11.0411%
BV:								
Min	(8.8619%)	(10.1799%)	(8.8619%)	(8.5110%)	(6.7726%)	(5.8450%)	(6.8954%)	(10.1799%)
Mean	(0.4689%)	(0.3919%)	(0.3864%)	(0.3451%)	(0.2164%)	(0.2765%)	(0.7084%)	(0.4966%)
Standard deviation	2.2450%	2.7458%	2.0497%	1.6668%	2.3571%	1.5952%	2.4174%	4.1405%
Median	(0.1245%)	(0.2930%)	(0.0742%)	(0.0340%)	(0.0890%)	(0.1047%)	(0.3877%)	(0.8605%)
Max	9.8949%	22.4800%	6.2668%	6.6671%	4.2472%	3.0909%	9.8949%	22.4800%
Correlation MV_BV	0.5408	0.5978	0.6006	0.7012	0.6313	0.5198	0.4759	0.5767

Index 2—All narrowbody aircraft MoM summary statistics

Time scenario	*1*		*2*		*3*		*4*	
Time	*6.1996–6.2017*		*6.1996–6.2007*		*6.2007–6.2010*		*6.2010–6.2017*	
	Weighted ave	Simple ave	Weighted ave	Simple ave	Weighted ave	Simple ave	Weighted ave	Simple ave
n	21,684	21,684	6817	6817	3738	3738	11,129	11,129
MV:								
Min	(7.5487%)	(7.7940%)	(7.5487%)	(6.9535%)	(6.3823%)	(6.1923%)	(7.4076%)	(7.7940%)
Mean	(0.3375%)	(0.3858%)	(0.2752%)	(0.2163%)	(0.8314%)	(0.9820%)	(0.2381%)	(0.3932%)
Standard deviation	2.9181%	2.7268%	2.9154%	2.4050%	2.3214%	2.1042%	3.0994%	3.3071%
Median	(0.1830%)	(0.0839%)	(0.0219%)	0.1259%	(0.8414%)	(0.3793%)	(0.3739%)	(0.8072%)
Max	18.9991%	17.7589%	18.9991%	17.7589%	3.8151%	4.0366%	10.8101%	10.2017%
BV:								
Min	(11.1762%)	(11.5597%)	(11.1762%)	(9.6318%)	(9.0393%)	(8.1191%)	(7.6968%)	(11.5597%)
Mean	(0.1783%)	(0.3050%)	(0.1504%)	(0.2394%)	0.3198%	(0.3083%)	(0.4455%)	(0.4248%)
Standard deviation	2.9292%	3.2875%	2.7586%	1.9589%	4.0798%	2.8496%	2.4942%	4.7483%
Median	(0.0334%)	(0.3553%)	0.0118%	(0.0437%)	0.5892%	(0.4355%)	(0.3570%)	(0.7230%)
Max	18.9991%	22.1773%	15.5275%	11.2055%	8.9263%	6.0484%	8.9147%	22.1773%
Correlation MV_BV	0.5071	0.3779	0.6172	0.6322	0.6097	0.5624	0.3553	0.2472

(*continued*)

Table 5.4 (continued)

Index 3—All widebody aircraft MoM summary statistics

Time scenario	*1*		*2*		*3*		*4*	
Time	*6.1996–6.2017*		*6.1996–6.2007*		*6.2007–6.2010*		*6.2010–6.2017*	
	Weighted ave	Simple ave	Weighted ave	Simple ave	Weighted ave	Simple ave	Weighted ave	Simple ave
n	18,820	18,820	5199	5199	3506	3506	10,115	10,115
MV:								
Min	(11.9864%)	(13.9115%)	(11.9864%)	(13.9115%)	(5.1791%)	(4.3236%)	(8.8526%)	(11.4233%)
Mean	(0.8324%)	(0.5698%)	(0.6489%)	(0.4111%)	(0.8677%)	(0.6317%)	(1.0697%)	(0.7313%)
Standard deviation	3.0276%	3.6371%	2.8309%	2.4281%	1.9170%	1.4201%	3.5095%	5.3533%
Median	(0.5709%)	(0.1340%)	(0.1555%)	0.0482%	(0.6655%)	(0.3335%)	(1.3975%)	(1.5799%)
Max	13.8674%	17.7728%	8.1385%	3.7627%	3.7196%	1.4257%	13.8674%	17.7728%
BV:								
Min	(9.0328%)	(10.2729%)	(9.0328%)	(10.2729%)	(5.8542%)	(4.2397%)	(7.8620%)	(9.1632%)
Mean	(0.6348%)	(0.4581%)	(0.5358%)	(0.4249%)	(0.5373%)	(0.2492%)	(0.8168%)	(0.5513%)
Standard deviation	2.3332%	2.8982%	2.0459%	1.8016%	2.2498%	1.5230%	2.6986%	4.3063%
Median	(0.2503%)	(0.1328%)	(0.0958%)	(0.1325%)	(0.5292%)	0.1618%	(0.3710%)	(0.8063%)
Max	11.9264%	22.7030%	4.6610%	2.4313%	4.7332%	2.0859%	11.9264%	22.7030%
Correlation MV_BV	0.6039	0.7690	0.6794	0.8257	0.7413	0.7671	0.5204	0.7550

Note: Index 3 data starts from 2.1997

Index 4—All narrowbody classic aircraft MoM summary statistics

Time scenario	*1*		*2*		*3*		*4*	
Time	*6.1996–6.2017*		*6.1996–6.2007*		*6.2007–6.2010*		*6.2010–6.2017*	
	Weighted ave	Simple ave	Weighted ave	Simple ave	Weighted ave	Simple ave	Weighted ave	Simple ave
n	14,309	14,309	5154	5154	2575	2575	6580	6580
MV:								
Min	(10.4857%)	(11.8900%)	(8.6975%)	(8.4228%)	(9.4386%)	(10.5345%)	(10.4857%)	(11.8900%)
Mean	(0.6287%)	(0.5985%)	(0.3074%)	(0.4001%)	(1.7808%)	(1.3630%)	(0.6832%)	(0.5829%)
Standard deviation	3.8874%	3.6736%	3.8871%	3.4704%	3.1797%	2.9790%	4.0363%	4.1408%
Median	(0.1769%)	(0.2379%)	(0.0427%)	(0.0333%)	(0.4436%)	(0.4284%)	(0.5869%)	(0.9381%)
Max	27.6055%	27.5382%	27.6055%	27.5382%	2.3868%	3.4204%	11.9470%	12.2305%
BV:								
Min	(11.8045%)	(19.5453%)	(11.8045%)	(8.9755%)	(9.3996%)	(7.4767%)	(9.1816%)	(19.5453%)
Mean	(0.1705%)	(0.4434%)	(0.0701%)	(0.3918%)	0.4505%	(0.4211%)	(0.6101%)	(0.5512%)
Standard deviation	3.5723%	5.1710%	3.3171%	2.3003%	4.3606%	2.4066%	3.4892%	8.2787%
Median	0.0255%	(0.3053%)	0.0675%	0.0139%	0.4801%	(0.5750%)	(0.6027%)	(1.3671%)
Max	20.6740%	42.8158%	20.6740%	12.3700%	9.6904%	4.7192%	11.8196%	42.8158%
Correlation MV_BV	0.3207	0.2856	0.4906	0.5272	0.0571	0.2878	0.2307	0.2522

(*continued*)

Table 5.4 (continued)

Index 5—All narrowbody NG aircraft MoM summary statistics

Time scenario	*1*		*2*		*3*		*4*	
Time	*6.1996–6.2017*		*6.1996–6.2007*		*6.2007–6.2010*		*6.2010–6.2017*	
	Weighted ave	Simple ave	Weighted ave	Simple ave	Weighted ave	Simple ave	Weighted ave	Simple ave
n	8054	8054	1511	1511	1134	1134	5409	5409
MV:								
Min	(8.8196%)	(8.5707%)	(8.8196%)	(8.3554%)	(5.4493%)	(4.3414%)	(7.7395%)	(8.5707%)
Mean	(0.2426%)	(0.1775%)	(0.3512%)	(0.0287%)	(0.3842%)	(0.5844%)	(0.0708%)	(0.1528%)
Standard deviation	2.6168%	2.4663%	2.2615%	1.8520%	2.6127%	1.9534%	2.9245%	3.1208%
Median	(0.2028%)	(0.0623%)	(0.1081%)	0.3006%	(0.4164%)	(0.5290%)	(0.2082%)	(0.5708%)
Max	10.4915%	12.6543%	6.9616%	6.5970%	5.6921%	3.4005%	10.4915%	12.6543%
BV:								
Min	(9.2265%)	(8.9584%)	(9.2265%)	(8.7541%)	(9.0960%)	(8.9584%)	(8.5839%)	(7.1403%)
Mean	(0.3834%)	(0.1433%)	(0.4552%)	(0.0466%)	(0.2383%)	(0.1954%)	(0.3693%)	(0.2346%)
Standard deviation	2.6670%	2.3600%	2.4419%	2.0655%	3.6603%	2.9645%	2.3358%	2.3449%
Median	(0.1551%)	0.1091%	(0.3276%)	0.1901%	0.0180%	(0.2186%)	(0.1253%)	(0.1703%)
Max	7.9271%	7.8473%	7.8990%	7.8473%	7.9271%	7.6919%	7.5810%	7.0396%
Correlation MV_BV	0.6765	0.6010	0.8738	0.8247	0.8592	0.8319	0.4345	0.4389

Note: Index 5 data starts from 10.1999

Table 5.5 Index 1–5, time scenarios 1–4 YoY summary statistics

Index 1—All aircraft YoY summary statistics								
Time scenario	*1*		*2*		*3*		*4*	
Time	*6.1996–6.2017*		*6.1996–6.2007*		*6.2007–6.2010*		*6.2010–6.2017*	
	Weighted ave	*Simple ave*	*Weighted ave*	*Simple ave*	*Weighted ave*	*Simple ave*	*Weighted ave*	*Simple ave*
n	40,507	40,507	12,015	12,015	7243	7243	21,249	21,249
MV:								
Min	(17.3111%)	(18.8734%)	(17.3111%)	(18.8734%)	(16.2065%)	(17.8841%)	(11.8674%)	(16.6675%)
Mean	(4.6253%)	(6.9297%)	(3.1340%)	(4.4320%)	(5.3550%)	(7.6586%)	(6.3532%)	(9.9548%)
Standard deviation	4.8758%	6.1592%	5.4652%	6.7955%	6.0326%	6.9198%	1.6679%	1.9222%
Median	(5.2979%)	(8.1317%)	(1.2845%)	(3.8370%)	(3.7718%)	(8.0994%)	(6.3455%)	(9.9207%)
Max	9.0681%	10.3204%	9.0681%	10.3204%	2.7579%	2.0513%	(1.9148%)	(3.1984%)
BV:								
Min	(12.2702%)	(23.9187%)	(11.8897%)	(13.9475%)	(7.9701%)	(7.6087%)	(12.2702%)	(23.9187%)
Mean	(4.0264%)	(6.4089%)	(3.2444%)	(4.4503%)	(2.8184%)	(4.3712%)	(5.5289%)	(9.7813%)
Standard deviation	3.1222%	4.8583%	3.6353%	4.5709%	2.4031%	2.1161%	1.7758%	4.3018%
Median	(4.3978%)	(6.4455%)	(3.1484%)	(4.8020%)	(2.9180%)	(4.5624%)	(5.4899%)	(9.3150%)
Max	3.8214%	7.1757%	3.8214%	7.1757%	2.2831%	0.9983%	(1.3479%)	0.6955%
Correlation MV_BV	0.6583	0.6727	0.7074	0.8101	0.6540	0.6578	0.2135	0.1709

(*continued*)

Table 5.5 (continued)

Index 2—All narrowbody aircraft YoY summary statistics

Time scenario	*1*		*2*		*3*		*4*	
Time	*6.1996–6.2017*		*6.1996–6.2007*		*6.2007–6.2010*		*6.2010–6.2017*	
	Weighted ave	Simple ave	Weighted ave	Simple ave	Weighted ave	Simple ave	Weighted ave	Simple ave
n	21,684	21,684	6817	6817	3738	3738	11,129	11,129
MV:								
Min	(19.7040%)	(24.6568%)	(17.0013%)	(17.1006%)	(19.7040%)	(24.6568%)	(9.6532%)	(14.2838%)
Mean	(3.7641%)	(5.5932%)	(2.6685%)	(3.2420%)	(7.3208%)	(10.0338%)	(3.7271%)	(6.9795%)
Standard deviation	6.0675%	6.8598%	6.4345%	7.0816%	8.0421%	8.5649%	3.2106%	3.5688%
Median	(2.5346%)	(5.1346%)	(0.3461%)	(1.7752%)	(8.1164%)	(10.2096%)	(3.4241%)	(6.4820%)
Max	13.4889%	14.9066%	13.4889%	14.9066%	4.3860%	2.4284%	4.0900%	0.6943%
BV:								
Min	(12.2077%)	(27.2727%)	(12.2077%)	(15.4005%)	(10.1388%)	(12.9394%)	(10.2257%)	(27.2727%)
Mean	(3.6421%)	(5.5566%)	(2.6927%)	(3.3729%)	(3.5531%)	(5.6633%)	(4.9158%)	(8.5340%)
Standard deviation	3.7596%	5.2345%	4.3850%	4.7236%	3.2739%	3.1068%	2.3513%	5.1726%
Median	(3.6069%)	(5.7479%)	(1.1376%)	(2.3847%)	(3.1376%)	(4.9966%)	(4.9133%)	(7.4739%)
Max	6.8807%	6.6489%	6.8807%	6.6489%	2.0000%	0.8029%	1.5731%	3.3517%
Correlation MV_BV	0.5911	0.5628	0.7792	0.7925	0.5130	0.6018	(0.0129)	0.0968

Index 3—All widebody aircraft YoY summary statistics

Time scenario	*1*		*2*		*3*		*4*	
Time	*6.1996–6.2017*		*6.1996–6.2007*		*6.2007–6.2010*		*6.2010–6.2017*	
	Weighted ave	Simple ave	Weighted ave	Simple ave	Weighted ave	Simple ave	Weighted ave	Simple ave
n	18,820	18,820	5199	5199	3506	3506	10,115	10,115
MV:								
Min	(17.4847%)	(20.8570%)	(17.4847%)	(20.8570%)	(14.7894%)	(16.6772%)	(14.2527%)	(19.4176%)
Mean	(5.1748%)	(8.1521%)	(3.6372%)	(5.7133%)	(4.7038%)	(5.6780%)	(7.3515%)	(12.1135%)
Standard deviation	4.8201%	6.5133%	5.4278%	7.2330%	5.4965%	5.8825%	1.9476%	2.6131%
Median	(6.0310%)	(9.8114%)	(2.6659%)	(5.3409%)	(2.8035%)	(5.3173%)	(7.3089%)	(12.0341%)
Max	6.8679%	14.3537%	6.8679%	14.3537%	2.0680%	1.8426%	(1.1832%)	(2.8446%)
BV:								
Min	(14.1344%)	(21.4474%)	(10.7984%)	(17.6005%)	(7.3023%)	(8.0740%)	(14.1344%)	(21.4474%)
Mean	(4.2523%)	(7.2642%)	(3.5668%)	(5.6565%)	(2.5691%)	(3.3177%)	(5.7581%)	(10.6810%)
Standard deviation	2.9977%	5.0509%	3.3536%	4.9333%	2.5090%	2.0180%	1.8847%	4.1995%
Median	(4.8217%)	(7.0698%)	(4.1027%)	(6.5418%)	(2.2102%)	(3.8709%)	(5.6341%)	(10.5951%)
Max	3.0421%	10.2264%	3.0421%	10.2264%	2.3909%	1.1592%	(1.3296%)	(1.3703%)
Correlation MV_BV	0.6739	0.7711	0.6773	0.8267	0.7186	0.6777	0.3589	0.5235

Note: Index 3 data starts from 2.1997

(*continued*)

Table 5.5 (continued)

Index 4—All narrowbody classic aircraft YoY summary statistics

Time scenario	*1*		*2*		*3*		*4*	
Time	*6.1996–6.2017*		*6.1996–6.2007*		*6.2007–6.2010*		*6.2010–6.2017*	
	Weighted ave	Simple ave	Weighted ave	Simple ave	Weighted ave	Simple ave	Weighted ave	Simple ave
n	14,309	14,309	5154	5154	2575	2575	6580	6580
MV:								
Min	(28.3734%)	(35.1805%)	(22.5717%)	(28.3124%)	(28.3734%)	(35.1805%)	(13.4787%)	(23.1088%)
Mean	(5.1532%)	(8.3287%)	(3.7789%)	(5.3564%)	(9.7495%)	(13.8659%)	(5.0869%)	(10.0660%)
Standard deviation	7.9603%	9.8426%	8.1589%	10.6748%	11.1611%	11.8628%	4.3405%	4.8333%
Median	(3.3424%)	(7.3913%)	(1.2413%)	(3.8728%)	(12.8931%)	(14.4737%)	(4.6902%)	(10.3711%)
Max	18.4992%	22.0899%	18.4992%	22.0899%	8.1947%	4.8758%	7.0261%	(0.1473%)
BV:								
Min	(19.0100%)	(42.1959%)	(15.0495%)	(21.8834%)	(12.4543%)	(14.5092%)	(19.0100%)	(42.1959%)
Mean	(5.1085%)	(8.2198%)	(3.7419%)	(5.3438%)	(4.7390%)	(7.7013%)	(7.0454%)	(12.3878%)
Standard deviation	4.9318%	7.6476%	5.3426%	6.8665%	4.0469%	3.4019%	3.9145%	8.0722%
Median	(4.8938%)	(7.9038%)	(1.5747%)	(4.3447%)	(4.7949%)	(8.8994%)	(6.8089%)	(9.9513%)
Max	6.3555%	6.1150%	6.3555%	6.1150%	3.2876	(0.0445%)	0.4133%	1.1509%
Correlation MV_BV	0.5806	0.5515	0.7632	0.8253	0.5726	0.7304	0.2368	0.0381

Index 5—All narrowbody NG aircraft YoY summary statistics

Time scenario	*1*		*2*		*3*		*4*	
Time	*6.1996–6.2017*		*6.1996–6.2007*		*6.2007–6.2010*		*6.2010–6.2017*	
	Weighted ave	Simple ave	Weighted ave	Simple ave	Weighted ave	Simple ave	Weighted ave	Simple ave
n	8054	8054	1511	1511	1134	1134	5409	5409
MV:								
Min	(14.7959%)	(15.1182%)	(10.7362%)	(13.0318%)	(14.7959%)	(15.1182%)	(9.9627%)	(0.4946%)
Mean	(2.5340%)	(2.9162%)	(1.0661%)	(1.3456%)	(5.1985%)	(5.4933%)	(2.7204%)	(3.3270%)
Standard deviation	4.8176%	4.8490%	5.1684%	5.1209%	5.7332%	5.8341%	3.2119%	3.4254%
Median	(1.5050%)	(2.1652%)	(0.0150%)	(0.6411%)	(4.1046%)	(4.6045%)	(2.2565%)	(3.6227%)
Max	8.3003%	7.6125%	8.3003%	7.6125%	3.4902%	2.6689%	3.9466%	5.0818%
BV:								
Min	(7.9454%)	(9.8625%)	(7.3660%)	(8.7727%)	(7.9454%)	(9.8625%)	(6.7911%)	(9.7981%)
Mean	(2.3981%)	(2.7402%)	(1.2705%)	(1.3340%)	(2.1158%)	(2.7673%)	(3.5260%)	(4.0625%)
Standard deviation	2.7899%	3.3539%	3.2710%	3.7247%	2.5693%	2.7203%	1.7489%	2.6594%
Median	(2.8309%)	(3.1083%)	(1.4714%)	(1.9320%)	(1.5467%)	(2.6689%)	(3.8750%)	(4.0713%)
Max	7.1148%	7.1622%	7.1148%	7.1622%	3.8669%	3.2484%	1.8971%	6.2021%
Correlation MV_BV	0.4975	0.5186	0.7685	0.7159	0.4746	0.4449	(0.1611)	0.2092

Note: Index 5 data starts from 10.1999

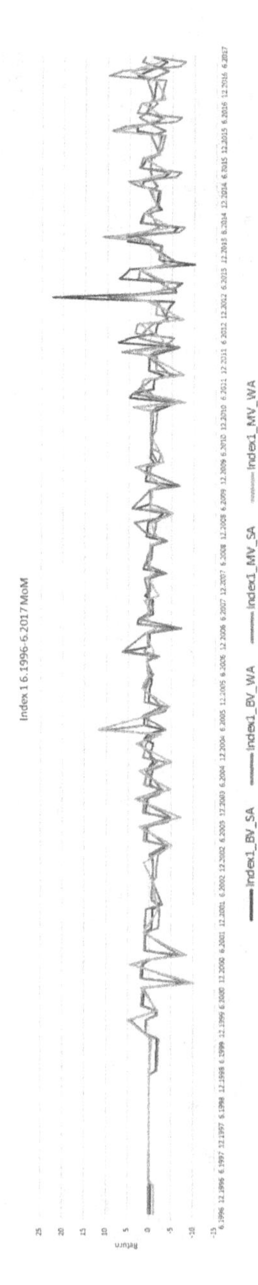

Fig. 5.9 Index 1 MoM—6.1996–6.2017

The monthly MV and BV mean from the resultant analysis of the data set are higher than compared with the baseline accounting book depreciation method of 0.283%. The weighted average MV median is higher but BV is lower and vice versa for simple average where MV median is lower and BV median is slightly higher. The accounting method is intended to approximate the depreciation but clearly, the data shows that for the mean figures the 0.283% is not enough but needs to be increased or faced with higher write downs during the term of the asset life.

4.2 *Cross Index Comparisons (Time 1—6.1996–6.2017): Index 2 (All Narrowbody Aircraft)*

Having the baseline Index 1 established, the analysis will move to cross index comparisons for the other Indexes 2–5 under the Time 1, all historical time scenarios. For Index 2 (all narrowbody aircraft) on an MoM basis, the MV mean for the weighted and simple averages are respectively −0.3375% and −0.3858% and the median is −0.1830% and −0.0839%, respectively. The corresponding standard deviations are 2.9181% and 2.7268% respectively for MV. For the BV, weighted and simple averages are respectively −0.1783% and −0.3050% and the median is −0.0334% and −0.3553%, respectively. The corresponding standard deviations are 2.992% and 3.2875% respectively.

Focusing on the weighted average case, narrowbody aircraft, represented by Index 2, have lower monthly mean and median compared with the total population in the Index 1 dataset for both MV and BV cases. The standard deviation is slightly higher for both BV and MV cases in Index 2 compared with Index 1. It can be deduced from the data that with the lower monthly mean and median numbers, narrowbody aircraft values depreciate less than the entire population including both narrowbody and widebody aircraft in Index 1. The standard deviation is slightly higher in Index 2 compared with Index 1 which means higher volatility in price deviations for narrowbody aircraft alone compared with a population set of both narrowbody and widebody aircraft.

The monthly MV and BV mean from the resultant analysis of the data set are all higher than compared with the baseline accounting book depreciation method of 0.283% except for weighted average BV mean. The median is all lower except for the weighted average BV case. Even so, it seems from this that for narrowbody aircraft, the accounting method needs to be adjusted upward or faced with higher write downs during the term of the asset life.

4.3 Cross Index Comparisons (Time 1—6.1996–6.2017): Index 3 (All Widebody Aircraft)

For Index 3 (all widebody aircraft) on an MoM basis, the weighted and simple average MV means are respectively −0.8324% and −0.5698% and the median is −0.5709% and −0.1340%, respectively for MV. The corresponding standard deviations are 3.0276% and 3.6371% respectively. For the BV, weighted and simple averages are respectively −0.6348% and −0.4581% and the median is −0.2503% and −0.1328%, respectively. The corresponding standard deviations are 2.3332% and 2.8982% respectively.

Focusing on the weighted average case, widebody aircraft, represented by Index 3, have higher monthly MV and BV mean and median compared with the total population Index 1 dataset. The standard deviation is higher for BV and MV cases with Index 3 compared with Index 1 as well. It can be deduced from the data that with the higher monthly mean and median numbers, widebody aircraft values depreciate faster on average than the population of aircraft in Index 1. The standard deviation is slightly higher in Index 3 compared with Index 1 which means higher volatility in price deviations compared to Index 1.

Comparing widebody aircraft Index 3 with the narrowbody Index 2 case, Index 3 has significantly higher MV and BV mean and median which results in higher price differences. The standard deviation differences in MV are higher with Index 3 while BV standard deviation is higher with Index 2. This means that widebody aircraft represented by Index 3 have slightly higher volatility in the MV which takes into account market demand considerations compared with the BV definition which has the opposite effect.

The monthly MV and BV mean from the resultant analysis of the data set are all higher than compared with the baseline accounting book depreciation method of 0.283%. Like the narrowbody Index 2 case, the MV median is higher except for the BV median. With the differences that are significant for widebody aircraft, again the accounting method needs to be adjusted upward or faced with higher write downs during the term of the asset life.

4.4 Cross Index Comparisons (Time 1—6.1996–6.2017): Index 4 (All Narrowbody Classic Aircraft)

For Index 4 (all narrowbody classic aircraft) on an MoM basis, the weighted and simple average MV means are respectively −0.6287% and −0.5985% and the MV median is −0.1769% and −0.2379%, respectively. The corresponding standard deviations are 3.8874% and 3.6736% respectively. For the BV, weighted and simple averages are respectively −0.1705% and −0.4434% and the median is 0.0255% and −0.3053% respectively. The corresponding standard deviations are 3.5723% and 5.1710%, respectively.

Focusing on the weighted average case, older narrowbody classic aircraft, represented by Index 4, have lower monthly MV and BV mean and median compared with the total population in Index 1. The standard deviation is significantly higher for BV and MV cases with Index 4 compared with Index 1. It can be deduced from the data that with the lower monthly mean and median numbers, older narrowbody classic aircraft values depreciate slower than the population of aircraft types in Index 1. This is a surprising result given that the age of the aircraft types in Index 4 is higher than the average of those in the population Index 1. The standard deviation is significantly higher in Index 4 compared with Index 1 which means higher volatility in the price deviations. This is logical as expected given the older age and technological profile of the Index compared with the aircraft types included in the other Indexes.

Compared with the all narrowbody aircraft Index 2, the data analysis is a bit more mixed. Older narrowbody classics Index 4 MV has a significantly higher mean but the median figures slightly lower. Similarly, for the BV comparisons, Index 4 mean is slightly lower and the median is higher than Index 2. These results in mixed signals for older narrowbody classic aircraft compared with the subset of narrowbody aircraft and conclusions cannot be easily attributed. There exists a large difference in MV mean figures which infers that market demand dynamic differences are very large for Index 4 versus Index 2. The standard deviations are higher for Index 4 than Index 2 for MV and BV which means higher volatility in the price movements.

Compared with the all widebody aircraft Index 3, older narrowbody classics Index 4 MV and BV mean and median are all lower. The standard deviations are higher for Index 4 than Index 3 for MV and BV which means higher volatility in the price movements. While the lower mean and median figures for Index 4 comparatively is not surprising, what is interesting is that the standard deviations for older classic narrowbody aircraft are even greater than those of widebody aircraft.

The monthly MV mean is higher than compared with the baseline accounting book depreciation method of 0.283% but all other metrics including BV median and BV mean and median are lower. With the different mixed indicators, it is hard to infer any adjustments to the accounting methods in this case.

4.5 *Cross Index Comparisons (Time 1—6.1996–6.2017): Index 5 (All Narrowbody NG Aircraft)*

For Index 5 (all narrowbody NG aircraft) on an MoM basis, the MV mean for the weighted and simple averages are respectively −0.2426% and −0.1775% and the median is −0.2028% and −0.0623%, respectively for MV. The corresponding standard deviations are 2.6168% and 2.4663% respectively. For the BV, weighted and simple averages are respectively −0.3834% and −0.1433% and the median is −0.1551% and 0.1091%, respectively. The corresponding standard deviations are 2.6670% and 2.3600%, respectively.

Focusing on the weighted average case, newer narrowbody NG aircraft Index 5 has lower monthly MV and BV mean compared with the total Index. The Index 5 median figure is lower for MV and slightly higher for BV. The standard deviation is higher for the BV case and slightly lower in the MV case for Index 5 compared with Index 1. It can be deduced from the data that with the lower monthly mean numbers, newer age and technology profiled narrowbody NG aircraft values depreciate slower than the population in Index 1. The standard deviation results are mixed comparing Index 5 with Index 1, which means lower market volatility in price deviations from a BV definition point of view.

Compared with all narrowbody aircraft Index 2, newer narrowbody NG Index 5 has lower MV mean, higher BV mean, and higher MV and BV median figures. These have conflicting signals to make any significant inferences in terms of price depreciation. The standard deviations for MV and BV are all lower with narrowbody aircraft NG Index 5. The volatility for newer profiled narrowbody aircraft can be inferred to be lower than the all narrowbody aircraft Index 2.

Comparing Index 5 with older classic types of narrowbody aircraft in Index 4, Index 5 has much lower MV mean and slightly higher BV mean. Similarly, Index 5 has higher MV median figures and the opposite for the BV case. The standard deviations are also much lower with Index 5 for the MV and BV cases. While the conflicting information is difficult to infer in terms of a price depreciation view, it can be inferred that Index 4 has higher volatility which is in line with current thinking that older assets are more volatile than newer assets controlled by aircraft type.

Finishing the comparison analysis with widebody aircraft Index 3, Index 5 has much lower MV and BV mean and median cases. The standard deviations are lower for MV and higher for BV for Index 5. It means that narrowbody NG aircraft Index 5 has slower depreciation than widebody aircraft Index 3 but standard deviation under the MV market dynamic case is in line with the industry thinking but the increased volatility in the BV case is surprising.

The monthly MV mean and median and BV median are lower than compared with the baseline accounting book depreciation method of 0.283%. The majority is lower than the accounting depreciation standard. It is recommended that the accounting method needs to be adjusted downward or faced with a higher RV at the terminal of the asset life.

4.6 Cross Time Comparisons Groups: Index 1 (All Aircraft Types)

There are some interesting observations that arise from further analyzing the time subsets data. Comparing the different periods, Time 2–4, with the entire period, Index 1's MV mean and median shows a lower figure than Time 1 in Time 2 (6.1996–6.2007) before the GFC and higher one for Time 3 during the GFC before increasing still for Time 4 (6.2010–6.2017). The standard deviation in decreased significantly in Time 3 and increased in Time 4 above Time 2. This infers that pre-GFC Time 2 had lower volatility than the historical figure in Time 1 and also had increased volatility after the GFC in Time 4.

BV moved differently where the MV mean decreased in Time 3 and increased significantly in Time 4 above Time 1 and 2. The standard deviation only slightly increased for Time 3 and Time 4. While there is no one clear signal, this does show that increased volatility during the GFC in BV and the increased MV and BV volatility in Time 4 compared to pre-GFC Time 2 and historical Time 1. In addition, the depreciation increased post the GFC in Time 4 than the period before GFC in Time 2.

4.7 Cross Time Comparisons Groups: Index 2 (All Narrowbody Aircraft)

Comparing across the 4 time subsets for Index 2's MV and BV mean, median, and standard deviation for Time 2 are all less than the entire historical period (Time 1). In Time 3, all the metrics increased significantly except for standard deviation MV where it decreased slightly. In Time 4, the mean and median decreased for all scenarios except for BV mean which decreased. Standard deviation in Time 4 increased for MV but decreased for BV. During GFC Time 3, price depreciation increased significantly. Volatility only increased for MV while BV decreased. Post GFC, there is a price depreciation and volatility divergence between MV and BV.

Compared to Index 1 in Time 3, both MV mean and median increased in Time 2 but divergence in Time 4 where Index 1 increased while Index 2 decreased. Volatility acted similarly in Index 1 and 2 with a decrease in Time 3 and increase in Time 4. BV acted similarly between Index 1 and 2 with BV mean decreased and increased in Time 3 and 4 respectively. Volatility diverged where after the increase in Time 3, Index 1 increased while Index 2 decreased in Time 4.

4.8 Cross Time Comparisons Groups: Index 3 (All Widebody Aircraft)

Comparing across the 4 time subsets for Index 3's MV and BV mean, median, and standard deviation for Time 2 are all less than the entire historical period (Time 1). In Time 3, all the metrics increased significantly except for standard deviation MV where it decreased. In Time 4, the mean and median increased for all except for BV median which decreased. Standard deviation in the Time 4 increased for MV and BV. During GFC Time 3, price depreciation increased significantly. Volatility decreased for MV while BV increased slightly. Post GFC, there is increased price depreciation and increased volatility as well.

The trends are very similar in Index 3 compared to Index 1 including mean, median, and volatility. Compared to Index 2, the MV price deviation is similar to Index 3 in Time 3 but while Index 3 increased further in

Time 4, Index 2 decreased. Volatility is less for Index 3 than Index 2 in Time 3 and it is reversed in Time 4. BV behaved differently compared to MV.

4.9 *Cross Time Comparisons Groups: Index 4 (All Narrowbody Classic Aircraft)*

Comparing across the 4 time subsets for Index 4's MV and BV mean, median, and standard deviation for Time 2 are all less than the entire historical period (Time 1). In Time 3, all the metrics increased significantly except for standard deviation MV where it decreased. In Time 4, MV median, BV mean, and median increased while the MV mean decreased. Standard deviation in Time 4 increased for MV but decreased for BV. During GFC Time 3, price depreciation increased significantly. Volatility decreased for MV while BV increased. Post GFC, there is a price depreciation and volatility divergence between MV and BV as MV price depreciation slowed and volatility increased and vice versa for BV.

Compared to others, the price depreciation and volatility is the highest for Index 4 in Time 3 as well as the price depreciation difference between Time 2. During Time 4, price depreciation contracted towards the mean in Time 1 but volatility increased similarly to Index 1. The trends are similar to Index 2 but all the figures are more pronounced.

4.10 *Cross Time Comparisons Groups: Index 5 (All Narrowbody NG Aircraft)*

Comparing across the 4 time subsets for Index 5's MV and BV mean, median, and standard deviation for Time 2 are all higher than the entire historical period (Time 1) except for MV median and standard deviation. In Time 3, MV mean and median increased while BV mean and median decreased. Volatility in Time 3 increased for MV and BV. In Time 4, MV mean and median decreased while BV mean and median increased. Standard deviation in Time 4 increased for MV but decreased for BV. During GFC Time 3, price depreciation increased only slightly. Volatility also only increased slightly for MV and more for BV. Post GFC, there is a price depreciation and volatility divergence between MV and BV as MV price depreciation slowed and volatility increased and vice versa for BV. Compared to others, the price depreciation is the lowest for Index 4 in Time 3 but for volatility, the lowest is Index 1.

4.11 *Correlations Between MV and BV*

The correlation between MV and BV varies between the different index and time groups. For the Time 1 case, the correlation is 0.5408 for the Index 1 weighted average case. Index 2 decreases slightly to 0.5071 and widebody aircraft Index 3 shows 0.6039. Breaking down the narrow-body subsets, classic narrowbody aircraft Index 4 is low at 0.3207 while Index 5 is higher at 0.6765. Comparing Index 1 among the time subsets Time 2–4, it increases to 0.6006, 0.6313, and 0.4759 respectively. This similar trend of slightly increased correlation in Time 3 while a decrease in Time 4 is exhibited in Index 1–5 but Index 3 has a higher peak with its correlation increasing to 0.7413 in Time 3. This peak is similar to Index 5 which exhibits the highest correlation among the Indexes both in Time 1 and also the highest among the Indexes in Time 2–4, where Time 2 correlation is 0.8738, Time 3 is 0.8592, and Time 4 falls to 0.4345. Index 4 is the lowest correlations among the Indexes in each of the Time scenarios.

Index 5 is the new and the newer generation of aircraft where the MV and BV tend to be similar in many cases which partially explain some of these observations. The opposite is viewed in regards to Index 4's older aircraft and low correlation between MV and BV. While there is some correlation among the different Indexes in the entire term, they do not exhibit high degrees of correlation between the two value definitions. Time 3 exhibits the highest correlations compared to the other Time groups.

5 Conclusion

Through these analyses, there are many observations. To establish the baseline case, this consists of all aircraft mean and median numbers means entire time frame, Time 1 (6.1996–6.2017). The MoM weighted average MV mean is −0.6327%, the median is −0.3724%, and the standard deviation is 2.6313%. The MoM weighted average BV mean is −0.4689%, the median is −0.1245%, and the standard deviation is 2.2450%.

5.1 *Cross Index Comparisons (Time 1—6.1996–6.2017)*

Index 2 MoM weighted average MV mean is −0.3375%, the median is −0.1830%, and the standard deviation is 2.9181%. The MoM weighted average BV mean is −0.1783%, median is −0.0334%, and standard

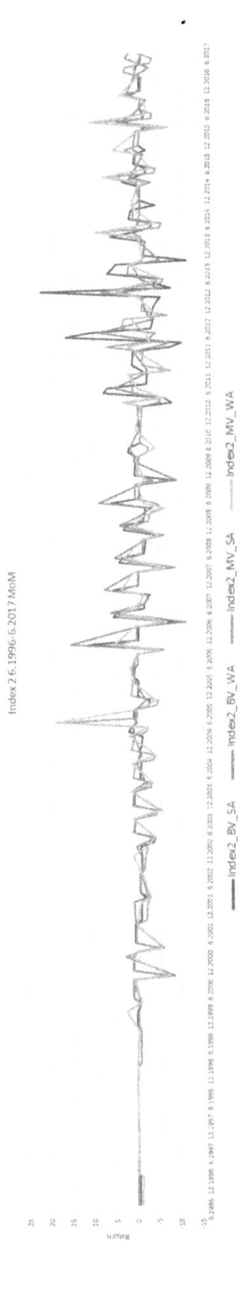

Fig. 5.10 Index 2 MoM—6.1996–6.2017

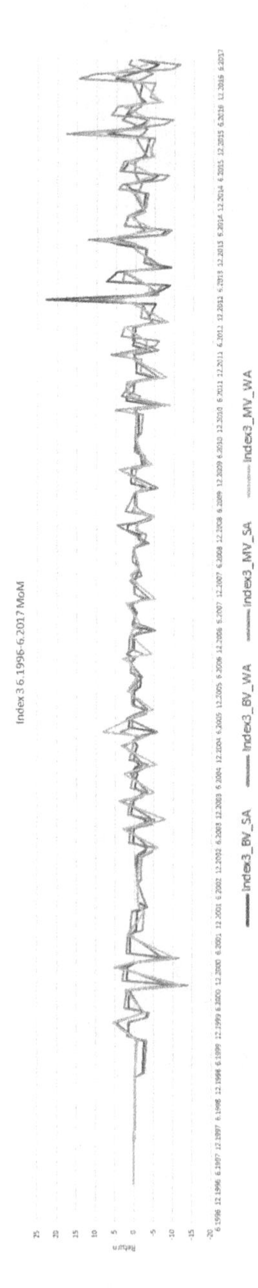

Fig. 5.11 Index 3 MoM—6.1996–6.2017. (Note: Index 3 data starts from 2.1997)

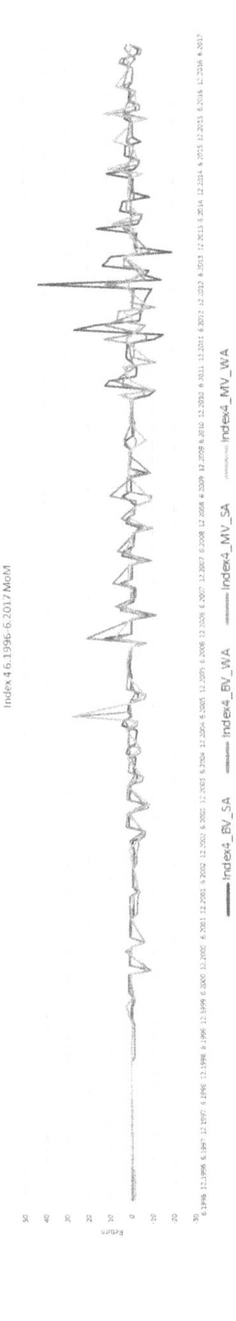

Fig. 5.12 Index 4 MoM—6.1996–6.2017

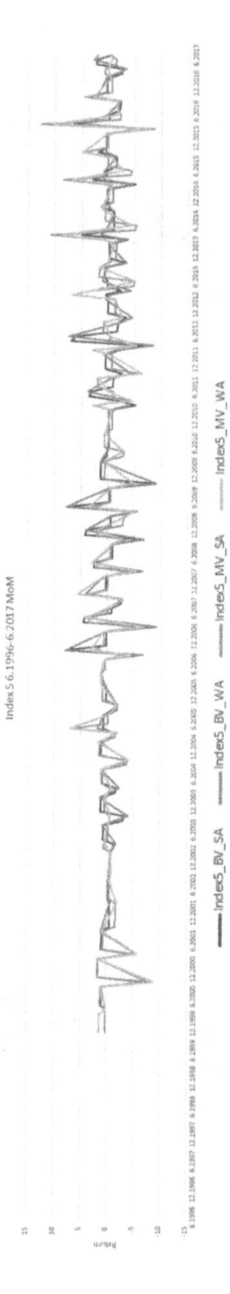

Fig. 5.13 Index 5 MoM—6.1996–6.2017. (Note: Index 5 data starts from 10.1999)

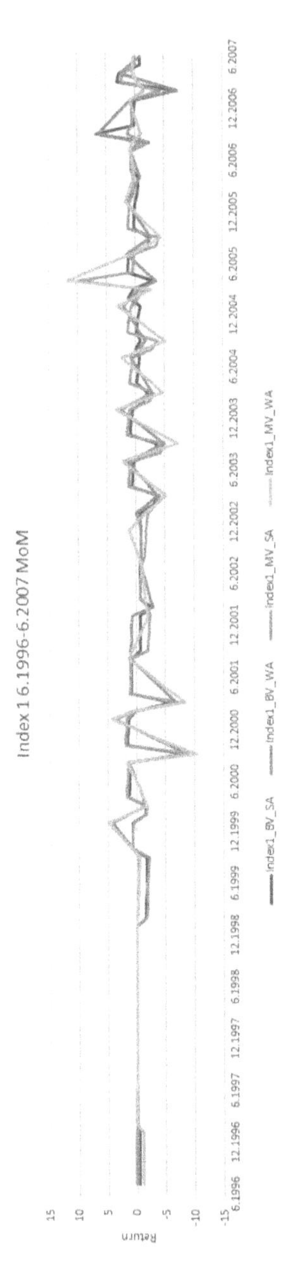

Fig. 5.14 Index 1 MoM—6.1996–6.2007

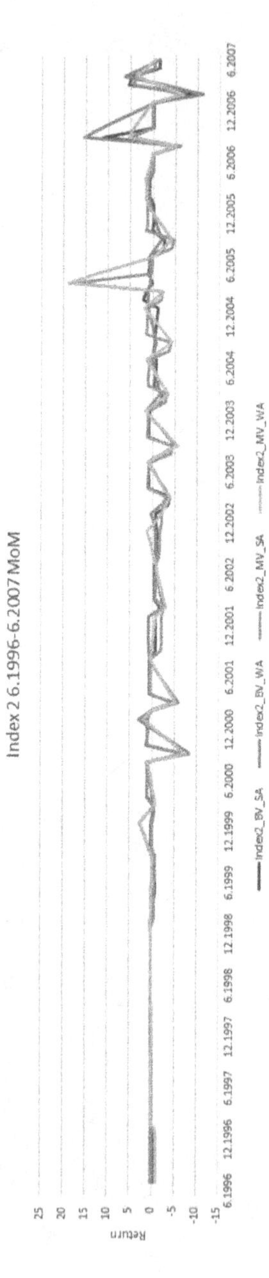

Fig. 5.15 Index 2 MoM—6.1996–6.2007

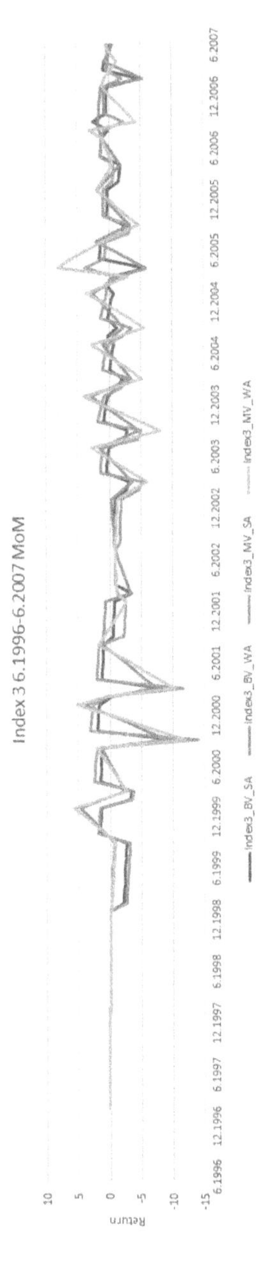

Fig. 5.16 Index 3 MoM—6.1996–6.2007. (Note: Index 3 data starts from 2.1997)

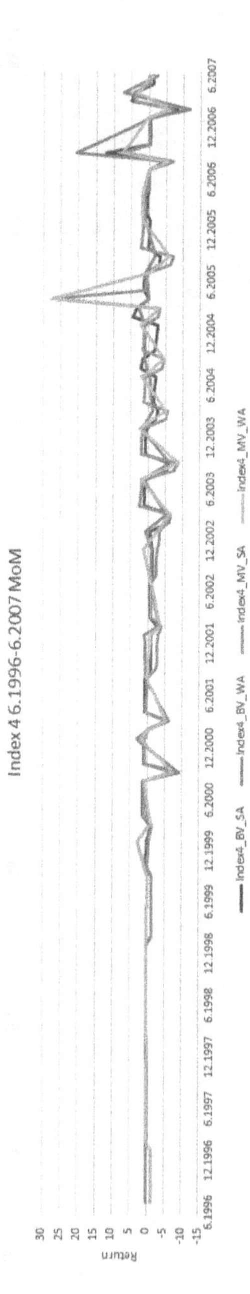

Fig. 5.17 Index 4 MoM—6.1996–6.2007

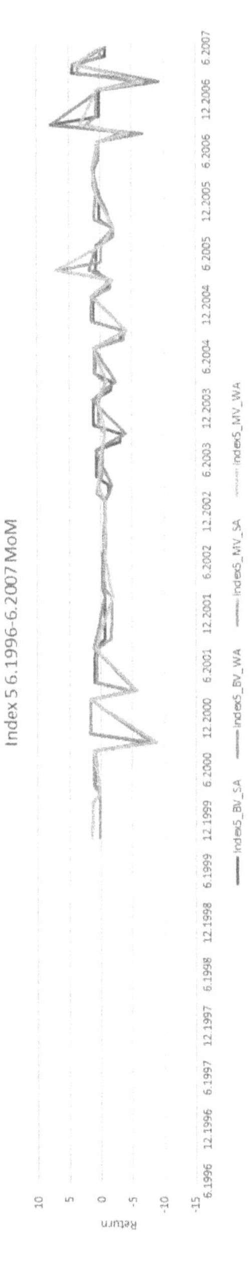

Fig. 5.18 Index 5 MoM—6.1996–6.2007. (Note: Index 5 data starts from 10.1999)

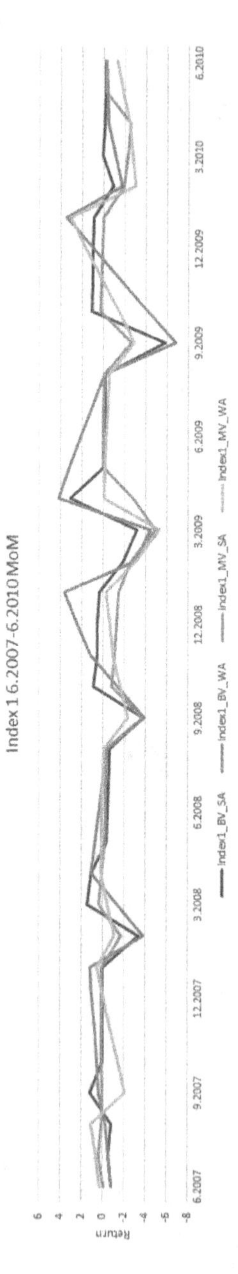

Fig. 5.19 Index 1 MoM—6.2007–6.2010

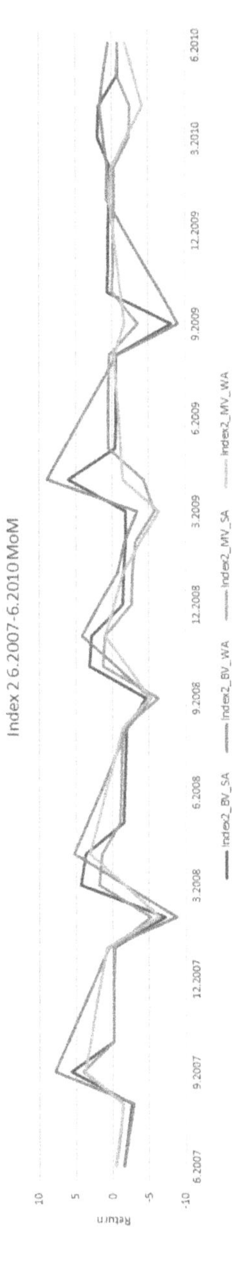

Fig. 5.20 Index 2 MoM—6.2007–6.2010

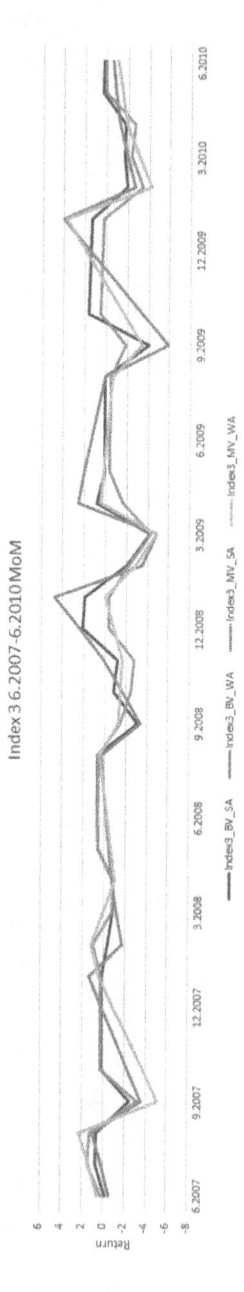

Fig. 5.21 Index 3 MoM—6.2007–6.2010. (Note: Index 3 data starts from 2.1997)

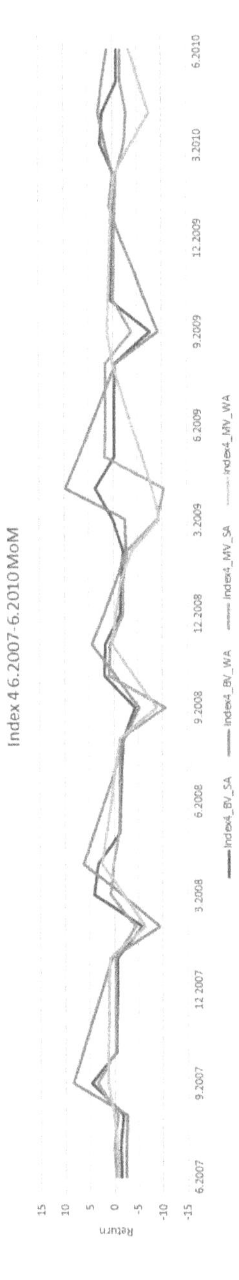

Fig. 5.22 Index 4 MoM—6.2007–6.2010

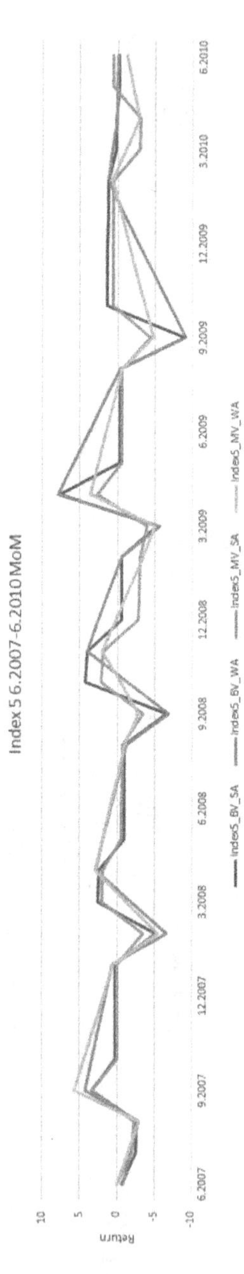

Fig. 5.23 Index 5 MoM—6.2007–6.2010. (Note: Index 5 data starts from 10.1999)

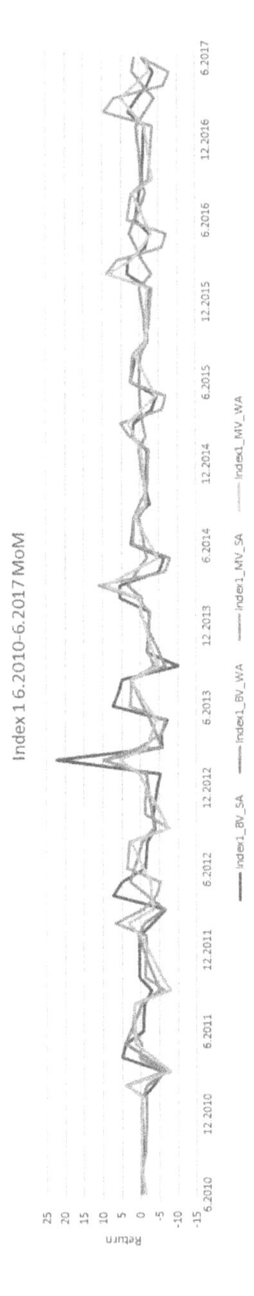

Fig. 5.24 Index 1 MoM—6.2010–6.2017

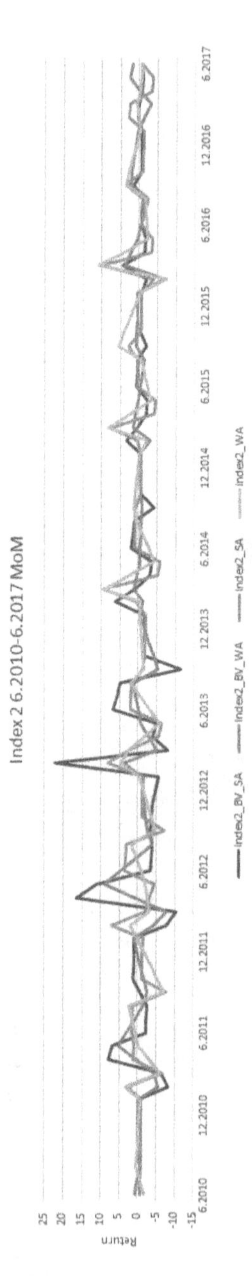

Fig. 5.25 Index 2 MoM—6.2010–6.2017

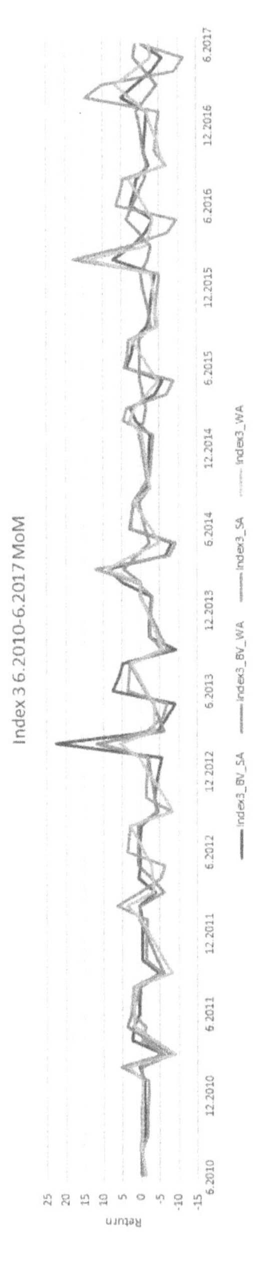

Fig. 5.26 Index 3 MoM—6.2010–6.2017. (Note: Index 3 data starts from 2.1997)

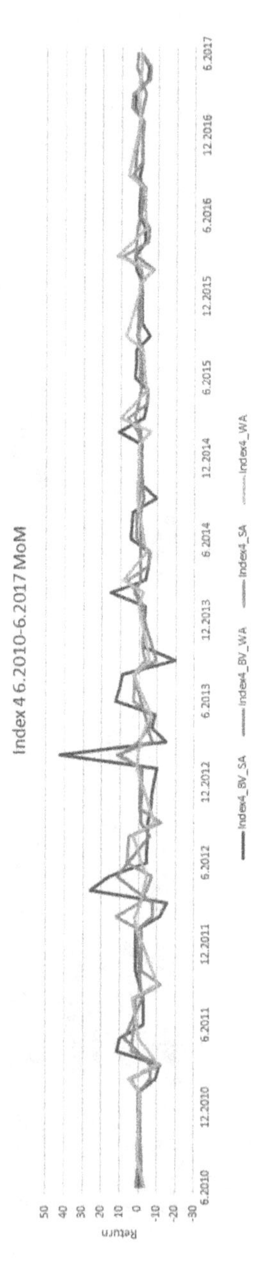

Fig. 5.27 Index 4 MoM—6.2010–6.2017

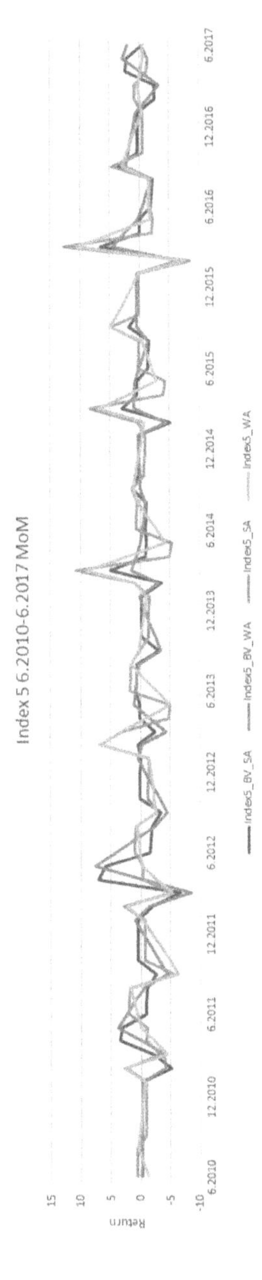

Fig. 5.28 Index 5 MoM—6.2010–6.2017

deviation is 2.992%. The data suggests the lower monthly mean and median numbers means that narrowbody aircraft values depreciate less than the entire population including both narrowbody and widebody aircraft in Index 1. The standard deviation is slightly higher in Index 2 compared with Index 1 which means higher volatility in price deviations for narrowbody aircraft alone compared with a population set of both narrowbody and widebody aircraft.

Index 3 MoM weighted average MV mean is −0.8324%, the median is −0.5709%, and the standard deviation is 3.0276%. The MoM weighted average BV mean is −0.6348%, median is −0.2503%, and standard deviation is 2.3332%. It can be deduced from the data that with the higher monthly mean and median numbers, widebody aircraft values depreciate faster on average than the population of aircraft in Index 1. In addition, volatility in price deviations is slightly higher in Index 3 compared with Index 1. Compared with the narrowbody Index 2 case, widebody aircraft Index 3 also depreciate faster while there is a divergence in volatility which is higher in MV while for BV it is lower than Index 2. The MV which takes into account market demand considerations compared with the BV definition which has the opposite effect.

Index 4 MoM weighted average MV mean is −0.6287%, the median is −0.1769%, and the standard deviation is 3.8874%. The MoM weighted average BV mean is −0.1705%, median is 0.0255%, and standard deviation is 3.5723%. It can be deduced from the data that with the older narrowbody classic aircraft Index 4 values depreciate slower and have higher volatility than the population of aircraft types in Index 1. This is a surprising result given the age of the aircraft types in Index 4 is higher than the average of those in the population Index 1 while its logical result that standard deviation is higher.

When Index 4 is compared with the all narrowbody aircraft Index 2, the data analysis is a bit more mixed. Index 4 MV mean depreciates significantly faster while with median depreciates slower and the vice versa for the BV case. These result in mixed signals for older narrowbody classic aircraft compared with the subset of narrowbody aircraft and conclusions cannot be easily attributed. It can be inferred that market demand dynamic differences are very large for Index 4 versus Index 2 due to the large differentials in MV mean figures. Volatility is found to be higher for narrowbody classic aircraft Index 4 than all narrowbody aircraft Index 2.

Index 4 has slower depreciation along with higher volatility compared with the all widebody aircraft Index 3. While the lower mean and median figures for Index 4 comparatively is not surprising, what is interesting is that the even greater volatility for older classic narrowbody aircraft compared to widebody aircraft.

Index 5 MoM weighted average MV mean is −0.2426%, median is −0.2028%, and standard deviation is 2.6168%. The MoM weighted average BV mean is −0.3834%, median is −0.1551%, and standard deviation is 2.6670%. It can be deduced from the data that with the lower monthly mean numbers, newer age and technology profiled narrowbody NG aircraft values depreciate slower than the all aircraft types in Index 1. The volatility figures have mixed results with Index 5 having a higher BV figure and slightly lower MV volatility.

Compared with all narrowbody aircraft Index 2, newer narrowbody NG Index 5 has lower MV mean, higher BV mean, and higher MV and BV median figures. These have conflicting signals to make any significant inferences in terms of price depreciation. The volatility for newer profiled narrowbody aircraft Index 5 can be inferred to be lower than the all narrowbody aircraft Index 2.

Index 5 has mixed results of lower MV mean and higher BV mean compared with older classic types of narrowbody aircraft in Index 4. While the conflicting information is difficult to make price depreciation conjectures with, it can be inferred that Index 5 has lower volatility than Index 4 which is in line with the thinking that older assets are more volatile than newer assets controlled by aircraft type. It means that narrowbody NG aircraft Index 5 has slower depreciation than widebody aircraft Index 3 but volatility under the MV market dynamic case is in line with the industry thinking but the increased volatility in the BV case is surprising.

The accounting definition of aircraft depreciation is also used as a baseline for comparison. The definition represents 3.4% depreciation per year or 0.283% deprecation per month on a straight line depreciation basis. For all aircraft types Index 1, MV and BV value depreciation is mostly higher than this standard and needs to be adjusted higher depreciation or there is a higher likelihood of write downs during the term of the asset life. This is the case for all narrowbody aircraft Index 2 and all widebody aircraft Index 3 as well. For narrowbody classic aircraft Index 4, the results provided mixed indicators that could provide any recommendations on adjustments while Index 5 results showed a need to adjust downward the monthly depreciation rate. Overall, Index 1–3 show the need to have a higher depreciation standard for aircraft assets to better match the actual movements of pricing over time. This data suggests that depreciation is higher overall and that the useful age of aircraft assets are less than 25 years.

The results suggest that the main hypothesis is correct that the aircraft market represented by All Aircraft Types Index 1 MoM WA MV has higher value depreciation compared to the accounting depreciation of 0.283% per month. Index 1's respective MoM WA MV mean in Time 1–4 is −0.6327%, −0.4730%, −0.7875%, and −0.8094%. In terms of the second part of the main hypothesis, the data also suggests this is correct as the standard deviation is less than 3% throughout the four time segments.

The data suggests some mixed results for the second hypothesis. Looking at the MoM WA MV case, narrowbody aircraft Index 2 has lower value depreciation throughout the four time segments than widebody aircraft. In analyzing the standard deviation, narrowbody aircraft is lower for all the time segments except for the Time 3 case. While there is one deviation in Time 3 standard deviation, the data suggests that most of the second hypothesis is correct.

The data suggests mixed results for the third hypothesis. Looking at the MoM WA MV case, narrowbody NG aircraft has lower value depreciation than narrowbody classic aircraft as well as narrowbody aircraft in all the periods except for Time 2. For standard deviation, narrowbody NG aircraft is lower in all time segments compared with narrowbody classic aircraft. In comparison to narrowbody aircraft, narrowbody NG has a lower standard deviation in all the periods except for Time 3. While there are deviations in Time 2 for the value depreciation for both narrowbody classic and narrowbody aircraft and Time 3 standard deviation for narrowbody aircraft, the data suggests that most of the third hypothesis is correct.

The data suggests mixed results for the fourth hypothesis. For the first part looking at the MoM WA MV case, the five aircraft market segments have lower value depreciation in the pre-GFC period Time 2 than during and post the GFC in Time 3 and Time 4, respectively for Indexes 1, 3, and 4 while for Index 2 and 5, Time 4 is lower than Time 3. As for standard deviation, Time 2 is the lowest for Index 5 while for Index 1–4, Time 3 is lower than Time 2. This suggests the data partially support the first part of the fourth hypothesis. Looking at the Index 1 results, the overall aircraft market has lower value depreciation in Time 2 than the other times while for most Indexes, in Time 3 the standard deviation is the lowest.

For the second part of the fourth hypothesis, Indexes 2, 4, and 5 have higher value depreciation during the GFC in Time 3 than the other periods but Time 4 is higher for Index 1 and 3. For standard deviation, Time 3 is not the highest relative figure as Time 4 is the highest for all Index 1–5. The data does not support the second part of the fourth hypothesis. Looking at the Index 1 results, the overall aircraft market has a higher value depreciation in Time 4 than the other times while for all Indexes, Time 4 also has the highest standard deviation.

5.2 *Cross Time Comparisons Groups: Index 1 (All Aircraft Types)*

There are some interesting observations that arise from further analyzing the time subsets data. Before the GFC, represented by Time 2, is generally a period of lower price depreciation and lower volatility. During the GFC as represented by Time 3, there is increased price depreciation for MV and lower for BV while volatility decreased significantly for MV and increased slightly for BV. Post GFC, represented by Time 4, price depreciation increased significantly especially compared to Time 2 and 1. Volatility increased during Time 4. This infers that pre-GFC Time 2 had lower volatility than the historical figure in Time 1 and also had increased volatility after the GFC in Time 4.

5.3 *Cross Time Comparisons Groups: Index 2 (All Narrowbody Aircraft)*

Comparing across the time subsets for Index 2 in Time 3, price depreciation increased while volatility decreased slightly while rebounding in Time 3. Post GFC there is a price depreciation and volatility divergence between MV and BV. Compared to Index 1 in Time 3, MV price depreciation increased in Time 2 but diverged in Time 4 where Index 1 increased while Index 2 decreased. Volatility acted similarly in Index 1 and 2 with a decrease in Time 3 and increase in Time 4. BV acted similarly between Index 1 and 2 but volatility diverged where after the increase in Time 3, Index 1 increased while Index 2 decreased in Time 4.

5.4 *Cross Time Comparisons Groups: Index 3 (All Widebody Aircraft)*

Index 3 cross time comparison results are similar to Index 2. During GFC Time 3, price depreciation increased significantly. Volatility decreased for MV while BV increased slightly. Post GFC there is increased price depreciation and increased volatility as well. The trends are very similar in Index 3 compared to Index 1 including mean, median, and volatility. Compared to Index 2, the MV price deviation is similar to Index 3 in Time 3 but while Index 3 increased further in Time 4, Index 2 decreased. Volatility is less for Index 3 than Index 2 in Time 3 and it is reversed in Time 4. BV exhibited and behaved differently compared to MV.

5.5 *Cross Time Comparisons: Index 4 (All Narrowbody Classic Aircraft)*

Index 4 cross time comparison results are similar to Index 2. During GFC Time 3, price depreciation increased significantly. Volatility decreased for MV while BV increased. Post GFC there is a price depreciation and volatility divergence between MV and BV as MV price depreciation slowed and volatility increased and vice versa for BV. Compared to others, the price depreciation and volatility are the highest for Index 4 in Time 3 as well as the price depreciation difference between Time 2. During Time 4, price depreciation contracted towards the mean in Time 1 but volatility increased similarly to Index 1. The trends are similar to Index 2 but all the figures are more pronounced.

5.6 *Cross Time Comparisons Groups: Index 5 (All Narrowbody NG Aircraft)*

Comparing across the time subsets for Index 5's MV and BV mean, median, and standard deviation for Time 2 are all higher than the entire historical period Time 1 except for MV median and standard deviation. In Time 3, MV mean and median increased while BV mean and median decreased. Volatility in Time 3 increased for MV and BV. In Time 4, MV mean and median decreased while BV mean and median increased. Standard deviation in Time 4 increased for MV but decreased for BV. During GFC Time 3, price depreciation increased only slightly. Volatility also only increased slightly for MV and more for BV. Post GFC there is a price depreciation and volatility divergence between MV and BV as MV price depreciation slowed and volatility increased and vice versa for BV. Compared to others, the price depreciation is the lowest for Index 4 in Time 3 but Index 1 volatility is the lowest.

5.7 *Correlation Between MV and BV*

The correlation between MV and BV varies between the different index and time groups. While there is some correlation among the different Indexes in the entire period, the data does not exhibit high degrees of correlation between the two value definitions in general except for Index 5—the highest correlation among the indexes. Time 3 exhibits the highest correlations compared to the other Time groups while a decreased correlation is exhibited in Index 1–5 in Time 4. Index 4 exhibited the lowest correlation between the Indexes in the Time groups. Index 5 is the new and the newest aircraft where the MV and BV tend to be similar in many cases which partially explain some of these observations. The opposite is viewed in regards to Index 4's older aircraft and low correlation between MV and BV.

The data suggests mixed results for the fifth hypothesis. For the first part of the hypothesis that there is slight to moderate correlation between MV and BV for the aircraft types throughout the four time segments is found to be supported for all the Index and Time scenarios except for the high correlation found in Index 5's Time 2 and 3. The data mostly support the first part of the hypothesis.

For the second and third parts of the hypothesis, the highest relative correlation occurs during pre-GFC Time 2 while the lowest is during the GFC Time 3. The data suggests that Time 2 is the highest relative correlation for all Indexes except for Index 3, where Time 3 is the highest relative correlation. The data suggests that Time 4 is the lowest relative correlation for all Indexes except for Index 4 where Time 3 is the lowest relative correlation. The data does not support the third part of the hypothesis. The lowest correlation between MV and BV is after the GFC Time 4.

Bibliography

Airbus. (2017). *Orders and Deliveries*. Retrieved from https://www.airbus.com/aircraft/market/orders-deliveries.html

Beechy, T. H. (1969). Quasi-Debt Analysis of Financial Leases. *The Accounting Review, 44*(2), 375–381.

Boeing Corporation. (2016). *Current Market Outlook 2016–2035*. Retrieved from Seattle, WA: www.boeing.com/resources/boeingdotcom/commercial/about-our-market/assets/downloads/cmo_print_2016_final_updated.pdf

Boeing Corporation. (2017). *Orders and Deliveries*. Retrieved from Seattle, WA: http://www.boeing.com/commercial/#/orders-deliveries

Bower, R. S. (1973). Issues in lease financing. *Financial Management*, 25–34.

Federal Aviation Administration. (2018). Glossy of Terms. Retrieved from www.faa.gov/air_traffic/flight_info/avn/maintenanceoperations/programstandards/webbasedtraining/CBIChg29/AVN300WEB/Glossarytems.htm

Froot, K. A. (1995). Hedging portfolios with real assets. *Journal of Portfolio Management, 21*(4), 60.

Gibson, W., & Morrell, P. (2004). Theory and practice in aircraft financial evaluation. *Journal of Air Transport Management, 10*(6), 427–433.

Gibson, W., & Morrell, P. (2005). *Airline finance and aircraft financial evaluation: evidence from the field*. Paper presented at the Conference conducted at the 2005 ATRS World Conference.

Hallerstrom, N. (2013). *Modeling Aircraft Loan & Lease Portfolios*. Retrieved from www.pkair.com/common/docs/modeling-aircraft-loan-and-lease-portfolios.pdf

Harris, R. S., & Pringle, J. J. (1985). Risk adjusted discount rates extensions from the average risk case. *Journal of Financial Research, 8*(3), 237–244.

International Society of Transport Aircraft Trading. (2009). *The Principles Of Appraisal Practice And Code Of Ethics*. Retrieved from https://members.istat.org/p/do/sd/sid=140&fid=168&req=direct

International Society of Transport Aircraft Trading. (2013). *Appraisers' Program Handbook Revision 7*. Retrieved from www.istat.org/d/do/382

Johnson, R. W., & Lewellen, W. G. (1972). Analysis of the Lease or Buy Decision. *The Journal of Finance, 27*(4), 815-823.

Myers, S. C. (1974). Interactions of corporate financing and investment decisions—implications for capital budgeting. *The Journal of Finance, 29*(1), 1–25.

Myers, S. C., Dill, D. A., & Bautista, A. J. (1976). Valuation of financial lease contracts. *The Journal of Finance, 31*(3), 799–819.

Oum, T. H., Zhang, A., & Zhang, Y. (2000). Socially optimal capacity and capital structure in oligopoly: the case of the airline industry. *Journal of Transport Economics and Policy*, 55–68.

Quigley, J. M. (1995). A simple hybrid model for estimating real estate price indexes. *Journal of Housing Economics*, *4*(1), 1–12.

CHAPTER 6

Comparative Examination Between the Aircraft and Other Asset Classes

1 Introduction and Research Questions

In Chaps. 3 and 4, the drivers and dynamics of aviation finance on aircraft asset pricing has been discussed. In Chap. 5, the construction of the indexes representing the aircraft asset and was empirically tested to show its characteristics. The extension to the main hypotheses is to test the similarities and differences between the baseline aircraft asset class indexes and the comparable asset classes to fill in gaps in the conversation found through the academic literature review.

These comparable asset classes include publicly listed aircraft lessors, other real asset subclasses such as shipping, infrastructure, real estate, commodities, agricultural land, timberland, as well as general interest rates, and representative indexes such as airline, transport, and the broader debt and equity markets. Analyzing the relationships between the aircraft asset classes and these benchmarks will be able to deduce many observations in the context of portfolio construction and risk along with the risk adjusted returns of the asset classes in a portfolio context.

Analyzing the comparative asset classes to the aircraft asset class segments will yield many interesting observations. The main emphasis of the analysis is on the aircraft asset classes represented by Index 1–5 and 20 comparative asset classes including the public aircraft lessor index. The listed aircraft lessors are the first investors who have the most exposure to the aircraft asset class. While they can take on leverage through financing

D. Yu, *Aircraft Valuation*,
https://doi.org/10.1007/978-981-15-6743-8_6

structures as well as buy and sell aircraft, the first empirical study is to analyze their performance over time compared to the performance of the overall aircraft asset classes. The second part is to analyze the aircraft asset classes to other real asset subclasses. Third, the correlation and covariance between each of the main investigative and various comparative indexes are derived along with a regression model and significance testing. With these derived statistic characteristic information, the inclusion, or not, of the aircraft asset class and public aircraft lessor asset class in portfolio construction and management can be determined. The time period in the analysis is similar to the Chap. 4 analysis (Times 1–4) and is as follows: (1) all available historical sample set information 6.1996–6.2017; (2) pre-GFC period, 6.1996–6.2007; (3) GFC period, 6.2007–6.2010; and (4) post-GFC period, 6.2010–6.2017. A more in-depth discussion of the index and time periods construction are found in Data and Methodology Sect. 5.4. The main question is how do the aircraft asset class segments compare in terms of various returns and volatility metrics against other real asset-backed classes and benchmarks in the four time segments?

The main hypothesis is that aircraft asset class as represented by Index 1–5 MoM WA MV have a lower standalone risk in terms of standard deviation and variance over the historical time periods compared with public aircraft lessors and other real asset-backed segments (such as shipping, real estate, commodities, precious metals, agricultural land, timberland, airlines, transportation index), specific commodities (such as gold, crude oil), leading indicators, rates (such as USD 3M LIBOR, USD 1–3M treasuries, USD ten-year swap rates, US Consumer Price Index (CPI)) and broad debt and equity indexes.

Making an extension of the main research question, the second hypothesis is that the aircraft asset classes have a lower relative volatility in terms of beta or β, covariance and correlation compared with the other comparative asset classes including publicly listed aircraft lessors in the four time segments.

Continuing in the same line of questioning, the third hypothesis is that the publicly listed aircraft leasing companies generate higher excess returns or α relative to the risk free rate, Real Assets and MSCI World compared to the aircraft asset class in the four time segments.

Combining return and volatility concepts, the fourth hypothesis is that the publicly listed aircraft leasing companies generate higher Sharpe ratios than the aircraft asset class in the four time segments.

2 Comparative Asset Classes Discussion

2.1 *Publicly Listed Aircraft Lessors*

One of the main comparative asset classes is the public aircraft lessor index. While many aircraft lessors are privately held entities or subsidiaries of larger groups, there is a handful of leasing companies that are publicly listed. Most of these are listed in 2006 and after. Publicly listed pure-play or aircraft leasing companies are the most directly comparable set to the aircraft asset class. The companies primarily own and lease aircraft while some have a small amount of aircraft that is managed on behalf of third party investors and owners. The public pure-play large commercial aircraft leasing companies that are still active as public entities include AerCap Holdings N.V. (AerCap), Aircastle Limited (Aircastle), Air Lease Corporation (ALC), Avation PLC (Avation), Avolon Holdings Limited (Avolon), China Aircraft Leasing Group Holdings Limited (CALC), FLY Leasing Limited (FLY), AviaAM Leasing (AviaAM), China Development Bank Financial Leasing Co., Ltd (CDB Leasing), and ALAFCO Aviation Lease and Finance Company K.S.C.P. (ALAFCO). While Genesis Lease Limited (GLS) and Avolon were once publicly listed and subsequently merged into Aercap and taken private by HNA Group, respectively. The pricing data available from the period when the companies were listed are included in the comparative dataset.

While CDB Leasing is not a pure-play aircraft lessor, it is included due to the significant percentage of the company's revenues and assets are aircraft which stand at 51.5% and 37.7%, respectively, for the 2016 figures. Also by that logic, GECAS, one of the top two lessors globally and an indirect subsidiary of General Electric, as well as other bank-owned Chinese leasing subsidiary companies are not included as they do not represent a significant amount of the listed company's total assets and revenues. There are other similar companies who are also included such as regional aircraft leasing players such as AeroCentury Corp. (ACY) and specialized public engine leasing players such as Willis Lease Finance Corporation (WLFC).

The majority of the companies is or was listed on the New York Stock Exchange while CDB Leasing and CALC are listed on the Hong Kong Stock Exchange, AviaAM is listed on the Warsaw Stock Exchange, Avation is listed on the London Stock Exchange, and ALAFCO is listed on Boursa Kuwait. There are also special purpose vehicle aircraft leasing fund corporations which are listed on the Specialized Funds Market including DP Aircraft 1 Limited (DP1), Doric Nimrod Air One Limited (DNA1), Doric Nimrod Air Two Limited (DNA2), Doric Nimrod Air Three Limited (DNA3), and Amedeo Air Four Plus (AAFP). These companies are all included in the data set index. These funds are special purpose vehicles and its size is smaller than the other public aircraft leasing companies listed above. One example is DNA1, which consists of a single Airbus A380-800 leased to Emirates while DNA2 contains three aircraft, while the other securities are larger in composition. The index created is an equal weighted total return index (including dividends) consisting of these companies. The listed public lessors generally utilize debt financing while the aircraft asset class is an asset price or 100% equity and contains no debt.

2.2 *Real Assets Overview*

Real assets are a broad range of hard, physical assets that derive value from their substance and properties. They have intrinsic value and can both be tangible or intangible which includes precious metals, commodities, real estate, land agricultural land, timberland, infrastructure, mining land and rights, machinery, and intellectual property (copyrights, trademarks, patents, trade secrets, etc). Real assets are not financial assets such as stocks and bonds. Some investors have slightly different definitions of the asset class such as including sustainable investments. Aircraft is a subset of this asset class and used as one of the market approximations in calculating excess return and beta. (More in the data collection and methodology in Sect. 5.4).

There are many benefits of the asset class. These include having many subclasses that have lower volatility compared to equities. They also have low correlations to traditional financial assets such as equity and fixed-income products. When combined with traditional financial assets, this low correlation makes them ideal for a diversified portfolio according to modern portfolio theory. See Table 6.1 for correlations between the asset classes during the 2006–2016 timeframe. Figure 6.1 shows the various subclasses' Sharpe ratios (more on the definition in Sect. 5.4 Data Collection and Methodology), which are higher compared to global bonds and stocks over the period 2006–2016.

Many of the elements of the real assets class are inputs into end-consumer products which often are leading indicators for inflation. Also,

Table 6.1 2006–2016 correlations among real asset subclasses

	Real estate	*Infrastructure*	*Timberland*	*Agriculture*	*Global bonds*	*Global stocks*
Real estate	1.00	0.55	0.23	0.12	0.19	(0.10)
Infrastructure	0.55	1.00	0.17	0.08	0.64	0.37
Timberland	0.23	0.17	1.00	0.64	(0.11)	0.07
Agriculture	0.12	0.08	0.64	1.00	0.09	(0.15)
Global bonds	0.19	0.64	(0.11)	0.09	1.00	0.23
Global stocks	(0.10)	0.37	0.07	(0.15)	0.23	1.00

Source: NCREIF 2017; Cambridge Associates 2017; Bloomberg 2017; Brookfield Asset Management 2017

Note: Data as of June 30, 2016. Real Estate is represented by the NCREIF Property Index, Infrastructure is represented by the Cambridge Associates Infrastructure Index (available only through 3/16), Timberlands is represented by the NCREIF Timberland Index, Agriculture is represented by the NCREIF Farmland Index, Global Bonds is represented by the Barclays Global Aggregate Bond Index, and Global Stocks is represented by MSCI World Index

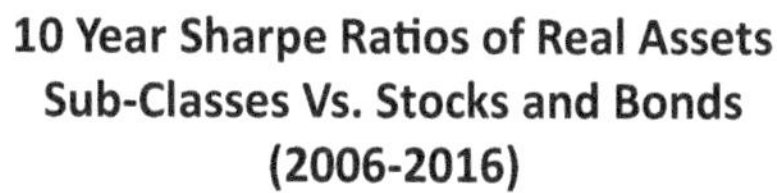

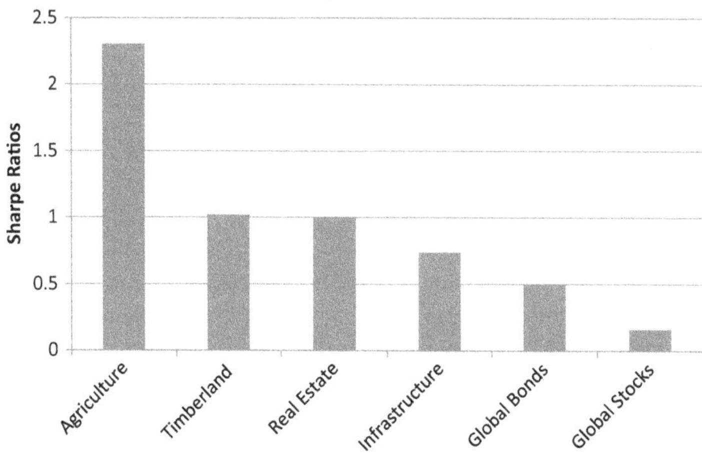

Fig. 6.1 Ten-year 2006–2016 sharpe ratios of real assets subclasses vs. stocks and bonds. Note: Data as of June 30, 2016. Sharpe Ratio based upon ten-year average annualized total returns and standard deviations of performance; assumes a risk-free rate of 1.7%. Agriculture is represented by the NCREIF Farmland Index, Timberlands is represented by the NCREIF Timberland Index, Real Estate is represented by the NCREIF Property Index, Infrastructure is represented by the Cambridge Associates Infrastructure Index, Global Bonds is represented by the Barclays Global Aggregate Bond Index, and Global Stocks is represented by the MSCI World Index. (MSCI 2017; Bloomberg 2017; NCREIF 2017; S&P Dow Jones Indexes 2017; Brookfield Asset Management 2017)

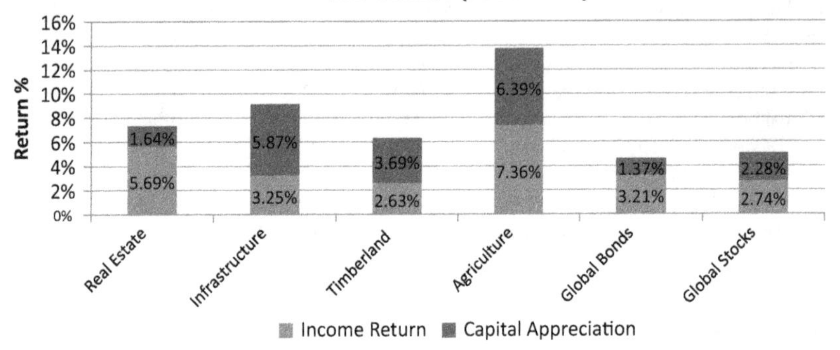

Fig. 6.2 Ten-year income and capital appreciation returns of real assets sub-classes. (Source: MSCI 2017; Bloomberg 2017; NCREIF 2017; S&P Dow Jones Indexes 2017; Brookfield Asset Management 2017. Note: Data as of June 30, 2016. Index data represent ten-year average annual returns for each index. Real Estate is represented by the NCREIF Property Index, Infrastructure is represented by the Dow Jones Brookfield Infrastructure Index, Timberlands is represented by the NCREIF Timberland Index, Agriculture represented by the NCREIF Farmland Index, Global Bonds is represented by the Barclays Global Aggregate Bond Index, and Global Stocks is represented by the MSCI World Index.)

there is potential for capital appreciation as many of these are supply constrained such as limited in quantity and scarce by nature (Fig. 6.2).

Inelastic demand properties of the asset class create stable and predictable income streams and growth. The duration and viewpoints of these assets are long term and are generally characterized as international by nature. Other macroeconomic drivers that contribute to the asset class include a growing global population, demographic change, rising wealth in emerging countries, climate change, and lack of resources. For the purpose of this study, the real asset class is represented by the S&P Real Assets Total Return Index.

Along with the main drivers, there are also risk elements that are associated with the asset class. These include economic downturns, and political- and weather-related risks. These risks are also found in some financial assets as well. Exposure in the asset class can come in different forms as there are different types of entities involved. An overview of these

Table 6.2 Various types of investment vehicles and attributes

Various Types of Investment Vehicles and Attributes

	Direct Investment	Club Deal	Closed-End Fund	Open-End Mutual Fund	Public Equity	ETF
Control Over the Asset						
Customization						
Lever of In-House Expertise Required						
Liquidity						
Min. Investment						
Diversification Within Vehicle						

High — Scale — Low

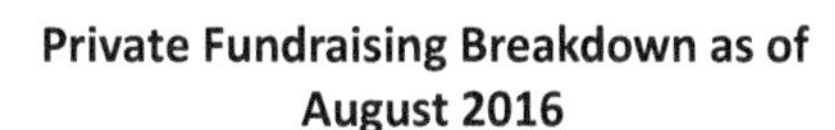

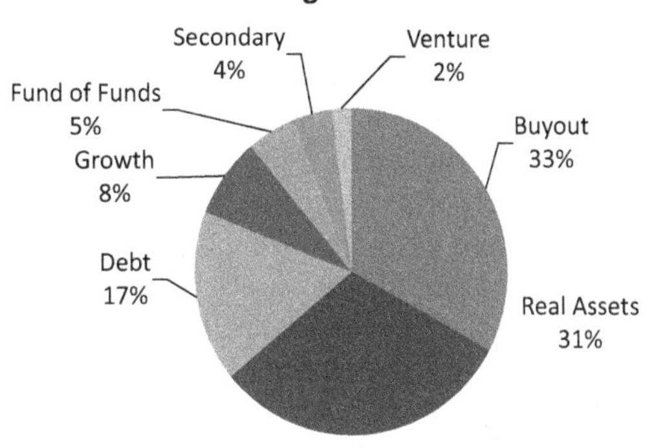

Fig. 6.3 Private fundraising breakdown as of August 2016. (Source: Bloomberg 2017; Brookfield Asset Management 2017)

different types of funds include private and public funds and investment vehicles, direct investments, public securities, real estate investment trusts (REITs), master limited partnerships (MLPs), public and private operating companies, exchanged listed futures, commodity futures and exchange traded funds (ETFs). See Table 6.2 for a more detailed breakdown by investment vehicle type. With these properties, real assets as an asset class represent 31% of all private funds raised as of August 2016 which represents the second largest category after buyouts (Fig. 6.3).

2.3 *Shipping*

A ship is a large ocean going vessel used to carry goods or passengers. The comparative context to large commercial aircraft is large commercial ships. These large commercial ships include cargo bulk carriers, container ships, roll on roll off (RoRo) carriers, tankers, and passenger and off-shore vessels. There are many similarities and differences between shipping and aircraft as an asset class. This is considered the most direct comparison to aircraft as an asset class. In both instances, the asset can transport passengers or cargo (the "Carriage") from one point to another and is a movable asset. It also creates value for the owner from the operations of the operator either itself or a third party. Similarities end in terms of the market characteristics including the larger volatility of the shipping asset class compared to aircraft. In terms of the supply base, shipping has multiple shipyards and new shipyards enter the market as there are lower barriers to entry. Large commercial aviation, on the other hand, has two main OEM (original equipment manufacturer) suppliers Boeing and Airbus with larger barriers to entry.

There are three main types of charters or leases. The first type is time charter, which is chartered for a specific time and analogous to the spot market. The length is usually shorter than the second type, a bareboat charter. The owner, or the lessor, still manages the ship but the charterer, or the lessee, has to pay for all the fuel, port charges, commissions, and the daily hire rate. The second type of charter is a bareboat charter or demise charter. This is similar to an aircraft operating lease where no administration or technical maintenance is carried out by the lessor. The charterer, or the lessee, has to pay for all the operating expenses including fuel, crew, port expenses, and insurance. This type of charter has a typically long lease period. The third type of charter is a voyage charter. This is for a single voyage between ports and the cost is determined on a lump-sum basis or more precise per-ton basis. With a voyage charter, the owner has to pay for the port, fuel, and crew costs. This is typically the shortest charter period.

The Carriage is dependent on the type of ship but can be general commodities, crude oil, liquid products, or others. The cost of moving the Carriage from one point to another is based on the charter type discussed above. Freight Forward Agreements (FFA) are financial derivative instruments that are used to measure the cost of a time charter for a specific route or an average of routes on a particular type of ship. FFAs are constructed as swaps, are net cash flow settled as there are no physical deliveries and are generally used by end-users and suppliers for hedging and/or investment speculation. Currently, these are traded over the counter or exchange-cleared contracts (such as SGX, LCH, Nasdaq, CME, ICE,

EEX). As FFAs are instruments, there is no need to own a ship, employ staff and crew, fuel, food, and so on.

Clean and dirty tankers are liquid carriers. Dirty tankers contain products with heavy or dirty residuals such as crude oil, vacuum gas oil, heavy fuel oil, and unrefined condensates. Clean tankers are products that do not contain those heavy or dirty residuals such as product or chemical tankers. While a tanker can be converted from clean to dirty or vice versa to take financial spread advantage of pricing between the two, it is best to note that the cargo hold is not thoroughly cleaned in between cargo products and remaining residuals can affect the next product. This can produce combustible conditions so it is best not to be repeated.

Baltic Dry Index represents dry bulk carrier costs including multiple routes of three component ship types including 40% Capesizes' five routes, 30% Panamaxes' five routes, and 30% Supramaxes' ten routes with a multiplier starting March 2018 after historically equal weighting between Capesizes, Panamaxes, Supramaxes, and Handymaxes with another multiplier for consistency purposes. This change to more weighting for bigger ships is due to a low volume of Handymax derivatives being traded and the desire for ETFs to be created. The specific vessel definitions used by the Baltic Exchange for the various indexes can be found in Table 6.3. It is noted that the routes and vessels that are contained in the index have changed over time and long consistency of the data is cautioned.

For this study, an equal weighted shipping index is created between dry bulk, liquid carriers, and containers. Dry bulk is represented by the Baltic Dry Index including Capesize, Panamax, and Supramax vessels and covering 20 global routes. Liquid carriers are an equal weight between Baltic Dirty Tanker Index covering 15 major routes and Baltic Clean Tanker Index covering nine major routes. Liquid carriers indexes include four ship classes including VLCC, Suezmax, Aframax, and Panamax. Container freight cost is represented by an equal weight of 12 Freightos Index routes which are the costs to ship 40-foot containers on ocean freight for those respective routes.

2.4 Real Estate

The real estate asset class is one of the largest segments with approximately $217 trillion of value with 13% in commercial real estate according to Savills.com's 2016 Around The World in Dollars and Cents. It is an

Table 6.3 Baltic exchange vessel definitions

- Capesize 2014 vessel:
  - Basis 180,000 mt dwt on 18.2 m SSW draft
  - Max age ten yrs
  - LOA 290 m, beam 45 m, TPC 121
  - 198,000 cbm grain
  - 14 knots laden/15 knots ballast on 62 mt fuel oil (380 cst), no diesel at sea
- Panamax vessel:
  - 74,000 mt dwt on 13.95 m SSW draft
  - Max age 12 yrs
  - LOA 225 m, beam 32.2 m
  - 89,000 cbm grain
  - 14 knots on 32 mt fuel oil (380 cst) laden/28 mt fuel oil (380 cst) ballast, no diesel at sea
- Supramax vessel:
  - Standard "Tess 58" type
  - 58,328 mt dwt on 12.80 m SSW draft
  - Max age 15 yrs
  - LOA 189.99 m, beam 32.26 m
  - 72,360 cbm grain, 70,557 cbm bale
  - Five holds, five hatches
  - 4 × 30 mt cranes with 12 cbm grabs
  - 14 knots laden on 33 mt fuel oil (380 cst), no diesel at sea
  - 14 knots ballast on 32 mt fuel oil (380 cst), no diesel at sea
  - 12 knots lade on 24 mt fuel oil (380 cst) no diesel at sea
  - 12.5 knots ballast on 23 mt fuel oil (380 cst), no diesel at sea
- Supramax vessel:
  - Standard "Tess 58"
  - 58,328 mt dwt on 12.80 m SSW draft
  - Max age 15 yrs
  - LOA 189.99 m, beam 32.26 m
  - 72,360 cbm grain, 70,557 cbm bale
  - Five holds, five hatches
  - 4 × 30 mt cranes with 12 cbm grabs
  - 4 knots laden on 33 mt fuel oil (380 cst), no diesel at sea
  - 14 knots ballast on 32 mt fuel oil (380 cst), no diesel at sea
  - 12 knots laden on 24 mt fuel oil (380 cst) no diesel at sea
  - 12.5 knots ballast on 23 mt fuel oil (380 cst), no diesel at sea

Source: Baltic Exchange Information Services Ltd, 2018a

established global asset class on its own and includes many subsegments. These include office, multifamily residential, retail, hospitality, and industrial real estate. The office segment has grades from AAA in major cities downward. The length of the leases is normally 5 to 20 years. The key driver is job creation and growth. Multifamily residential are rental buildings with multiple units. The main drivers are main economic factors such as employment, demographics, competitive supply of rental housing, and availability of single-family housing available for sale. The typical length of the lease is one year and as such is more volatile and sensitive to economic factors. Industrial segment includes warehouses, logistics, distribution, and light manufacturing facilities. The leases are generally long. The demand is mainly derived from overall global commercial growth and activity. This facilitates the need for product inventory and the flow of goods from one location to another. Hospitality segment is more operational in nature. It includes hotels, motels, vacation rentals, and executive rental housing and the drivers are business and leisure travel where business travel accounts for more of the traffic but leisure activity has grown. To represent this asset class, an equal weighted of three indexes, the S&P Global REIT USD Index, S&P Corelogic Case Shiller U.S National Price SA Index, and NCREIF Property Index is used.

2.5 *Infrastructure*

The infrastructure asset class tends to provide essential products or services. These include transportation and energy which can be divided into utilities and renewable energy. These assets are also long term in nature with associated stable cash flows and tend to benefit from inelastic demand. This is an area where public-private partnerships (PPPs) are common, where the government partners with private capital investors to build and operate an infrastructure asset.

The transportation subclass includes infrastructure networks utilized to move passengers or cargo such as toll roads, railroads, ports, airports, bridges, and tunnels. Machinery such as aircraft, ships, and railcars are also sometimes included in this category. Generally, these concessions or leases are 10–99 years or more and pricing is linked to inflation or other volume measures and maintenance capital expenditures that are required for upkeep. Given the long-term concession structure of the projects, these tend to have high barriers of entry.

The energy subclass consists of power generation, distribution and support systems, and renewable power sources. Utilities represent the electric power generation aspect of infrastructure and also include water generation and wastewater treatment systems. Renewable power includes solar, wind, hydro, geothermal, and biomass sources which are both environmentally friendly and seek to reduce the carbon footprint. The substitutes are more traditional coal, gas, and nuclear power plants. To represent this asset class, an equal weight index consisting of the Dow Jones Brookfield Global Infrastructure Composite Index and S&P Global Listed Infrastructure Index is used.

2.6 *Commodities*

The definition of a commodity is a primary agricultural or raw material product which can be bought and sold in a transaction. There are many types of commodities and they can be categorized as soft and hard commodities. Hard commodities are mined such as precious materials like gold and oil. Examples of soft commodities would be agricultural products. Commodities can be used for hedging by suppliers and end-users as well as investors and speculators. Some commodities such as gold and oil have been indicators of the economy, while others are used for hedging purposes. There are many types of instruments used to get exposure to the space including physical or financial delivery. Some of these investment instruments include ETFs (exchange traded funds), futures, forwards, swaps, or listed or private companies. Many types of commodities have both industrial uses and intrinsic values such as those commodities used in the manufacturing of electronics. To represent the Commodities asset class, the Thomson Reuters/CoreCommunity CRB Commodity Index is used.

Specific commodities are used as comparative set by themselves. Specific commodities of interest are crude oil as it is a major direct input for aircraft operations and also gold. Gold has also been classically used as an inflationary hedging tool and is represented by the gold spot price per troy ounce. Crude oil is represented by an equal weighted average of WTI spot crude oil price (representing the US market) and Brent crude oil spot price (representing the European market). Precious metals is another subclass of general commodities and is represented by the S&P GSCI Precious Metals Spot Index.

2.7 *Agricultural Land and Timberland*

Agricultural or farmland and timberland are the biggest category for sustainable resources. These assets are naturally self-reproducing. With proper management, the harvest rate must be below the regeneration rate or else the resource risks depletion over time. Agricultural land is essential land for the production of food and is scarce in nature. Investments focus on productive land and returns are based on the growth of the crops, livestock, and land appreciation. This became an investable class by institutional investors starting in the 1970s. US Farmland as a major asset class has over $1.9 trillion in total asset value according to the United States Department of Agriculture (USDA). Of that, approximately $300–400 billion is institutional-quality farmland according to Hancock Agricultural Investment Group. National Council of Real Estate Investment Fiduciaries (NCREIF) Farmland Index stands at $2 billion as of December 31, 2009. These include three types, namely, (a) row crops such as annual crops like corn, soybeans; (b) permanent or perennial crops such as fruits trees and nuts; and (c) livestock which includes pasture land for grazing, dairy, and livestock farms for meat. The main type of farmland usage is bulk commodity crops that have efficiencies of scale, storable crops, and livestock which enable large scale production.

Timberland produces wood-based raw materials used for many industries. With the growing population, there is increasing demand for these essential raw material assets. The export markets are an important source of output. Timberland became an investable class by institutional investors starting in the 1980s. The assets under management have doubled in the last ten years according to the Timberland Investment Management Organizations, who are invested in the asset class. There is some debate as to the actual investable universe. On the high end, Professor Michael Clutter of the University of Georgia estimated in 2004, the investable timberland asset size to be US$400 billion while Brookfield Asset Management estimated the value to be US$120 billion. While others think the overall market size is higher. The International Woodland Company in its 2009 Timberland Investable Universe paper noted that there is over US$120 billion invested in timberland as an asset class for institutional investors (Acquila Capital 2015).

Both agricultural land and timberland can either be owned and directly managed or leased to operators. Different risk profiles exist as direct operations versus leasing the land. Some of the risks for direct operations include both skill and nature related such as weather and disease. The rewards, on the other hand, can be higher for direct operations than leasing the land to an operator as the annual crop and livestock sales can be based on managing supply and demand elements optimally. That said, institutional investors are not the best suited to manage the direct operational end of the business. From this perspective, the answer is no from the author's perspective. Capital, management, and technological improvements and advancement can also improve the yield from the land. On the other hand, if the land is leased to an operator, the lease rate is normally a fixed amount over a moderate to long tenor and the main risk entailed is the credit risk of the operator. Either way, specialty property management is needed to look after timberland and farmland assets like all asset-backed asset classes. To represent the Agricultural land asset class, the NCREIF Farmland Total Return Index is used while Timberland is represented by an equal weighted average of the S&P Global Timber & Forestry Index and NCREIF Timberland Index.

2.8 *Interest Rate Products*

Interest rates are very important to any financial asset and hedging tool. Aircraft are predominately denominated in US dollars. Also, aircraft have long residual, useful lives and as such, similar financing can match the longer length tenors. Usually, the longer tenor floating debt is set to match its floating nature and the fixed nature of the rental cash flow stream. This mismatch is fixed by implementing a USD swap rate. The rental stream is usually a function of the competitive supply and demand environment, the base risk-free and benchmark financing rates allowing for various premia including credit, geographic, and structural, among others. The airline or end-user does have the option to evaluate whether leasing or owning outright with debt is more appropriate. In addition, both rental stream and debt is benchmarked on the risk-free rate and three-month USD LIBOR rate. As such 1–3M US Treasuries is the set risk-free rate and represented by Bloomberg-Barclays US 1–3 Month T-Bills Total Return Index. USD 3M LIBOR rates and USD ten-year interest rate swap rates are included in the comparative data sets. As aviation travel demand is correlated to GDP, US CPI is also used as a comparative set as well.

2.9 *Sector-Specific and Broad Equity and Debt Indexes*

Airline- and transportation-specific industry indexes are also examined to get a better comparison viewpoint for similar asset profiles as aircraft values. To represent the Airlines asset class, the S&P 500 Airlines Total

Return Index is used while an equal weighted average of the Dow Jones Transportation Average and Nasdaq Transportation Index is used to represent the Transportation asset class. Both broader equity and debt indexes are included in the comparative set. For equity, S&P500 and MSCI World Index individually are used to represent equity indexes in USA and worldwide. The broader global debt index is represented by the Bloomberg-Barclays Global Aggregate Bond Total Return Index Value Unhedged.

3 Data and Methodology

The baseline empirical data sets and index construction are derived and extended upon from the methodologies described in Chap. 4. Again, all data and calculated returns are calculated and monthly values and return figures are derived. A subset of the sample set is calculated through various Index 1–5 to represent the various aircraft asset types. Index 1–5 represents the value of the aircraft assets by type and its associated depreciation profile. They do not have associated lease or cash flow thus the comparison directly to other comparative asset classes as later specified is not a likewise comparison. Given the lack of actual return cash flow data, the monthly negative returns or depreciation of aircraft value, as depicted by market value (MV), is sign inversed and considered positive returns for comparative purposes. The associated historic aircraft lease cash flows data by type and vintage from each of the various sources are ideally available. Unfortunately, this is not publicly available nor are there any other suitable substitutes.

Similar to Chap. 4, Index 1–5 are used as the baseline comparative indexes to represent the different segments and these indexes are as follows: (1) All aircraft types, (2) all narrowbody aircraft, (3) all widebody, (4) all narrowbody classic aircraft, and (5) all narrowbody next generation aircraft. For this comparison with other asset classes, only the weighted-average approach on MV for each index is considered and sign inversed as discussed previously. This weighted average approach takes into account the differences between the smaller values such as the older and smaller aircraft and the larger values with larger weighting for the newer or larger aircraft. The simple arithmetic-average approach treats all aircraft as equal weighting. Again, month-over-month calculation is the focus instead of the year-over-year approach given the lack of seasonality in the figures.

3.1 *Month over Month Simple or Equal Weighted Average Index Total Return Calculations*

$$\text{Asset Class Simple Average Return}_{i,t} = \left(\frac{1}{n}\right) * \sum_{n}^{j=0} \text{Asset Class Return}_{i,j,t}$$

where

$$\text{Asset Class Return}_{i,j,t} = \frac{\text{Asset Class Value}_{i,j,t} + \text{Asset Class Dividend Value}_{i,j,t}}{\text{Asset Class Value}_{i,j,t-1}} - 1$$

i = specific asset class	j = subset member of asset class i	t = time, month, and year of the valuation date (in months)

The overall time frame of the dataset is 6.1996–6.2017, in addition, segmentation of the time periods which are the same as in Chap. 4 are analyzed to see if there are observations that can be postulated due to the major economic event, the Global Financial Crisis, in Time 3. The time scenarios (Times 1–4) selected are as follows: (1) all available historical information 6.1996–6.2017; (2) pre-GFC period, 6.1996–6.2007; (3) GFC period, 6.2007–6.2010; and (4) post-GFC period, 6.2010–6.2017. Again, these are the same time segments as in Chap. 4 and chosen for ease of comparison purposes. For each index under the different conditions, after the sign inversion, statistical analyses are performed including mean, median, standard deviation, variance, correlation, and covariances with the comparative data sets. Utilizing these figures, both return and volatility metrics are calculated. The return metrics include the following: excess return to the risk-free rate, α relating to Real Assets, α relating to MSCI World, volatility metrics (including β relating to the market as Real Assets, β relating to the market such as MSCI World) and Sharpe ratios (which combines both excess return and volatility concepts).

α is the excess return to the market returns. α and β's market is calculated under two market scenarios—Real Assets and MSCI World—as aviation is within the real assets market and is global in nature so the S&P 500 is not used. The risk-free rate throughout the study is set as the US

Treasuries 1–3M Index's respective time scenario mean. Excess return compared to the risk-free rate, Real Assets and MSCI World returns, is used to observe the comparative performance of the asset classes.

$$(\text{Alpha})\alpha_i = (r_i - r_m)$$

r_i= return of index i (our case mean)	r_m = mean return of market index	i = index number	m = real assets or MSCI world index

$$\text{Excess Return Over } r_f = (r_i - r_f)$$

r_i = return of index i (our case mean)	r_f = risk free rate (represented by US Treasuries 1–3M)	i = index number

There are several measurements for volatility that are analyzed such as standard deviation, the variance, and beta. While standard deviation for each asset class is focused on each on a standalone basis, variance is the square of standard deviation and beta measures the volatility on a comparative basis relative to the market index proxies, like alpha, include Real Assets and MSCI World.

$$Var(x) = E\left[(x-\mu)^2\right] = \sigma^2$$

$$cov(x,y) = E\left[(x - E[x])(y - E[y])\right]$$

$$\beta_i = \frac{cov(r_i, r_m)}{Var(r_m)} = \frac{\sigma_{i,m}}{\sigma_m^2}$$

$\sigma_{i,m}$ = covariance of between index i and market index	σ_m^2 = variance of index i	i = index number

β is also part of the capital asset pricing model (CAPM) which uses risk-free rate, excess market returns, and beta volatility to calculate the return expectations for an asset.

3.2 *Capital Assets Pricing Model*

$$\text{Expected return } E(r_i) = r_f + \beta_i * \left(E(r_m) - r_f\right)$$

r_f = risk-free rate (represented by US Treasuries 1–3M)	$E(r_m)$= expected market return (mean)	β_i = beta of index i	i = index number

Combining the concepts of asset class excess return over the risk-free rate and adjusted for the asset standalone volatility or standard deviation, is the Sharpe ratio.

$$\text{Sharpe Ratio}_i = \frac{\overline{r}_p - r_f}{\sigma_p}$$

$\overline{r}_p$ = expected mean portfolio return of index i	r_f = risk-free rate (represented by US Treasuries 1–3M)	σ_p= standard deviation of index i	i = index number

The comparative sets are the monthly returns and compared against the baseline Indexes 1–5 MoM WA MV as well as to each other. The comparative data includes historical trading data from Bloomberg, Reuters, respective company information. and other publicly available information. Each of the different comparative data sets represents a different asset class and thus different indexes are created.

For the public pure-play aircraft leasing company index, the components include AAFP, ACY, AerCap, Aircastle, ALAFCO, ALC, Avation, AviaAM, Avolon, BOC Aviation, CALC, CDB Leasing, DP1, DNA1, DNA2, DNA3, FLY, GLS, and WLFC. See Publicly Listed Aircraft Lessors Section 5.2.1 for a fuller discussion. The overall time frame for this index is from 6.2006 to 6.2017 while companies are added to the index when they are taken public and when companies are acquired or taken private, they are removed from the index. As such, GLS was acquired and Avolon was taken private after their initial public offering; only the data from their time as public companies are used. For the other companies, all data while the respective companies are public are used which do not encompass the entire duration of the sample historical data time frame 6.1996–6.2017 and noted as such.

The comparative data set contains 20 asset classes including Aircraft Lessors that are compared with the five aircraft type indexes from Chap. 4. Some of the data members and asset classes' historical pricing do not go as far as 6.1996. These are noted and all available historical figures are used for computation. These comparative asset classes and their respective members are the following:

1. **Aircraft Lessors**—(a) AAFP; (b) ACY; (c) AerCap; (d) Aircastle; (e) ALAFCO; (f) ALC; (g) Avation; (h) AviaAM; (i) Avolon; (j) BOC Aviation; (k) CALC; (l) CDB Leasing; (m) DP1; (n) DNA1; (o) DNA2; (p) DNA3; (q) FLY; (r) GLS; and (s) WLFC (Source: Bloomberg)
2. **Shipping**—(a) Baltic Dry Bulk Index; (b) Liquid Carriers (Dirty and Clean) Route Indexes; and (c) Average of 12 Freightos Baltic Container Route Indexes (Source: Bloomberg; Baltic Exchange Information Services Ltd)
3. **Real Estate**—(a) S&P Global REIT USD Index; (b) S&P Corelogic Case Shiller U.S National Price SA Index; and (c) NCREIF Property Index (Source: Bloomberg)
4. **Agricultural Land**—NCREIF Farmland Total Return Index (Source: Bloomberg; NCREIF)
5. **Timberland**—(a) S&P Global Timber & Forestry Index; and (b) NCREIF Timberland Index (Source: Bloomberg; NCREIF)
6. **Commodities**—Thomson Reuters/CoreCommunity CRB Commodity Index (Source: Bloomberg)

3.2.1 Commodities—Specific Individual

7. **Crude Oil**—(a) WTI spot crude oil price; and (b) Brent crude oil spot price for Europe market (Source: Bloomberg)
8. **Gold**—Gold Spot Price per Troy Ounce (Source: Bloomberg)

9. **Precious Metals**—S&P GSCI Precious Metals Spot Index (Source: Bloomberg)
10. **Airlines**—S&P 500 Airlines Total Return Index (Source: Bloomberg)
11. **Transportation**—(a) Dow Jones Transportation Average; and (b) Nasdaq Transportation Index (Source: Bloomberg)
12. **Infrastructure**—(a) Dow Jones Brookfield Global Infrastructure Composite Index; and (b) S&P Global Listed Infrastructure Index (Source: Bloomberg)
13. **Real Assets**—S&P Real Assets Total Return Index (Source: Bloomberg)

3.2.2 *Specific Interest Rates*

14. **US Treasuries 1–3M**—Bloomberg-Barclays US 1–3 Month T-Bills Total Return Index (Source: Bloomberg)
15. **USD Swap Rate10Y**—USD Swap Rate Ten Year (Source: Bloomberg)
16. **USD 3M LIBOR**—USD Three-Month LIBOR (Source: Bloomberg)
17. **US CPI**—US CPI Index (Source: Bloomberg)

3.2.3 *Specific Broad Equity Indexes*

18. **MSCI World—MSCI World (Equity) Index** (Source: Bloomberg; MSCI)
19. **S&P 500–S&P 500 Equity Index** (Source: Bloomberg; S&P Dow Jones Indexes)
20. **Broad Debt**—Bloomberg-Barclays Global Aggregate Bond Total Return Index Value Unhedged (Source: Bloomberg, Barclays)

For the NCREIF Property Index, NCREIF Farmland Total Return Index, and NCREIF Timberland Index, the indexes' data is quarterly so for the immediate months in between are smoothed out and calculated on a straight line basis. The respective comparative indexes are created by calculating the simple or equal weighted average of each of the subsets' monthly total returns. These returns are total returns which include either the price appreciation (or depreciation) and any dividends if there are any. The results of the analysis of these 20 comparative indexes representing different asset classes are compared with the baseline aircraft data represented by Index 1–5 MoM WA MV for the same time period (Time 1–4) and conclusions are postulated from the analysis.

4 Results and Analyses

4.1 *Return and Volatility Comparative Analysis*

Summary statistics of Index 1–5 and the comparative asset classes for Time 1–4 can be found in Table 6.4. Figures 6.4, 6.5, and 6.6 show the ranking summaries by asset class for Alpha (Real Assets), Alpha (MSCI World), and Sharpe Ratios while Fig. 6.7 shows the mean and standard deviation breakdown for Time 1–4.

4.2 *Return and Volatility Comparative Analysis: Time 1 (6.1996–6.2017)*

In the all historical period Time 1, return and volatility of Index 1–5 MoM WA MV is compared to the 20 comparative asset class benchmarks. The return analysis is focused on mean figures for the asset class and comparative excess returns over the risk-free rate and market proxies, Real Assets and MSCI World. By these return measures, Index 1–5 is ranked 12, 13, 7, 19, and 20, respectively, out of 25. Aircraft Lessors index is ranked in second in 25 comparative sets. The top ranked is the Shipping asset class. The mean for all datasets is positive except for the USD 3M LIBOR and USD Swap Rate Ten Yr.

In terms of excess return over the risk-free rate are shown to be all positive for all the indexes except for Commodities index and rates data (US CPI, USD 3M LIBOR, and USD Swap Rate Ten Yr) with the US Treasuries 1–3M as the risk-free benchmark, ranked 21. Index 1 MoM WA MV excess return over risk-free rate is 0.4526% and ranked 12. For alpha or excess returns over MSCI World, MSCI World is ranked 11 so more than half of the asset sets are negative. Index 1 MoM WA MV alpha over MSCI World is -0.0157%. Real Assets is ranked below MSCI World at 17 with Index 1 and 4, while Precious Metals, Real Estate, and Gold is ranked in between. Index 1 MoM WA MV alpha over Real Assets is 0.0529%.

The Sharpe ratio tells a bit of a different story, where Indexes 3, 1, and 4 moved up and are ranked third, sixth, and 11th respectively. Indexes 2 and 5 did not move in rankings and Aircraft Lessors moved down from 2 to 12. Sharpe ratios adjust the excess return to the risk-free rate to take into account the individual volatility and standard deviation. Ahead of Index 3 in the ranking are Agricultural Land and Real Estate asset classes. Index 1 MoM WA MV's Sharpe ratio is 0.1720%.

In terms of standalone volatility with standard deviation and variance, Index 1–5 is ranked in the third quartile. In terms of relative volatility or

Table 6.4 Time 1–4 comparative indexes' summary statistics.

Time scenario 1—6.1996–6.2017 summary statistics

	n	*Min (%)*	*Max (%)*	*Median (%)*	*Mean (%)*	*Std. Dev*	*Var*	*β (Real asset)*	*β (MSCI world)*	*Mean over RF (%)*[1]	*α (Real assets) (%)*	*α (MSCI world) (%)*	*Sharpe ratios (%)*[1]
Index 1 MoM WA MV	12,015	(11.8712)	10.2745	0.3724	0.6327	2.631	6.9239	(0.0788)	(0.0262)	0.4526	0.0529	(0.0157)	0.1720
Index 2 MoM WA MV	6817	(18.9991)	7.5487	0.1830	0.3375	2.924	8.5493	(0.1430)	(0.0582)	0.1574	(0.2423)	(0.3109)	0.0538
Index 3 MoM WA MV[2]	5199	(8.1385)	11.9864	0.5709	0.8324	3.034	9.2039	(0.0490)	(0.0136)	0.6524	0.2527	0.1840	0.2150
Index4 MoM WA MV	5154	(27.6055)	10.4857	0.1769	0.6287	3.895	15.1720	(0.1470)	(0.0418)	0.4487	0.0490	(0.0197)	0.1152
Index 5 MoM WA MV[3]	1511	(6.9616)	8.8196	0.2028	0.2426	2.623	6.8799	(0.1345)	(0.0860)	0.0625	(0.3372)	(0.4058)	0.0238
Aircraft lessors[4]	2508	(35.7439)	43.4109	0.9766	1.2407	9.781	95.6756	1.4329	0.8995	1.0606	0.6609	0.5922	0.1084
Real estate	396	(11.4848)	6.2830	0.7373	0.6105	1.526	2.3272	0.5035	0.2072	0.4305	0.0308	(0.0379)	0.2822
Shipping	132	(46.6789)	52.9171	1.6946	1.5879	13.081	171.1027	0.8431	0.3615	1.4079	1.0082	0.9395	0.1076
Agr. land	132	(0.0033)	7.5933	0.6524	0.9759	1.070	1.1439	(0.0148)	0.0246	0.7958	0.3961	0.3275	0.7441
Timberland	264	(14.2696)	13.6359	0.7384	0.6513	2.599	6.7528	0.7738	0.4059	0.4712	0.0715	0.0029	0.1813
Commodities	132	(22.3251)	13.7866	0.2053	0.1149	4.791	22.9495	1.1257	0.4343	(0.0651)	(0.4648)	(0.5335)	(0.0136)
Crude oil	264	(33.2646)	40.2353	1.4434	0.8397	9.760	95.2632	1.5859	0.5784	0.6596	0.2599	0.1913	0.0676
Gold	132	(16.8896)	16.8453	0.1383	0.5862	4.841	23.4387	0.6108	0.1070	0.4062	0.0065	(0.0622)	0.0839
Precious metals	132	(18.6736)	15.1079	(0.0375)	0.6166	5.120	26.2124	0.7248	0.1679	0.4366	0.0369	(0.0318)	0.0853
Airlines	132	(26.5725)	24.2940	0.4430	0.7020	8.768	76.8696	0.9667	1.0175	0.5219	0.1222	0.0536	0.0595
Transportation	264	(20.8783)	17.0595	1.2737	0.9241	5.616	31.5438	1.0068	0.9232	0.7441	0.3444	0.2757	0.1325

Infrastructure[5]	67	(14.4816)	10.6156	1.1467	0.8671	3.811	14.5251	1.1409	0.7216	0.6870	0.2873	0.2187	0.1803
Real assets[6]	26	(16.8700)	8.6055	0.9022	0.5797	3.188	10.1623	0.9863	0.6417	0.3997	0.0000	(0.0687)	0.1254
US treasuries 1-3M	132	(0.7639)	0.5636	0.1097	0.1800	0.193	0.0372	0.0014	(0.0003)	0.0000	(0.3997)	(0.4684)	0.0000
USD swap rate 10Y	132	(30.1781)	31.4864	(0.6147)	(0.1501)	7.712	59.4683	0.1384	0.3738	(0.3302)	(0.7299)	(0.7985)	(0.0428)
USD 3M LIBOR	132	(35.7205)	54.7351	(0.0457)	(0.0791)	10.020	100.3959	(1.0004)	(0.2430)	(0.2592)	(0.6589)	(0.7275)	(0.0259)
US CPI	132	(3.6850)	1.3768	0.1844	0.1610	0.374	0.1397	0.0206	0.0073	(0.0191)	(0.4188)	(0.4875)	(0.0511)
MSCI world	132	(18.9312)	11.3076	1.1843	0.6484	4.395	19.3202	1.2200	0.9960	0.4684	0.0687	0.0000	0.1066
S&P 500	132	(16.7929)	10.9193	1.1789	0.7866	4.347	18.8930	1.0308	0.9418	0.6065	0.2068	0.1381	0.1395
Broad debt	396	(3.9732)	6.2131	0.4137	0.3908	1.607	2.5831	0.3062	0.0869	0.2108	(0.1889)	(0.2576)	0.1312

[1]Risk free rate is US treasuries 1-3M mean

[2]Starts from 2.1997

[3]Starts from 10.1999

[4]Starts from 6.2006

[5]Starts from 12.2001

[6]Starts from 5.2005

(*continued*)

Table 6.4 (continued)

Time scenario 2—6.1996–6.2007 summary statistics

	n	*Min (%)*	*Max (%)*	*Median (%)*	*Mean (%)*	*Std. Dev*	*Var*	*β (Real asset)*	*β (MSCI world)*	*Mean over RF (%)*[1]	*α (Real assets) (%)*	*α (MSCI world) (%)*	*Sharpe ratios (%)*[1]
Index 1 MoM WA MV	12,015	(11.8712)	10.2745	0.0241	0.4730	2.615	6.8381	(0.2282)	(0.0443)	0.1675	(0.8592)	(0.3192)	0.0640
Index 2 MoM WA MV	6817	(18.9991)	7.5487	0.0219	0.2752	2.926	8.5642	(0.2835)	(0.0456)	(0.0304)	(1.0570)	(0.5171)	(0.0104)
Index 3 MoM WA MV[2]	5199	(8.1385)	11.9864	0.1555	0.6489	2.842	8.0784	(0.2031)	(0.0482)	0.3433	(0.6833)	(0.1433)	0.1208
Index4 MoM WA MV	5154	(27.6055)	8.6975	0.0427	0.3074	3.902	15.2251	(0.4268)	(0.0355)	0.0018	(1.0248)	(0.4848)	0.0005
Index 5 MoM WA MV[3]	1511	(6.9616)	8.8196	0.1081	0.3512	2.274	5.1701	(0.1167)	(0.0857)	0.0456	(0.9810)	(0.4410)	0.0201
Aircraft lessors[4]	2508	(35.7439)	43.4109	1.4964	1.5074	11.380	129.4978	1.4820	0.5986	1.2018	0.1752	0.7152	0.1056
Real estate	396	0.3294	1.4261	0.7777	0.8025	0.249	0.0619	0.0046	0.0030	0.4969	(0.5297)	0.0102	1.9978
Shipping	132	(19.8428)	38.8833	1.6946	1.6218	10.606	112.4882	(1.7228)	0.0582	1.3162	0.2896	0.8296	0.1241
Agr. land	132	(0.0033)	7.5933	0.5683	0.9428	1.264	1.5968	(0.1853)	0.0541	0.6372	(0.3894)	0.1506	0.5043
Timberland	264	(3.0094)	6.8164	0.6407	0.8305	1.528	2.3347	0.3771	0.1269	0.5250	(0.5017)	0.0383	0.3436
Commodities	132	(9.1297)	13.1207	0.3799	0.5334	4.191	17.5671	1.3222	0.1344	0.2278	(0.7988)	(0.2588)	0.0544
Crude oil	264	(24.8600)	40.2353	2.3593	1.5049	10.183	103.7001	1.8852	0.0680	1.1993	0.1727	0.7126	0.1178
Gold	132	(9.3094)	16.8453	(0.0990)	0.4861	4.060	16.4822	1.5978	0.0862	0.1805	(0.8461)	(0.3062)	0.0445
Precious metals	132	(11.0819)	15.1079	(0.1829)	0.5525	4.113	16.9185	1.8302	0.1321	0.2469	(0.7797)	(0.2397)	0.0600

Airlines	132	(26.2798)	22.3267	0.2081	0.2267	8.716	75.9605	(0.8776)	1.1151	(0.0789)	(1.1055)	(0.5655)	(0.0090)
Transportation	264	(20.8783)	14.3578	1.3184	1.0792	5.610	31.4725	0.0499	0.9137	0.7736	(0.2530)	0.2870	0.1379
Infrastructure[5]	67	(7.5067)	6.5623	1.7221	1.4574	2.981	8.8839	1.1293	0.4250	1.1519	0.1252	0.6652	0.3865
Real assets[6]	26	(3.2100)	4.6231	1.4849	1.3322	1.655	2.7375	0.9615	0.1274	1.0266	0.0000	0.5399	0.6205
US treasuries 1-3M	132	(0.7639)	0.5636	0.3768	0.3056	0.174	0.0303	0.0014	0.0016	0.0000	(1.0266)	(0.4867)	0.0000
USD swap rate 10Y	132	(11.8870)	29.5720	(0.7257)	(0.0169)	5.534	30.6279	(0.8563)	0.1998	(0.3225)	(1.3492)	(0.8092)	(0.0583)
USD 3M LIBOR	132	(25.1986)	22.4335	0.0000	0.1127	5.489	30.1328	(0.1181)	0.1208	(0.1929)	(1.2195)	(0.6796)	(0.0351)
US CPI	132	(3.6850)	1.3768	0.1869	0.1829	0.417	0.1739	0.0162	0.0055	(0.1227)	(1.1493)	(0.6094)	(0.2942)
MSCI world	132	(13.3108)	8.9797	1.2204	0.7923	4.058	16.4664	0.7661	0.9924	0.4867	(0.5399)	0.0000	0.1199
S&P 500	132	(14.4436)	9.7768	1.1789	0.8869	4.325	18.7061	0.5083	1.0007	0.5813	(0.4454)	0.0946	0.1344
Broad debt	396	(3.6564)	4.8062	0.2923	0.4584	1.526	2.3298	0.4338	0.0126	0.1529	(0.8738)	(0.3338)	0.1001

[1]Risk free rate is US treasuries 1-3M mean

[2]Starts from 2.1997

[3]Starts from 10.1999

[4]Starts from 6.2006

[5]Starts from 12.2001

[6]Starts from 5.2005

(*continued*)

Table 6.4 (continued)

Time scenario 3—6.2007–6.2010 summary statistics

	n	*Min (%)*	*Max (%)*	*Median (%)*	*Mean (%)*	*Std. Dev*	*Var*	*β (Real asset)*	*β (MSCI world)*	*Mean over RF (%)*[1]	*α (Real assets) (%)*	*α (MSCI world) (%)*	*Sharpe ratios (%)*[1]
Index 1 MoM WA MV	7243	(3.1365)	5.2569	0.4361	0.7875	1.5366	2.3613	(0.0025)	0.0043	0.6647	0.8203	1.5287	0.4326
Index 2 MoM WA MV	3738	(3.8151)	6.3823	0.8414	0.8314	2.3534	5.5386	(0.0731)	(0.0552)	0.7086	0.8641	1.5726	0.3011
Index 3 MoM WA MV	3506	(3.7196)	5.1791	0.6655	0.8677	1.9435	3.7770	0.0317	0.0319	0.7449	0.9004	1.6089	0.3833
Index 4 MoM WA MV	2575	(2.3868)	9.4386	0.4436	1.7808	3.2236	10.3913	(0.0371)	(0.0087)	1.6580	1.8135	2.5220	0.5143
Index 5 MoM WA MV	1134	(5.6921)	5.4493	0.4164	0.3842	2.6487	7.0157	(0.0954)	(0.0860)	0.2613	0.4169	1.1254	0.0987
Aircraft lessors	703	(21.0529)	28.6740	(0.1267)	0.1248	12.1500	147.6225	1.6956	1.3258	0.0020	0.1575	0.8660	0.0002
Real estate	111	(11.4848)	6.2830	0.2778	(0.4176)	3.2185	10.3588	0.5587	0.4178	(0.5404)	(0.3849)	0.3236	(0.1679)
Shipping	37	(46.6789)	52.9171	4.2564	2.0756	19.0199	361.7573	1.7928	1.0997	1.9528	2.1084	2.8168	0.1027
Agr. land	37	0.2223	2.6400	0.6871	0.9184	0.7744	0.5997	(0.0417)	(0.0244)	0.7955	0.9511	1.6596	1.0273
Timberland	74	(14.2696)	13.6359	0.7601	(0.0064)	4.8729	23.7453	0.8556	0.6774	(0.1292)	0.0264	0.7348	(0.0265)
Commodities	37	(22.3251)	13.7866	1.4600	(0.2506)	7.0390	49.5477	1.0676	0.6522	(0.3734)	(0.2178)	0.4907	(0.0530)
Crude oil	74	(33.2646)	31.0283	2.5407	1.0363	11.5926	134.3894	1.5944	1.0284	0.9135	1.0691	1.7775	0.0788
Gold	37	(16.8896)	13.0137	2.2551	1.9184	6.3206	39.9500	0.4550	0.1533	1.7956	1.9511	2.6596	0.2841
Precious metals	37	(18.6736)	13.6739	2.4609	1.8424	6.7977	46.2088	0.5269	0.1919	1.7195	1.8751	2.5836	0.253[illegible]
Airlines	37	(26.5725)	24.2940	(0.3039)	(0.0682)	10.9579	120.0758	1.2431	0.9975	(0.1910)	(0.0354)	0.6730	(0.0174)
Transportation	74	(15.9992)	17.0595	0.7613	(0.5351)	7.2704	52.8581	1.1055	0.9177	(0.6579)	(0.5023)	0.2061	(0.0905)

Infrastructure	37	(14.4816)	10.6156	0.1469	(0.4818)	5.6275	31.6688	1.0903	0.8147	(0.6047)	(0.4491)	0.2594	(0.1074)
Real assets	37	(16.8700)	8.6055	0.2376	(0.0327)	4.7674	22.7277	0.9730	0.6853	(0.1556)	0.0000	0.7085	(0.0326)
US treasuries 1-3M	37	(0.0052)	0.4931	0.0261	0.1228	0.1479	0.0219	(0.0045)	(0.0039)	0.0000	0.1556	0.8640	0.0000
USD swap rate 10Y	37	(30.1781)	20.4611	(0.6905)	(1.1272)	9.6657	93.4253	0.2944	0.1999	(1.2500)	(1.0945)	(0.3860)	(0.1293)
USD 3M LIBOR	37	(35.7205)	54.7351	(2.3114)	(4.3719)	18.6447	347.6242	(1.1088)	(0.7622)	(4.4947)	(4.3391)	(3.6307)	(0.2411)
US CPI	37	(1.7705)	1.0478	0.1931	0.1344	0.4886	0.2387	0.0250	0.0096	0.0116	0.1672	0.8756	0.0238
MSCI world	37	(18.9312)	11.3076	(0.5233)	(0.7412)	6.4175	41.1844	1.2419	0.9730	(0.8640)	(0.7085)	0.0000	(0.1346)
S&P500	37	(16.7929)	9.5607	0.1997	(0.7020)	5.8968	34.7718	1.0776	0.8707	(0.8249)	(0.6693)	0.0392	(0.1399)
Broad debt	111	(3.7627)	6.2131	0.4698	0.5453	2.1623	4.6754	0.2494	0.1551	0.4224	0.5780	1.2865	0.1954

[1]Risk free rate is US Treasuries 1-3M mean

(*continued*)

Table 6.4 (continued)

Time scenario 4—6.2010–6.2017 summary statistics

	n	*Min (%)*	*Max (%)*	*Median (%)*	*Mean (%)*	*Std. dev*	*Var*	*β (Real asset)*	*β (MSCI world)*	*Mean over RF (%)*[1]	*α (Real assets) (%)*	*α (MSCI world) (%)*	*Sharpe ratios (%)*[1]
Index 1 MoM WA MV	21,249	(9.1773)	8.0354	0.8896	0.8094	2.991	8.9450	(0.1342)	(0.0283)	0.7988	0.2246	(0.1570)	0.2671
Index 2 MoM WA MV	11,129	(10.8101)	7.4076	0.3739	0.2381	3.118	9.7204	(0.1713)	(0.0676)	0.2275	(0.3468)	(0.7284)	0.0730
Index 3 MoM WA MV	10,115	(13.8674)	8.8526	1.3975	1.0697	3.621	13.1110	(0.1166)	(0.0124)	1.0591	0.4848	0.1033	0.2925
Index 4 MoM WA MV	6580	(11.9470)	10.4857	0.5869	0.6632	4.060	16.4859	(0.1706)	(0.0535)	0.6527	0.0784	(0.3032)	0.1607
Index 5 MoM WA MV	5409	(10.4915)	7.7395	0.2082	0.0708	2.942	8.6543	(0.1698)	(0.0711)	0.0602	(0.5141)	(0.8956)	0.0205
Aircraft lessors	1615	(11.4832)	11.8357	0.9033	1.1893	4.788	22.9297	1.1876	1.0206	1.1787	0.6045	0.2229	0.2462
Real estate	255	(3.6164)	4.1223	0.6971	0.7416	1.378	1.8989	0.5152	0.3148	0.7310	0.1567	(0.2249)	0.5305
Shipping	85	(40.6584)	34.5437	0.0137	1.0382	13.630	185.7674	0.7355	0.6785	1.0276	0.4533	0.0718	0.0754
Agr. land	85	0.1628	3.1867	0.6933	1.0409	0.819	0.6708	0.0450	0.0540	1.0303	0.4561	0.0745	1.2580
Timberland	170	(5.8907)	5.8097	1.1077	0.6333	2.488	6.1907	0.7740	0.6169	0.6227	0.0485	(0.3331)	0.2503
Commodities	85	(12.9667)	10.4144	(0.2376)	(0.3453)	4.409	19.4364	1.4085	0.8491	(0.3559)	(0.9302)	(1.3117)	(0.0807)
Crude oil	170	(19.3238)	23.4352	(0.1860)	(0.1855)	8.015	64.2453	1.9535	1.2562	(0.1961)	(0.7704)	(1.1519)	(0.0245)
Gold	85	(11.0520)	12.2000	0.0505	0.1540	5.130	26.3127	1.0022	0.3361	0.1434	(0.4309)	(0.8125)	0.0280
Precious metals	85	(14.0686)	12.1650	(0.0771)	0.1615	5.608	31.4538	1.1959	0.4692	0.1510	(0.4233)	(0.8049)	0.0269
Airlines	85	(13.4083)	18.3223	1.5083	1.6831	7.737	59.8639	1.1376	1.0683	1.6725	1.0982	0.7166	0.2162
Transportation	85	(9.2305)	13.8789	1.0469	1.1819	4.732	22.3955	1.1644	1.0493	1.1713	0.5970	0.2155	0.2475

Infrastructure	85	(6.4879)	9.1862	1.0870	0.9610	3.256	10.6014	1.2880	0.7909	0.9504	0.3761	(0.0054)	0.2919
Real assets	85	(6.8932)	6.9889	0.8330	0.5849	2.567	6.5906	1.1035	0.6549	0.5743	0.0000	(0.3816)	0.2237
US treasuries 1-3M	85	(0.0052)	0.0777	0.0052	0.0106	0.015	0.0002	0.0019	0.0011	0.0000	(0.5743)	(0.9559)	0.0000
USD swap rate 10Y	85	(21.6955)	31.4864	0.5666	(0.0113)	9.539	90.9912	0.6003	1.2965	(0.0219)	(0.5962)	(0.9778)	(0.0023)
USD 3M LIBOR	85	(34.8474)	47.2129	0.2147	1.4884	9.662	93.3586	(0.1724)	0.2021	1.4778	0.9035	0.5219	0.1529
US CPI	85	(0.6069)	0.5810	0.1713	0.1369	0.207	0.0428	0.0242	0.0161	0.1263	(0.4480)	(0.8296)	0.6103
MSCI world	85	(8.5761)	10.3647	1.3117	0.9664	3.692	13.6294	1.3543	1.0857	0.9559	0.3816	0.0000	0.2589
S&P500	85	(7.0240)	10.9193	1.2859	1.1792	3.425	11.7335	1.1497	0.9766	1.1686	0.5943	0.2128	0.3412
Broad debt	255	(3.9732)	3.3980	0.4710	0.2230	1.434	2.0561	0.4686	0.1947	0.2124	(0.3618)	(0.7434)	0.1482

[1]Risk free rate Is US treasuries 1-3M mean

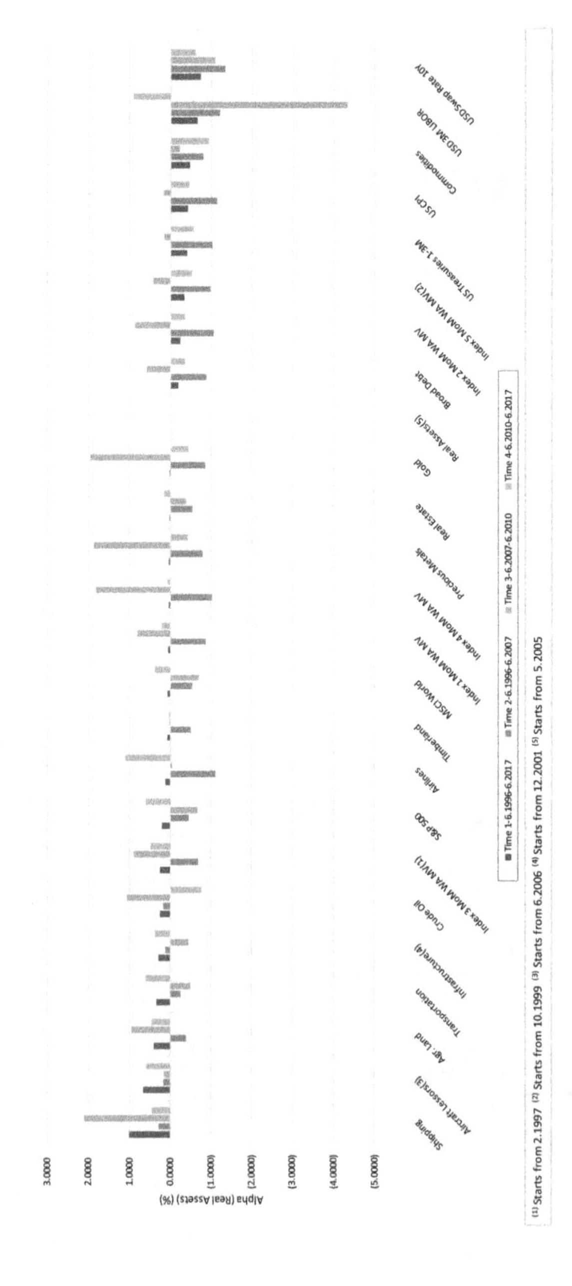

Fig. 6.4 Asset class rankings by alpha (real assets) over four time segments

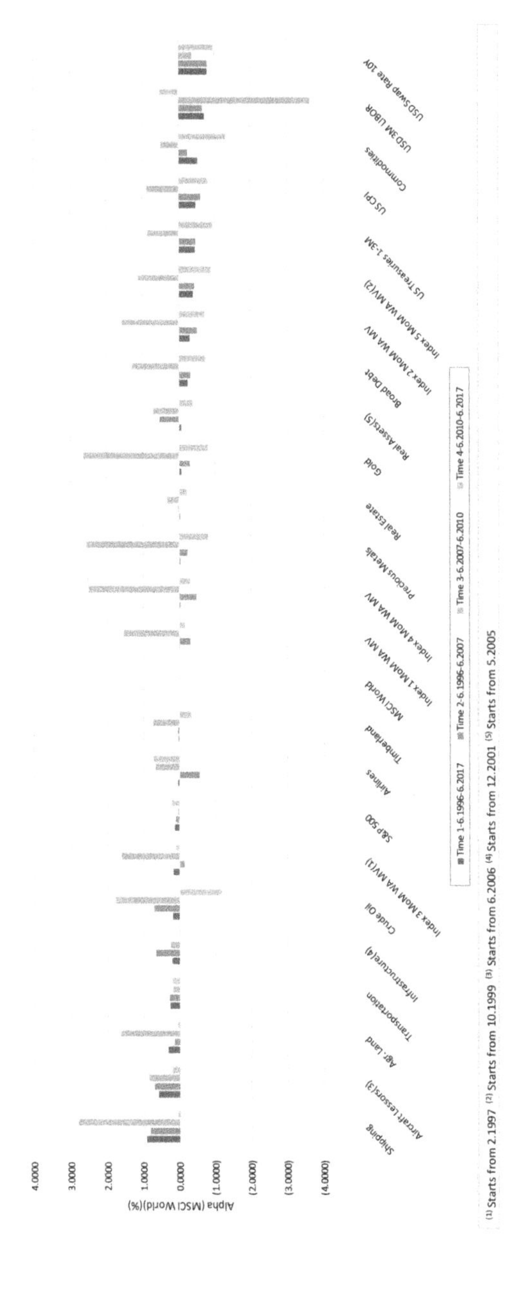

Fig. 6.5 Asset class rankings by alpha (MSCI world) over four time segments

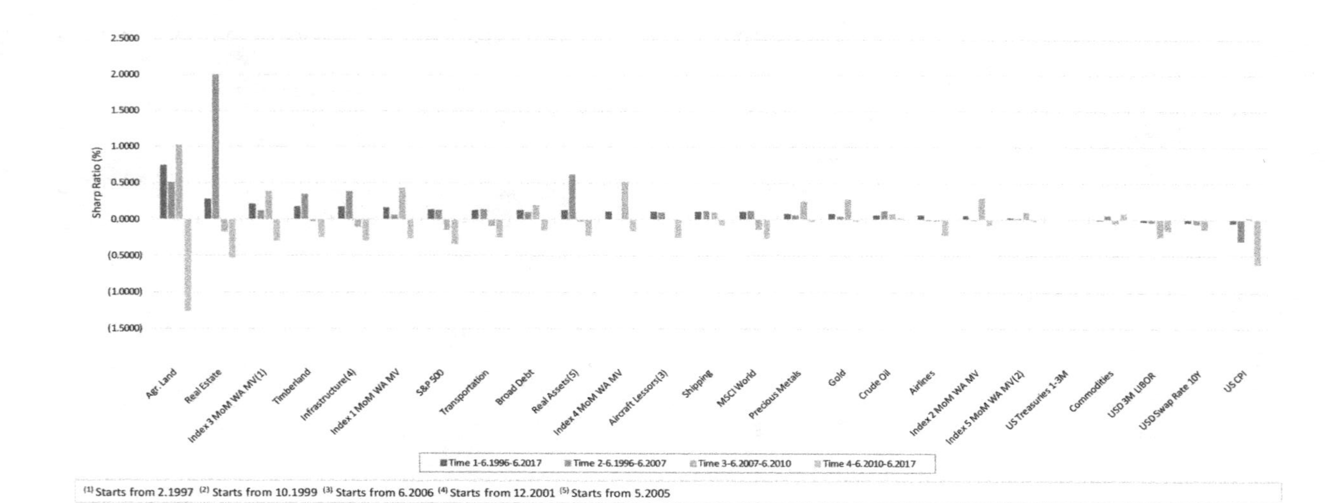

(1) Starts from 2.1997 (2) Starts from 10.1999 (3) Starts from 6.2006 (4) Starts from 12.2001 (5) Starts from 5.2005

Fig. 6.6 Asset class rankings by sharpe ratios over four time segments

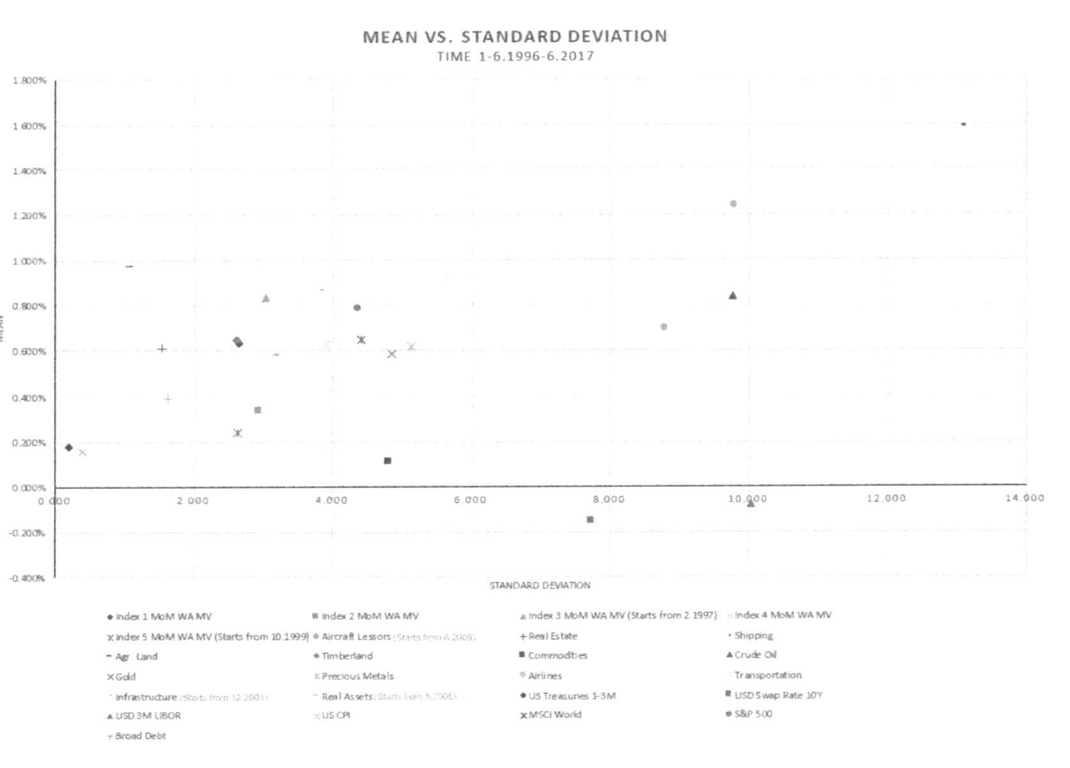

Fig. 6.7 Risk and return of aircraft and comparative asset classes time 1–4

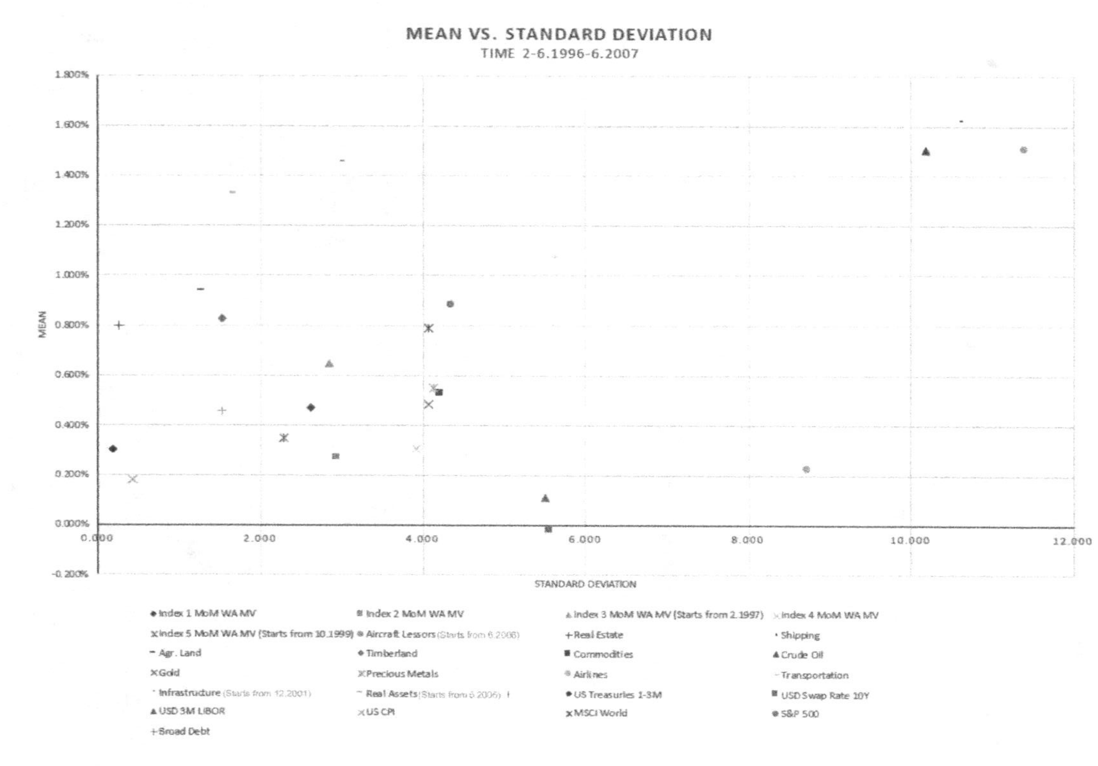

Fig. 6.7 (continued)

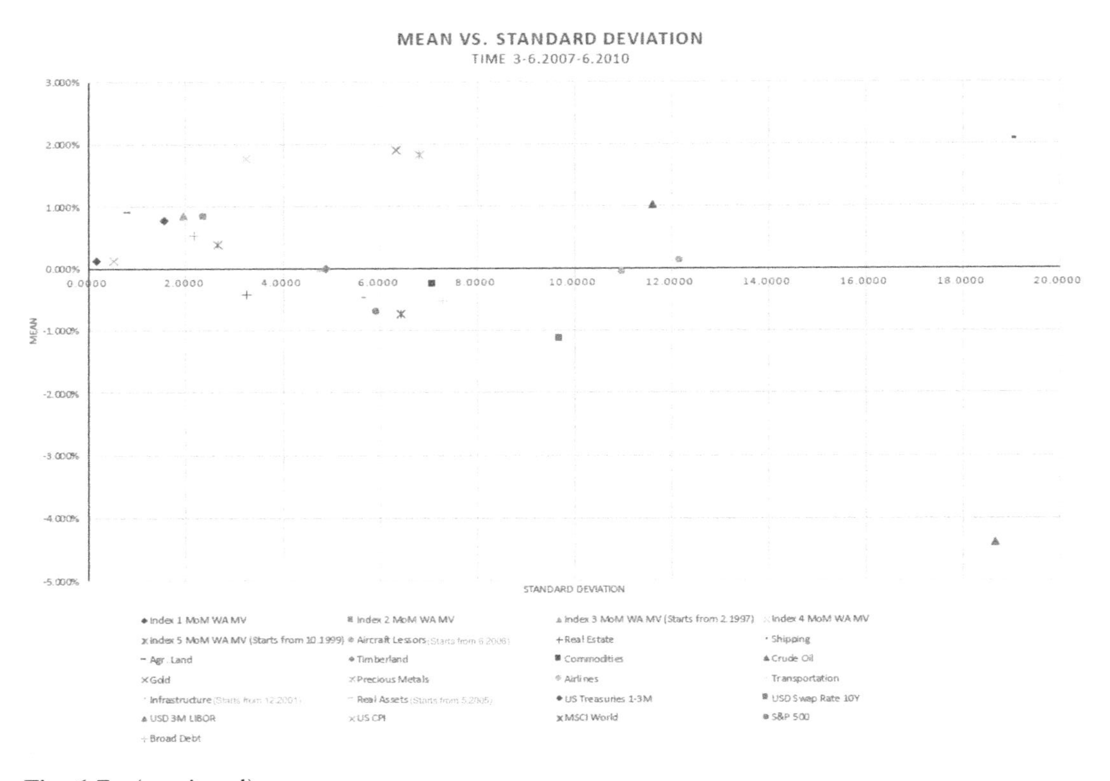

Fig. 6.7 (continued)

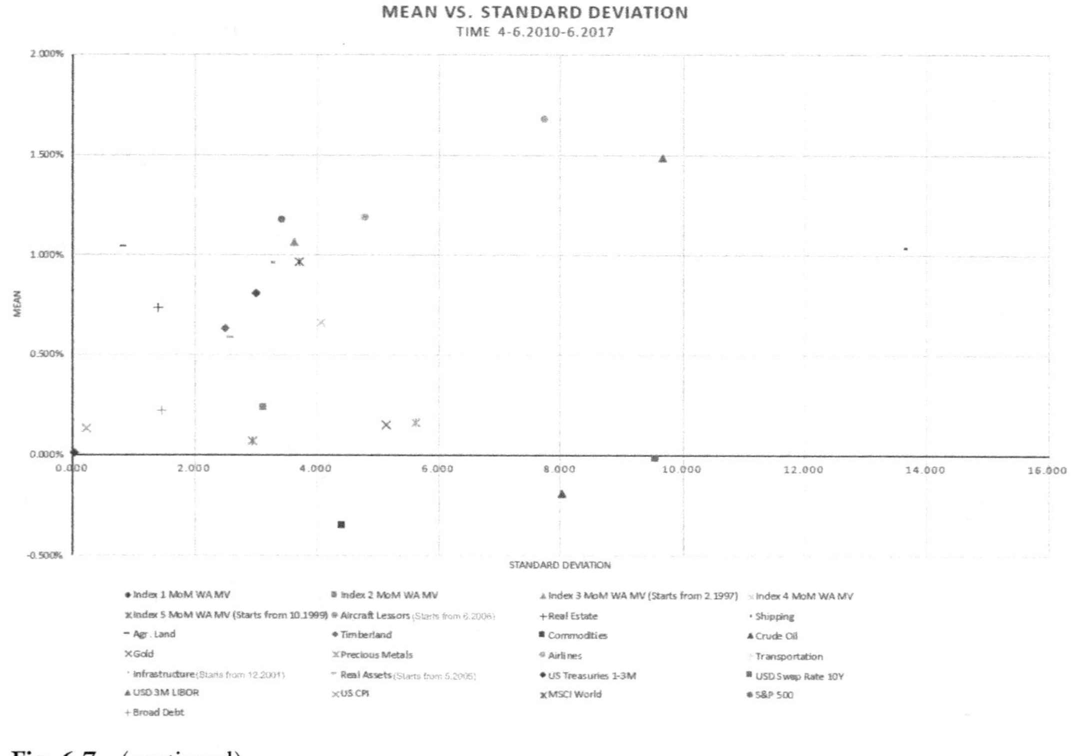

Fig. 6.7 (continued)

beta with respect to Real Assets and MSCI World, Index 1–5 are mostly in the bottom quartile of the rankings. Aircraft Lessors are found to be near the top with standard deviation, variance, beta to Real Asset and MSCI World at three, three, two, and five, respectively.

4.3 *Return and Volatility Comparative Analysis: Time 2 (6.1996–6.2007)*

In the pre-GFC Time 2, return and volatility of Index 1–5 MoM WA MV are compared to the 20 comparative asset class benchmarks. For Time 2, the aircraft asset classes have moved down slightly. Index 3 is ranked 12 while Index 1, 5, 4, and 2 are in the third and fourth quartiles ranked 16, 18, 19, and 21, respectively. Like Time 1 by return measures, Aircraft Lessors is ranked second and the Shipping asset class is top ranked. The mean return for all datasets is positive except for the USD Swap Rate Ten Yr.

In terms of excess return over the risk-free rate are all positive for all the indexes except for Index 2 MoM WA MV, Airlines, US CPI, USD 3M LIBOR, and USD Swap Rate Ten Yr. Again, US Treasuries 1–3M is the risk-free benchmark and ranked 20. Index 1 MoM WA MV excess return over risk-free rate is 0.1675% and ranked 16. MSCI World is ranked 11 and for alpha over MSCI World, one more than half the asset classes are negative. Index 1 MoM WA MV excess return over MSCI World is -0.3192%. With the Real Assets benchmark ranked fifth, there are several additions to the negative alpha over Real Assets compared to negative alpha over MSCI World. Index 1 MoM WA MV excess return over Real Assets index is -0.8592%.

The Sharpe ratio tells a slightly different story, where Index 1–4 moved down compared to Time 1. Indexes 3 and 1 moved down to nine and 14, respectively, Index 4 and 2 are in the lower quartile at 18 and 22, while Index 5 increased two rankings. Aircraft Lessors index stayed at the same rank at 12. The highest Sharpe ratios are Real Estate, Real Assets, and Agricultural Land asset classes for Time 2. Index 1 MoM WA MV's Sharpe ratio is 0.0640%.

In terms of standalone volatility with standard deviation and variance, Index 1–5 is ranked in the third quartile like Time 1. In terms of relative volatility or beta with respect to MSCI World, Index 1–5 are at the bottom of the rankings. In terms of beta with respect to Real Assets, Index 1–5 is in the third and mostly fourth quartiles. Aircraft Lessors are found to be at the top in terms of standard deviation and variance and four and five with respect to beta to Real Asset and MSCI World, respectively.

4.4 *Return and Volatility Comparative Analysis: Time 3 (6.2007–6.2010)*

In the period GFC Time 2, return and volatility of Index 1–5 MoM WA MV are compared to the 20 comparative asset class benchmarks. During this period, Index 1–5 are ranked generally in the upper quartile with Index 4 ranked fourth out of 25 while Index 3, 2, 1, and 5 ranking seven–nine and 11, respectively. In Time 3, Aircraft Lessors is ranked right under the median at 13 while Shipping, Gold, and Precious Metals are top-ranked asset classes. A little less than half of the datasets for the mean return are negative.

In terms of excess return over the risk-free rate, again a little less than half the group is negative as US Treasuries 1–3M is the risk-free benchmark and ranked 14. Index 1 MoM WA MV excess return over risk-free rate is 0.6647% and ranked ninth. During Time 3, MSCI World performed poorly and had a mean of -0.7412% and ranked 23. That means for alpha or excess returns over MSCI World the only negative excess returns are USD Swap Rate 10Y and USD 3M LIBOR. All other indexes exhibit positive alpha to MSCI World. Index 1 MoM WA MV excess return over MSCI World is 1.5287%. Real Assets ranked 16, while still negative, performed much better than the MSCI World. Given that Real Assets is slightly under the median, a bit less than half have negative alpha. Index 1 MoM WA MV alpha over Real Assets is 0.8203%.

The Sharpe ratio tells a slightly different story, with Index 1–4 in the top two–five rank while Index 5 is at ten. Index 1–5 all moved significantly higher compared to Time 2 while Aircraft Lessors decreased one to 13. The highest Sharpe ratio is the Agricultural Land asset class for Time 3. Index 1 MoM WA MV's Sharpe ratio is 0.4326%.

Again, in terms of volatility with respect to standard deviation, variance, or beta to Real Assets or MSCI World, Index 1–5 are in the lower half of the rankings for each. Aircraft Lessors still near the top such as in Time 2 with ranking of third for standard deviation and variance while second and third for the beta to Real Assets or MSCI World, respectively.

4.5 *Return and Volatility Comparative Analysis: Time 4 (6.2010–6.2017)*

In the post-GFC period Time 4, return and volatility of Index 1–5 MoM WA MV are compared to the 20 comparative asset class benchmarks. The return analysis is focused on mean figures for asset class and comparative excess returns over the risk-free rate and market proxies, Real Assets and MSCI World. By these return measures, Index 3 is ranked sixth out of 25 while Index 1 and 4 are at 11 and 13 with Index 2 and 5 are ranked 16 and 21 in the comparative set. Time 4 Index 1–5 rankings are similar to Time 1. Aircraft Lessors is ranked near the top at three. The top ranked is the Airlines asset class. For the mean return for all datasets are positive except for USD Swap Rate 10Y, Crude Oil, and Commodities.

In terms of excess return over the risk-free rate are all positive except for USD Swap Rate 10Y, Crude Oil, and Commodities with US Treasuries 1–3M as the risk-free benchmark ranked 22. Index 1 MoM WA MV excess return over risk-free rate is 0.7988% and ranked 11. For alpha or excess returns over MSCI World, more than half of the comparative set is negative as the benchmark is ranked nine. Index 1 MoM WA MV excess return over MSCI World is -0.1570%. Real Assets is ranked 14 and is higher than MSCI World which is similar as Times 1–3. Negative alpha over Real Assets occurs in slightly less than half the comparative sets. Index 1 MoM WA MV alpha over Real Assets is 0.2246%.

The Sharpe ratio for Time 4 tells a slightly different story, with most of the comparative sets are negative except for the top three ranked Commodities, Crude Oil, and USD Swap Rate 10Y. Index 5 moved to five on top of Index 1–4 which are spread out with a bigger range from 8 to 21. Aircraft Lessors decreased slightly to 15. Index 1 MoM WA MV's Sharpe ratio is -0.2671%.

In terms of volatility with respect to standard deviation, variance, or beta with respect to Real Assets or MSCI World, Index 1–5 is similar to Time 1 rankings with most in the third and fourth quartiles with the exception of Index 4 in standard deviation and variance ranked 11. Only USD 3M LIBOR is the lowest for the beta with respect to Real Assets. Aircraft Lessors are found to be near the middle.

4.6 *Correlation and Covariance Observations*

Correlation statistics of Index 1–5 and the comparative asset classes for Time 1–4 can be found in Table 6.5. The covariance statistics for Index 1–5 compared with other asset classes are found in Table 6.6. Figure 6.8 shows Time 1 case for correlation versus covariance of the various asset classes. Time 2–4 figures are found in the Appendix.

Table 6.5 Time 1–4 asset class summary correlation statistics

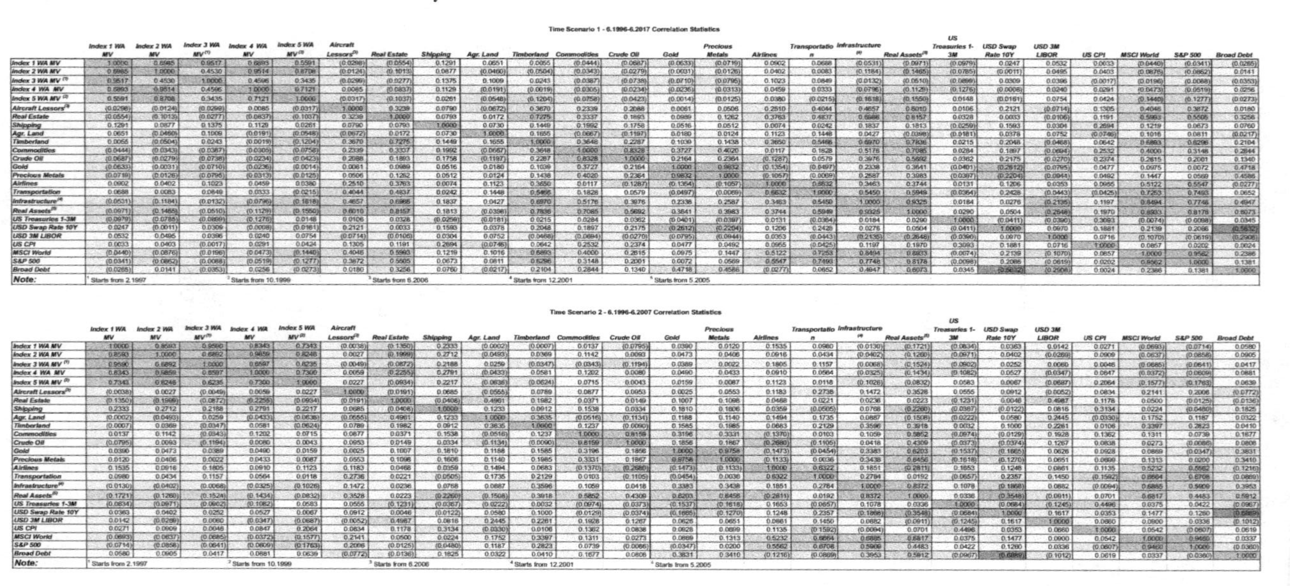

Time Scenario 1 - 6.1996-6.2017 Correlation Statistics

	Index 1 WA MV	Index 2 WA MV	Index 3 WA MV[1]	Index 4 WA MV	Index 5 WA MV[2]	Aircraft Lessors[3]	Real Estate	Shipping	Agr. Land	Timberland	Commodities	Crude Oil	Gold	Precious Metals	Airlines	Transportation	Infrastructure[4]	Real Assets[5]	US Treasuries 1-3M	USD Swap Rate 10Y	USD 3M LIBOR	US CPI	MSCI World	S&P 500	Broad Debt
Index 1 WA MV	1.0000	0.6985	0.9517	0.6893	0.5591	(0.0298)	(0.0554)	0.1291	0.0651	0.0055	(0.0444)	(0.0687)	(0.0633)	(0.0719)	0.0902	0.0688	(0.0531)	(0.0971)	(0.0979)	0.0247	0.0532	0.0033	(0.0440)	(0.0341)	(0.0265)
Index 2 WA MV	0.6985	1.0000	0.4530	0.9514	0.8708	(0.0124)	(0.1013)	0.0877	(0.0460)	(0.0504)	(0.0343)	(0.0279)	(0.0031)	(0.0126)	0.0402	0.0083	(0.1184)	(0.1465)	(0.0785)	(0.0011)	0.0495	0.0403	(0.0876)	(0.0862)	0.0141
Index 3 WA MV[1]	0.9517	0.4530	1.0000	0.4596	0.3435	(0.0299)	(0.0277)	0.1375	0.1009	0.0243	(0.0387)	(0.0738)	(0.0710)	(0.0795)	0.1023	0.0849	(0.0132)	(0.0510)	(0.0899)	0.0309	0.0396	(0.0017)	(0.0196)	(0.0088)	(0.0353)
Index 4 WA MV	0.6893	0.9514	0.4596	1.0000	0.7121	0.0085	(0.0837)	0.1129	(0.0191)	(0.0019)	(0.0305)	(0.0234)	(0.0236)	(0.0313)	0.0459	0.0333	(0.0796)	(0.1129)	(0.1276)	(0.0008)	0.0240	0.0291	(0.0473)	(0.0519)	0.0256
Index 5 WA MV[2]	0.5591	0.8708	0.3435	0.7121	1.0000	(0.0317)	(0.1037)	0.0261	(0.0548)	(0.1204)	(0.0758)	(0.0423)	(0.0014)	(0.0125)	0.0380	(0.0215)	(0.1818)	(0.1550)	0.0148	(0.0161)	0.0754	0.0424	(0.1440)	(0.1277)	(0.0273)
Aircraft Lessors[3]	(0.0298)	(0.0124)	(0.0298)	0.0085	(0.0317)	1.0000	0.3239	0.0790	(0.0672)	0.3670	0.2339	0.2088	0.0061	0.0506	0.2510	0.4044	0.4657	0.6010	0.0106	0.2121	(0.0714)	0.1305	0.4046	0.3872	0.0180
Real Estate	(0.0554)	(0.1013)	(0.0277)	(0.0837)	(0.1037)	0.3239	1.0000	0.0793	0.0172	0.7275	0.3337	0.1893	0.0989	0.1262	0.3763	0.4837	0.6986	0.8157	0.0328	0.0033	(0.0106)	0.1191	0.5993	0.5505	0.3256
Shipping	0.1291	0.0877	0.1375	0.1129	0.0261	0.0790	0.0793	1.0000	0.0730	0.1449	0.1992	0.1758	0.0516	0.0512	0.0074	0.0242	0.1837	0.1813	(0.0259)	0.1593	0.0304	0.2694	0.1219	0.0673	0.0760
Agr. Land	0.0651	(0.0460)	0.1009	(0.0191)	(0.0548)	(0.0672)	0.0172	0.0730	1.0000	0.1655	(0.0667)	(0.1197)	0.0180	0.0124	0.1123	0.1448	0.0427	(0.0398)	(0.0181)	0.0378	0.0752	(0.0746)	0.1016	0.0811	(0.0217)
Timberland	0.0055	(0.0504)	0.0243	(0.0019)	(0.1204)	0.3670	0.7275	0.1449	0.1655	1.0000	0.3648	0.2287	0.1039	0.1438	0.3650	0.5466	0.6970	0.7836	0.0215	0.2048	(0.0468)	0.0642	0.6893	0.6296	0.2104
Commodities	(0.0444)	(0.0343)	(0.0387)	(0.0305)	(0.0758)	0.2339	0.3337	0.1992	(0.0667)	0.3648	1.0000	0.8328	0.3727	0.4020	0.0117	0.1828	0.5176	0.7085	0.0284	0.1897	(0.0694)	0.2532	0.4000	0.3148	0.2844
Crude Oil	(0.0687)	(0.0279)	(0.0738)	(0.0234)	(0.0423)	0.2088	0.1893	0.1758	(0.1197)	0.2287	0.8328	1.0000	0.2164	0.2364	(0.1287)	0.0579	0.3976	0.5692	0.0362	0.2175	(0.0270)	0.2374	0.2615	0.2001	0.1340
Gold	(0.0633)	(0.0031)	(0.0710)	(0.0236)	(0.0014)	0.0061	0.0989	0.0516	0.0180	0.1039	0.3727	0.2164	1.0000	0.9632	(0.1354)	(0.0497)	0.2338	0.3641	(0.0401)	(0.2612)	(0.0795)	0.0477	0.0975	0.0072	0.4718
Precious Metals	(0.0719)	(0.0126)	(0.0795)	(0.0313)	(0.0125)	0.0506	0.1262	0.0512	0.0124	0.1438	0.4020	0.2364	0.9632	1.0000	(0.1057)	(0.0069)	0.2587	0.3983	(0.0397)	(0.2204)	(0.0944)	0.0492	0.1447	0.0569	0.4586
Airlines	0.0902	0.0402	0.1023	0.0459	0.0380	0.2510	0.3763	0.0074	0.1123	0.3650	0.0117	(0.1287)	(0.1354)	(0.1057)	1.0000	0.6632	0.3463	0.3744	0.0131	0.1206	0.0353	0.0955	0.5122	0.5547	(0.0277)
Transportation	0.0688	0.0083	0.0849	0.0333	(0.0215)	0.4044	0.4837	0.0242	0.1448	0.5466	0.1828	0.0579	(0.0497)	(0.0069)	0.6632	1.0000	0.5450	0.5949	(0.0364)	0.2428	(0.0443)	(0.0425)	0.7253	0.7493	0.0652
Infrastructure[4]	(0.0531)	(0.1184)	(0.0132)	(0.0796)	(0.1818)	0.4657	0.6986	0.1837	0.0427	0.6970	0.5176	0.3976	0.2338	0.2587	0.3463	0.5450	1.0000	0.9325	0.0184	0.0276	(0.2135)	0.1197	0.8494	0.7748	0.4947
Real Assets[5]	(0.0971)	(0.1465)	(0.0510)	(0.1129)	(0.1550)	0.6010	0.8157	0.1813	(0.0398)	0.7836	0.7085	0.5692	0.3641	0.3983	0.3744	0.5949	0.9325	1.0000	0.0290	0.0504	(0.2648)	0.1970	0.8903	0.8178	0.6073
US Treasuries 1-3M	(0.0979)	(0.0785)	(0.0899)	(0.1276)	0.0148	0.0106	0.0328	(0.0259)	(0.0181)	0.0215	0.0284	0.0362	(0.0401)	(0.0397)	0.0131	(0.0364)	0.0184	0.0290	1.0000	(0.0411)	(0.0390)	0.3093	(0.0074)	(0.0098)	0.0345
USD Swap Rate 10Y	0.0247	(0.0011)	0.0309	(0.0008)	(0.0161)	0.2121	0.0033	0.1593	0.0378	0.2048	0.1897	0.2175	(0.2612)	(0.2204)	0.1206	0.2428	0.0276	0.0504	(0.0411)	1.0000	0.0970	0.1881	0.2139	0.2086	(0.5832)
USD 3M LIBOR	0.0532	0.0495	0.0396	0.0240	0.0754	(0.0714)	(0.0106)	0.0304	0.0752	(0.0468)	(0.0694)	(0.0270)	(0.0795)	(0.0944)	0.0353	(0.0443)	(0.2135)	(0.2648)	(0.0390)	0.0970	1.0000	0.0716	(0.1070)	(0.0619)	(0.2906)
US CPI	0.0033	0.0403	(0.0017)	0.0291	0.0424	0.1305	0.1191	0.2694	(0.0746)	0.0642	0.2532	0.2374	0.0477	0.0492	0.0955	(0.0425)	0.1197	0.1970	0.3093	0.1881	0.0716	1.0000	0.0857	0.0202	0.0024
MSCI World	(0.0440)	(0.0876)	(0.0196)	(0.0473)	(0.1440)	0.4046	0.5993	0.1219	0.1016	0.6893	0.4000	0.2615	0.0975	0.1447	0.5122	0.7253	0.8494	0.8903	(0.0074)	0.2139	(0.1070)	0.0857	1.0000	0.9562	0.2386
S&P 500	(0.0341)	(0.0862)	(0.0088)	(0.0519)	(0.1277)	0.3872	0.5505	0.0673	0.0811	0.6296	0.3148	0.2001	0.0072	0.0569	0.5547	0.7493	0.7748	0.8178	(0.0098)	0.2086	(0.0619)	0.0202	0.9562	1.0000	0.1381
Broad Debt	(0.0265)	0.0141	(0.0353)	0.0256	(0.0273)	0.0180	0.3256	0.0760	(0.0217)	0.2104	0.2844	0.1340	0.4718	0.4586	(0.0277)	0.0652	0.4947	0.6073	0.0345	(0.5832)	(0.2906)	0.0024	0.2386	0.1381	1.0000
Note:	[1] Starts from 2.1997			[2] Starts from 10.1999				[3] Starts from 6.2006				[4] Starts from 12.2001			[5] Starts from 5.2005										

Time Scenario 2 - 6.1996-6.2007 Correlation Statistics

	Index 1 WA MV	Index 2 WA MV	Index 3 WA MV[1]	Index 4 WA MV	Index 5 WA MV[2]	Aircraft Lessors[3]	Real Estate	Shipping	Agr. Land	Timberland	Commodities	Crude Oil	Gold	Precious Metals	Airlines	Transportation	Infrastructure[4]	Real Assets[5]	US Treasuries 1-3M	USD Swap Rate 10Y	USD 3M LIBOR	US CPI	MSCI World	S&P 500	Broad Debt
Index 1 WA MV	1.0000	0.8593	0.9590	0.8343	0.7343	(0.0038)	(0.1350)	0.2333	(0.0002)	(0.0007)	0.0137	(0.0795)	0.0390	0.0120	0.1535	0.0980	(0.0130)	(0.1721)	(0.0834)	0.0363	0.0142	0.0271	(0.0693)	(0.0714)	0.0580
Index 2 WA MV	0.8593	1.0000	0.6892	0.9859	0.8246	0.0027	(0.1999)	0.2712	(0.0493)	0.0369	0.1142	0.0093	0.0473	0.0406	0.0916	0.0434	(0.0402)	(0.1260)	(0.0971)	0.0402	(0.0289)	0.0909	(0.0637)	(0.0858)	0.0905
Index 3 WA MV[1]	0.9590	0.6892	1.0000	0.6597	0.8235	(0.0049)	(0.0872)	0.2188	0.0259	(0.0347)	(0.0343)	(0.1194)	0.0389	0.0022	0.1805	0.1157	(0.0068)	(0.1524)	(0.0902)	0.0252	0.0060	0.0046	(0.0685)	(0.0641)	0.0417
Index 4 WA MV	0.8343	0.9859	0.6597	1.0000	0.7300	0.0059	(0.2255)	0.2791	(0.0433)	0.0581	0.1202	0.0080	0.0490	0.0433	0.0910	0.0564	(0.0325)	(0.1434)	(0.1082)	0.0527	(0.0347)	0.0847	(0.0372)	(0.0609)	0.0881
Index 5 WA MV[2]	0.7343	0.8246	0.6235	0.7300	1.0000	0.0227	(0.0934)	0.2217	(0.0636)	(0.0624)	0.0715	0.0043	0.0159	0.0087	0.1123	0.0118	(0.1026)	(0.0832)	0.0583	0.0067	(0.0687)	0.2064	(0.1577)	(0.1763)	0.0639
Aircraft Lessors[3]	(0.0038)	0.0027	(0.0049)	0.0059	0.0227	1.0000	(0.0191)	0.0685	(0.0555)	0.0789	0.0877	0.0953	0.0025	0.0553	0.1183	0.2736	0.1472	0.3528	0.0555	0.0912	(0.0052)	0.0834	0.2141	0.2006	(0.0772)
Real Estate	(0.1350)	(0.1999)	(0.0872)	(0.2255)	(0.0934)	(0.0191)	1.0000	(0.0408)	0.4961	0.1982	0.0371	0.0149	0.1007	0.1098	0.0468	0.0221	0.0236	0.0223	(0.1231)	0.0046	0.4987	0.1178	0.0500	(0.0125)	(0.0136)
Shipping	0.2333	0.2712	0.2188	0.2791	0.2217	0.0685	(0.0408)	1.0000	0.1233	0.0912	0.1538	0.0334	0.1810	0.1606	0.0359	(0.0505)	0.0768	(0.2260)	(0.0367)	(0.0122)	0.0816	0.3134	0.0224	(0.0480)	0.1825
Agr. Land	(0.0002)	(0.0493)	0.0259	(0.0433)	(0.0636)	(0.0555)	0.4961	0.1233	1.0000	0.3635	(0.0516)	(0.1134)	0.1188	0.1140	0.1494	0.1735	0.0887	(0.1508)	(0.0222)	0.0580	0.2445	(0.0330)	0.1752	0.1187	0.0322
Timberland	(0.0007)	0.0369	(0.0347)	0.0581	(0.0624)	0.0789	0.1982	0.0912	0.3635	1.0000	0.1237	(0.0090)	0.1585	0.1985	0.0683	0.2129	0.3596	0.3918	0.0032	0.1000	0.2261	0.0106	0.3397	0.2823	0.0410
Commodities	0.0137	0.1142	(0.0343)	0.1202	0.0715	0.0877	0.0371	0.1538	(0.0516)	0.1237	1.0000	0.8159	0.3196	0.3331	(0.1370)	0.0103	0.1059	0.5852	(0.0974)	(0.0129)	0.1928	0.1362	0.1311	0.0739	0.1677
Crude Oil	(0.0795)	0.0093	(0.1194)	0.0080	0.0043	0.0953	0.0149	0.0334	(0.1134)	(0.0090)	0.8159	1.0000	0.1856	0.1867	(0.2680)	(0.1105)	0.0418	0.4309	(0.0373)	(0.0374)	0.1267	0.0838	0.0273	(0.0086)	0.0806
Gold	0.0390	0.0473	0.0389	0.0490	0.0159	0.0025	0.1007	0.1810	0.1188	0.1585	0.3196	0.1856	1.0000	0.9758	(0.1473)	(0.0454)	0.3383	0.6203	(0.1537)	(0.1665)	0.0626	0.0928	0.0869	(0.0347)	0.3831
Precious Metals	0.0120	0.0406	0.0022	0.0433	0.0087	0.0553	0.1098	0.1606	0.1140	0.1985	0.3331	0.1867	0.9758	1.0000	(0.1133)	0.0036	0.3438	0.6456	(0.1618)	(0.1270)	0.0651	0.0699	0.1313	0.0200	0.3410
Airlines	0.1535	0.0916	0.1805	0.0910	0.1123	0.1183	0.0468	0.0359	0.1494	0.0683	(0.1370)	(0.2680)	(0.1473)	(0.1133)	1.0000	0.6322	0.1851	(0.2811)	0.1653	0.1248	0.0861	0.1135	0.5232	0.5562	(0.1216)
Transportation	0.0980	0.0434	0.1157	0.0564	0.0118	0.2736	0.0221	(0.0505)	0.1735	0.2129	0.0103	(0.1105)	(0.0454)	0.0036	0.6322	1.0000	0.2784	0.0192	(0.0657)	0.2357	0.1450	(0.1592)	0.8664	0.6708	(0.0869)
Infrastructure[4]	(0.0130)	(0.0402)	(0.0068)	(0.0325)	(0.1026)	0.1472	0.0236	0.0768	0.0887	0.3596	0.1059	0.0418	0.3383	0.3438	0.1851	0.2784	1.0000	0.8372	0.1078	(0.1868)	0.0882	(0.0094)	0.6885	0.5909	0.3953
Real Assets[5]	(0.1721)	(0.1260)	(0.1524)	(0.1434)	(0.0832)	0.3528	0.0223	(0.2260)	(0.1508)	0.3918	0.5852	0.4309	0.6203	0.6456	(0.2811)	0.0192	0.8372	1.0000	0.0336	(0.3548)	(0.0911)	0.0701	0.6817	0.4483	0.5912
US Treasuries 1-3M	(0.0834)	(0.0971)	(0.0902)	(0.1082)	0.0583	0.0555	(0.1231)	(0.0367)	(0.0222)	0.0032	(0.0974)	(0.0373)	(0.1537)	(0.1618)	0.1653	(0.0657)	0.1078	0.0336	1.0000	(0.0684)	(0.1245)	0.4496	0.0375	0.0422	(0.0967)
USD Swap Rate 10Y	0.0363	0.0402	0.0252	0.0527	0.0067	0.0912	0.0046	(0.0122)	0.0580	0.1000	(0.0129)	(0.0374)	(0.1665)	(0.1270)	0.1248	0.2357	(0.1868)	(0.3548)	(0.0684)	1.0000	0.1617	0.0353	0.1477	0.1260	(0.6889)
USD 3M LIBOR	0.0142	(0.0289)	0.0060	(0.0347)	(0.0687)	(0.0052)	0.4987	0.0816	0.2445	0.2261	0.1928	0.1267	0.0626	0.0651	0.0861	0.1450	0.0882	(0.0911)	(0.1245)	0.1617	1.0000	0.0660	0.0900	0.0336	(0.1012)
US CPI	0.0271	0.0909	0.0046	0.0847	0.2064	0.0834	0.1178	0.3134	(0.0330)	0.0106	0.1362	0.0838	0.0928	0.0699	0.1135	(0.1592)	(0.0094)	0.0701	0.4496	0.0353	0.0660	1.0000	0.0542	(0.0607)	0.0619
MSCI World	(0.0693)	(0.0637)	(0.0685)	(0.0372)	(0.1577)	0.2141	0.0500	0.0224	0.1752	0.3397	0.1311	0.0273	0.0869	0.1313	0.5232	0.8664	0.6885	0.6817	0.0375	0.1477	0.0900	0.0542	1.0000	0.9460	0.0337
S&P 500	(0.0714)	(0.0858)	(0.0641)	(0.0609)	(0.1763)	0.2006	(0.0125)	(0.0480)	0.1187	0.2823	0.0739	(0.0066)	(0.0347)	0.0200	0.5562	0.6708	0.5909	0.4483	0.0422	0.1260	0.0336	(0.0607)	0.9460	1.0000	(0.0360)
Broad Debt	0.0580	0.0905	0.0417	0.0881	0.0639	(0.0772)	(0.0136)	0.1825	0.0322	0.0410	0.1677	0.0806	0.3831	0.3410	(0.1216)	(0.0869)	0.3953	0.5912	(0.0967)	(0.6889)	(0.1012)	0.0619	0.0337	(0.0360)	1.0000
Note:	[1] Starts from 2.1997			[2] Starts from 10.1999				[3] Starts from 6.2006				[4] Starts from 12.2001			[5] Starts from 5.2005										

Table 6.5 (continued)

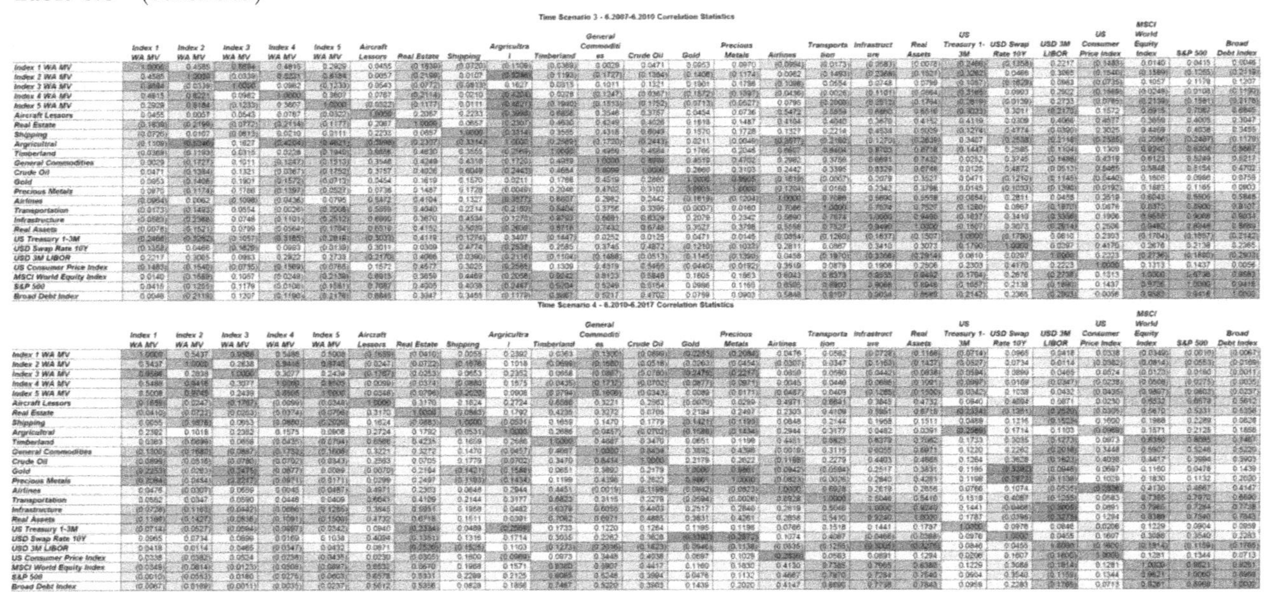

Time Scenario 3 - 6.2007-6.2010 Correlation Statistics

	Index 1 WA MV	Index 2 WA MV	Index 3 WA MV	Index 4 WA MV	Index 5 WA MV	Aircraft Lessors	Real Estate	Shipping	Argricultral	Timberland	General Commodities	Crude Oil	Gold	Precious Metals	Airlines	Transportation	Infrastructure	Real Assets	US Treasury 1-3M	USD Swap Rate 10Y	USD 3M LIBOR	US Consumer Price Index	MSCI World Equity Index	S&P 500	Broad Debt Index
Index 1 WA MV	1.0000	0.4585	0.8884	0.4815	0.2929	0.0455	(0.1839)	(0.0720)	(0.1109)	(0.0369)	0.0029	0.0471	0.0953	0.0970	(0.0954)	(0.0173)	(0.0683)	(0.0076)	(0.2466)	(0.1358)	0.2217	(0.1483)	0.0140	0.0415	0.0046
Index 2 WA MV	0.4585	1.0000	(0.0339)	0.8221	0.8184	0.0057	(0.2199)	0.0107	(0.5246)	(0.1193)	(0.1727)	(0.1384)	(0.1406)	(0.1174)	0.0062	(0.1493)	(0.2368)	(0.1521)	(0.3262)	0.0466	0.3065	(0.1540)	(0.1589)	(0.1255)	(0.2119)
Index 3 WA MV	0.8884	(0.0339)	1.0000	0.0982	(0.1233)	0.0543	(0.0772)	(0.0813)	0.1627	0.0315	0.1011	0.1321	0.1901	0.1788	(0.1098)	0.0554	0.0748	0.0799	(0.1057)	(0.1629)	0.0963	(0.0735)	0.1057	0.1179	0.1207
Index 4 WA MV	0.4815	0.8221	0.0982	1.0000	0.3607	0.0787	(0.2114)	0.0210	(0.4204)	0.0228	(0.1247)	(0.0367)	(0.1572)	(0.1397)	(0.0436)	(0.0026)	(0.1101)	(0.0564)	(0.3165)	0.0923	0.2922	(0.1569)	(0.0248)	(0.0108)	(0.1190)
Index 5 WA MV	0.2929	0.8184	(0.1233)	0.3607	1.0000	(0.0322)	(0.1177)	0.0111	(0.4621)	(0.1940)	(0.1513)	(0.1752)	(0.0713)	(0.0527)	0.0795	(0.2008)	(0.2512)	(0.1784)	(0.2818)	(0.0139)	0.2733	(0.0788)	(0.2139)	(0.1581)	(0.2178)
Aircraft Lessors	0.0455	0.0057	0.0543	0.0787	(0.0322)	1.0000	0.2067	0.2233	(0.3998)	0.6858	0.3548	0.3757	0.0454	0.0736	0.5472	0.5859	0.6990	0.6516	(0.3033)	0.3011	(0.2170)	0.1572	0.6915	0.7087	0.6846
Real Estate	(0.1839)	(0.2199)	(0.0772)	(0.2114)	(0.1177)	0.2067	1.0000	0.0857	(0.2307)	0.4630	0.4249	0.4026	0.1619	0.1487	0.4104	0.4040	0.3670	0.4152	0.4119	0.0309	0.4066	0.4577	0.3859	0.4005	0.3047
Shipping	(0.0726)	0.0107	(0.0813)	0.0210	0.0111	0.2233	0.0857	1.0000	(0.3314)	0.3565	0.4318	0.6049	0.1570	0.1728	0.1327	0.2214	0.4534	0.5029	(0.1274)	0.4774	(0.0390)	0.3025	0.4469	0.4038	0.3455
Argricultral	(0.1109)	(0.5246)	0.1627	(0.4204)	(0.4621)	(0.3998)	(0.2307)	(0.3314)	1.0000	(0.2569)	(0.1720)	(0.2443)	0.0211	(0.0049)	(0.3577)	(0.2180)	(0.1270)	(0.2639)	0.3407	(0.2538)	(0.2116)	(0.2585)	(0.2058)	(0.2487)	(0.1179)
Timberland	(0.0369)	(0.1193)	0.0315	0.0228	(0.1940)	0.6858	0.4630	0.3565	(0.2569)	1.0000	0.4959	0.4684	0.1788	0.2048	0.6607	0.8404	0.8703	0.8718	(0.1447)	0.2585	(0.1104)	0.1309	0.8242	0.6204	0.8667
General Commodities	0.0029	(0.1727)	0.1011	(0.1247)	(0.1513)	0.3548	0.4249	0.4318	(0.1720)	0.4959	1.0000	0.8099	0.4519	0.4702	0.2982	0.3756	0.6881	0.7432	0.0252	0.3745	(0.1488)	0.4319	0.8123	0.5249	0.5217
Crude Oil	0.0471	(0.1384)	0.1321	(0.0367)	(0.1752)	0.3757	0.4026	0.6049	(0.2443)	0.4684	0.8099	1.0000	0.2860	0.3103	0.2442	0.3395	0.8329	0.6748	0.0125	0.4872	(0.0513)	0.5465	0.5848	0.6154	0.4702
Gold	0.0953	(0.1406)	0.1901	(0.1572)	(0.0713)	0.0454	0.1619	0.1570	0.0211	0.1788	0.4519	0.2860	1.0000	0.9905	(0.1619)	(0.0007)	0.2079	0.3521	0.0471	(0.1210)	(0.1145)	(0.0440)	0.1606	0.0966	0.0759
Precious Metals	0.0970	(0.1174)	0.1788	(0.1397)	(0.0527)	0.0736	0.1487	0.1728	(0.0049)	0.2048	0.4702	0.3101	0.9905	1.0000	(0.1204)	0.0160	0.2342	0.3798	0.0145	(0.1033)	(0.1390)	(0.0192)	0.1683	0.1165	0.0903
Airlines	(0.0954)	0.0062	(0.1098)	(0.0436)	0.0795	0.5472	0.4104	0.1327	(0.3577)	0.6607	0.2982	0.2442	(0.1619)	(0.1204)	1.0000	0.7086	0.5890	0.5558	(0.0854)	0.2811	0.0458	0.3519	0.6043	0.5506	0.5848
Transportation	(0.0173)	(0.1493)	0.0554	(0.0026)	(0.2008)	0.5859	0.4040	0.2214	(0.2180)	0.8404	0.3756	0.3395	(0.0007)	0.0160	0.7086	1.0000	0.7674	0.7527	(0.1260)	0.0867	(0.1970)	0.0879	0.8373	0.8900	0.8107
Infrastructure	(0.0683)	(0.2368)	0.0748	(0.1101)	(0.2512)	0.6990	0.3670	0.4534	(0.1270)	0.8703	0.6881	0.8329	0.2079	0.2342	0.5890	0.7674	1.0000	0.9490	(0.1637)	0.3410	(0.3358)	0.1906	0.9658	0.9069	0.9034
Real Assets	(0.0076)	(0.1521)	0.0799	(0.0564)	(0.1784)	0.6516	0.4152	0.5029	(0.2639)	0.8718	0.7432	0.6748	0.3527	0.3798	0.5558	0.7527	0.9490	1.0000	(0.1507)	0.3073	(0.2914)	0.2506	0.9482	0.8946	0.8989
US Treasury 1-3M	(0.2466)	(0.3262)	(0.1057)	(0.3165)	(0.2818)	(0.3033)	0.4119	(0.1274)	0.3407	(0.1447)	0.0252	0.0125	0.0471	0.0145	(0.0854)	(0.1260)	(0.1637)	(0.1507)	1.0000	(0.1790)	0.0610	0.2303	(0.1704)	(0.1657)	(0.2142)
USD Swap Rate 10Y	(0.1358)	0.0466	(0.1629)	0.0923	(0.0139)	0.3011	0.0309	0.4774	(0.2538)	0.2585	0.3745	0.4872	(0.1210)	(0.1033)	0.2811	0.0867	0.3410	0.3073	(0.1790)	1.0000	0.0297	0.4170	0.2676	0.2138	0.2365
USD 3M LIBOR	0.2217	0.3065	0.0963	0.2922	0.2733	(0.2170)	0.4066	(0.0390)	(0.2116)	(0.1104)	(0.1488)	(0.0513)	(0.1145)	(0.1390)	0.0458	(0.1970)	(0.3358)	(0.2914)	0.0610	0.0297	1.0000	0.2223	(0.2736)	(0.1680)	(0.2903)
US Consumer Price Index	(0.1483)	(0.1540)	(0.0735)	(0.1569)	(0.0788)	0.1572	0.4577	0.3025	(0.2585)	0.1309	0.4319	0.5465	(0.0440)	(0.0192)	0.3519	0.0879	0.1906	0.2506	0.2303	0.4170	0.2223	1.0000	0.1313	0.1437	0.0056
MSCI World Equity Index	0.0140	(0.1589)	0.1057	(0.0248)	(0.2139)	0.6915	0.3859	0.4469	(0.2058)	0.8242	0.8123	0.5848	0.1605	0.1663	0.6043	0.8373	0.9655	0.9482	(0.1704)	0.2676	(0.2736)	0.1313	1.0000	0.9736	0.9583
S&P 500	0.0415	(0.1255)	0.1179	(0.0108)	(0.1581)	0.7087	0.4005	0.4038	(0.2487)	0.6204	0.5249	0.6154	0.0966	0.1160	0.6506	0.8900	0.9068	0.8946	(0.1657)	0.2138	(0.1680)	0.1437	0.9736	1.0000	0.9416
Broad Debt Index	0.0046	(0.2119)	0.1207	(0.1190)	(0.2178)	0.6846	0.3047	0.3455	(0.1179)	0.8667	0.5217	0.4702	0.0759	0.0903	0.5848	0.8107	0.9034	0.8989	(0.2142)	0.2365	(0.2903)	0.0056	0.9583	0.9416	1.0000

Time Scenario 4 - 6.2010-6.2017 Correlation Statistics

	Index 1 WA MV	Index 2 WA MV	Index 3 WA MV	Index 4 WA MV	Index 5 WA MV	Aircraft Lessors	Real Estate	Shipping	Argricultral	Timberland	General Commodities	Crude Oil	Gold	Precious Metals	Airlines	Transportation	Infrastructure	Real Assets	US Treasury 1-3M	USD Swap Rate 10Y	USD 3M LIBOR	US Consumer Price Index	MSCI World Equity Index	S&P 500	Broad Debt Index
Index 1 WA MV	1.0000	0.5437	0.9586	0.5486	0.5008	(0.1659)	(0.0410)	0.0055	0.2392	0.0363	(0.1300)	(0.0899)	(0.2253)	(0.2094)	0.0476	0.0562	(0.0726)	(0.1166)	(0.0714)	0.0965	0.0418	0.0338	(0.0349)	(0.0010)	(0.0067)
Index 2 WA MV	0.5437	1.0000	0.2838	0.9446	0.9745	(0.0247)	(0.0722)	(0.1876)	0.1018	(0.0699)	(0.1680)	(0.0518)	(0.0263)	(0.0454)	(0.0307)	0.0347	(0.1163)	(0.1437)	(0.0527)	0.0734	0.0114	(0.0382)	(0.0814)	(0.0553)	(0.0169)
Index 3 WA MV	0.9586	0.2838	1.0000	0.3277	0.2439	(0.1767)	(0.0253)	0.0653	0.2352	0.0658	(0.0887)	(0.0780)	(0.2476)	(0.2217)	0.0659	0.0590	(0.0442)	(0.0836)	(0.0594)	0.0899	0.0465	0.0524	(0.0123)	0.0160	(0.0011)
Index 4 WA MV	0.5486	0.9446	0.3277	1.0000	0.8505	(0.0099)	(0.0374)	(0.0880)	0.1575	(0.0435)	(0.1732)	(0.0702)	(0.0877)	(0.0971)	0.0045	0.0446	(0.0696)	(0.1091)	(0.0997)	0.0169	(0.0347)	(0.0238)	(0.0508)	(0.0275)	(0.0035)
Index 5 WA MV	0.5008	0.9745	0.2439	0.8505	1.0000	(0.0348)	(0.0796)	(0.2020)	0.0908	(0.0794)	(0.1606)	(0.0343)	0.0089	(0.0171)	(0.0487)	0.0409	(0.1285)	(0.1500)	(0.0342)	0.1038	0.0432	(0.0435)	(0.0897)	(0.0603)	(0.0237)
Aircraft Lessors	(0.1659)	(0.0247)	(0.1767)	(0.0099)	(0.0348)	1.0000	0.3170	0.1624	0.2724	0.6566	0.3221	0.2563	(0.0070)	0.0299	0.4971	0.6641	0.3845	0.4732	0.0940	0.4094	0.0871	0.0230	0.6532	0.6578	0.5612
Real Estate	(0.0410)	(0.0722)	(0.0253)	(0.0374)	(0.0796)	0.3170	1.0000	(0.0683)	0.1792	0.4235	0.3272	0.0705	0.2194	0.2497	0.2303	0.4109	0.5951	0.6718	(0.3334)	(0.1351)	(0.2526)	(0.0365)	0.5670	0.5331	0.5356
Shipping	0.0055	(0.1876)	0.0653	(0.0880)	(0.2020)	0.1624	(0.0683)	1.0000	(0.0531)	0.1659	0.1470	0.1779	(0.1421)	(0.1193)	0.0848	0.2144	0.1958	0.1511	0.0488	0.1316	(0.1524)	0.1600	0.1968	0.2289	0.0628
Argricultral	0.2392	0.1018	0.2352	0.1575	0.0908	0.2724	0.1792	(0.0531)	1.0000	0.2686	(0.0457)	(0.0702)	(0.1588)	(0.1434)	0.2944	0.3177	0.0482	0.0391	(0.2969)	0.1714	0.1103	(0.0969)	0.1571	0.2125	0.1898
Timberland	0.0363	(0.0699)	0.0658	(0.0435)	(0.0794)	0.6566	0.4235	0.1659	0.2686	1.0000	0.4687	0.3470	0.0651	0.1198	0.4451	0.6823	0.6379	0.7062	0.1733	0.3035	(0.1273)	0.0973	0.8360	0.8085	0.7467
General Commodities	(0.1300)	(0.1680)	(0.0887)	(0.1732)	(0.1606)	0.3221	0.3272	0.1470	(0.0457)	0.4687	1.0000	0.6434	0.3892	0.4398	(0.0019)	0.3115	0.6055	0.6971	0.1220	0.2262	(0.2018)	0.3448	0.5907	0.5246	0.5220
Crude Oil	(0.0899)	(0.0518)	(0.0780)	(0.0702)	(0.0343)	0.2563	0.0705	0.1779	(0.0702)	0.3470	0.6434	1.0000	0.2179	0.2622	(0.1198)	0.2279	0.4403	0.4665	0.1264	0.3628	(0.1623)	0.4038	0.4417	0.3994	0.3903
Gold	(0.2253)	(0.0263)	(0.2476)	(0.0877)	0.0089	(0.0070)	0.2194	(0.1421)	(0.1588)	0.0651	0.3892	0.2179	1.0000	0.9661	(0.0942)	(0.0594)	0.2517	0.3831	0.1195	(0.3392)	(0.0946)	0.0697	0.1160	0.0476	0.1439
Precious Metals	(0.2094)	(0.0454)	(0.2217)	(0.0971)	(0.0171)	0.0299	0.2497	(0.1193)	(0.1434)	0.1198	0.4398	0.2622	0.9661	1.0000	(0.0823)	(0.0026)	0.2840	0.4261	0.1198	(0.2873)	(0.1138)	0.1029	0.1830	0.1132	0.2020
Airlines	0.0476	(0.0307)	0.0659	0.0045	(0.0487)	0.4971	0.2303	0.0848	0.2944	0.4451	(0.0019)	(0.1198)	(0.0942)	(0.0823)	1.0000	0.6928	0.2619	0.2856	0.0766	0.1074	(0.0635)	(0.2836)	0.4130	0.4667	0.4147
Transportation	0.0562	0.0347	0.0590	0.0446	0.0409	0.6641	0.4109	0.2144	0.3177	0.6823	0.3115	0.2279	(0.0594)	(0.0026)	0.6928	1.0000	0.5046	0.5410	0.1518	0.4087	(0.1255)	0.0563	0.7385	0.7970	0.6690
Infrastructure	(0.0726)	(0.1163)	(0.0442)	(0.0696)	(0.1285)	0.3845	0.5951	0.1958	0.0482	0.6379	0.6055	0.4403	0.2517	0.2840	0.2619	0.5046	1.0000	0.9240	0.1441	(0.0466)	(0.3005)	0.0891	0.7965	0.7294	0.7738
Real Assets	(0.1166)	(0.1437)	(0.0836)	(0.1091)	(0.1500)	0.4732	0.6718	0.1511	0.0391	0.7062	0.6971	0.4665	0.3831	0.4261	0.2856	0.5410	0.9240	1.0000	0.1787	(0.0396)	(0.3278)	0.1294	0.8380	0.7540	0.7843
US Treasury 1-3M	(0.0714)	(0.0527)	(0.0594)	(0.0997)	(0.0342)	0.0940	(0.3334)	0.0488	(0.2969)	0.1733	0.1220	0.1264	0.1195	0.1198	0.0766	0.1518	0.1441	0.1787	1.0000	0.0976	0.0846	0.0206	0.1229	0.0904	0.0959
USD Swap Rate 10Y	0.0965	0.0734	0.0899	0.0169	0.1038	0.4094	(0.1351)	0.1316	0.1714	0.3035	0.2262	0.3628	(0.3392)	(0.2873)	0.1074	0.4087	(0.0466)	(0.0396)	0.0976	1.0000	0.0455	0.1607	0.3068	0.3540	0.2283
USD 3M LIBOR	0.0418	0.0114	0.0465	(0.0347)	0.0432	0.0871	(0.2526)	(0.1524)	0.1103	(0.1273)	(0.2018)	(0.1623)	(0.0946)	(0.1138)	(0.0635)	(0.1255)	(0.3005)	(0.3278)	0.0846	0.0455	1.0000	(0.1600)	(0.1814)	(0.1159)	(0.1765)
US Consumer Price Index	0.0338	(0.0382)	0.0524	(0.0238)	(0.0435)	0.0230	(0.0365)	0.1600	(0.0969)	0.0973	0.3448	0.4038	0.0697	0.1029	(0.2836)	0.0563	0.0891	0.1294	0.0206	0.1607	(0.1600)	1.0000	0.1281	0.1344	0.0713
MSCI World Equity Index	(0.0349)	(0.0814)	(0.0123)	(0.0508)	(0.0897)	0.6532	0.5670	0.1968	0.1571	0.8360	0.5907	0.4417	0.1160	0.1830	0.4130	0.7385	0.7965	0.8380	0.1229	0.3068	(0.1814)	0.1281	1.0000	0.9621	0.8261
S&P 500	(0.0010)	(0.0553)	0.0160	(0.0275)	(0.0603)	0.6578	0.5331	0.2289	0.2125	0.8085	0.5246	0.3994	0.0476	0.1132	0.4667	0.7970	0.7294	0.7540	0.0904	0.3540	(0.1159)	0.1344	0.9621	1.0000	0.8968
Broad Debt Index	(0.0067)	(0.0169)	(0.0011)	(0.0035)	(0.0237)	0.5612	0.5356	0.0628	0.1898	0.7467	0.5220	0.3903	0.1439	0.2020	0.4147	0.6690	0.7738	0.7843	0.0959	0.2283	(0.1765)	0.0713	0.8261	0.8968	1.0000

Table 6.6 Index 1–5 MoM WA MV summary covariance statistics

Index 1 - All Aircraft MoM MV vs. Comparative Metrics - Covariance Statistics								
Time Scenario	*1*		*2*		*3*		*4*	
Time	*6.1996-6.2017*		*6.1996-6.2007*		*6.2007-6.2010*		*6.2010-6.2017*	
	Weighted Ave	*Simple Ave*	*Weighted Ave*	*Simple Ave*	*Weighted Ave*	*Simple Ave*	*Weighted Ave*	*Simple Ave*
Aircraft Lessors[(1)]	(0.7676)	1.1047	(0.1144)	(1.2080)	(0.5206)	(1.2512)	(1.7860)	2.1662
Real Estate	(0.2217)	0.3197	(0.0871)	0.2864	(0.4595)	0.5233	(0.2348)	0.0704
Shipping	4.4275	0.4798	6.4214	(5.0428)	(3.1791)	1.5666	4.4990	7.9704
Agr. Land	0.1824	0.2682	(0.0008)	0.6590	(0.1283)	0.2491	0.5792	(0.3145)
Timberland	0.0374	(0.0164)	(0.0026)	(0.2234)	(0.2756)	(0.6883)	0.3074	0.4329
Commodities	(0.5577)	0.3038	0.1495	(0.7393)	0.0309	0.9696	(1.6935)	1.4492
Crude Oil	(1.7577)	0.8028	(2.0999)	0.0617	0.8164	1.3001	(2.0508)	1.5136
Gold	(0.8034)	0.9071	0.4112	(0.4213)	0.9001	(0.0601)	(3.4160)	3.5393
Precious Metals	(0.9644)	1.0798	0.1279	(0.3015)	0.9858	(0.2075)	(3.4540)	3.8937
Airlines	2.0734	(0.4411)	3.4713	1.0871	(1.5787)	(0.6057)	1.0880	(2.5880)
Transportation	1.0152	0.9373	1.4325	1.4774	(0.0689)	(0.5902)	0.8984	0.5490
Infrastructure[(2)]	(0.5465)	1.1269	(0.1077)	0.8306	(0.4895)	0.5689	(0.7145)	1.3322
Real Assets[(3)]	(0.8010)	0.7621	(0.6248)	(0.1087)	(0.0557)	0.0528	(0.8846)	1.1107
US Treasuries 1-3M	(0.0495)	0.0507	(0.0377)	0.0413	(0.0545)	0.0659	(0.0031)	(0.0009)
USD Swap Rate 10Y	0.4983	(0.6358)	0.5217	(0.1911)	(2.7051)	0.3868	1.8023	(1.8689)
USD 3M LIBOR	1.3969	(0.9272)	0.2023	0.6198	6.1807	(4.9515)	1.1935	(2.0514)
US CPI	0.0033	(0.0189)	0.0293	(0.0943)	(0.1083)	0.1480	0.0207	0.0055
MSCI World	(0.5065)	1.0572	(0.7300)	1.2129	0.1789	(0.2878)	(0.3862)	1.1766
S&P 500	(0.3883)	1.1015	(0.8016)	1.4654	0.3895	(0.5714)	(0.0359)	1.0503
Broad Debt	(0.1117)	0.8923	0.2304	1.3560	0.0455	(0.0238)	(0.6517)	0.3927
Note:	[1] Starts from 6.2006			[2] Starts from 12.2001			[3] Starts from 5.2005	

Index 2 - All Narrowbody Aircraft MoM MV vs. Comparative Metrics - Covariance Statistics								
Time Scenario	*1*		*2*		*3*		*4*	
Time	*6.1996-6.2017*		*6.1996-6.2007*		*6.2007-6.2010*		*6.2010-6.2017*	
	Weighted Ave	*Simple Ave*	*Weighted Ave*	*Simple Ave*	*Weighted Ave*	*Simple Ave*	*Weighted Ave*	*Simple Ave*
Aircraft Lessors[(1)]	(0.3551)	0.1301	0.0915	(1.7085)	0.2166	(1.5430)	-1.22094	0.7179
Real Estate	(0.4519)	0.3835	(0.1444)	0.2995	(0.7025)	0.4003	-0.57646	0.1311
Shipping	3.3484	(1.0488)	8.3539	(8.0177)	(6.9034)	1.0821	-0.70855	8.4575
Agr. Land	(0.1436)	0.3596	(0.1810)	0.6435	(0.9302)	0.4299	0.256818	(0.1239)
Timberland	(0.3832)	(0.1220)	0.1639	(0.7516)	(1.4517)	(1.4022)	-0.64082	1.1117
Commodities	(0.4808)	(0.0006)	1.3898	(1.5576)	(2.7843)	0.4209	-2.28179	2.0205
Crude Oil	(0.7959)	(0.0140)	0.2763	(1.1124)	(3.6597)	0.9461	-1.26097	1.1435
Gold	(0.0441)	(0.1064)	0.5582	(0.6452)	(2.0345)	(0.3577)	-0.41596	1.2050
Precious Metals	(0.1889)	(0.0283)	0.4856	(0.6980)	(1.8270)	(0.7609)	-0.78418	1.6556
Airlines	1.0259	(0.4532)	2.3188	0.9562	0.1563	(3.0160)	-0.73259	(1.5394)
Transportation	0.1366	0.6500	0.7099	0.6818	(2.5516)	0.2395	0.630481	0.4048
Infrastructure[(2)]	(1.4625)	1.3401	(0.4476)	0.7250	(3.0439)	0.8393	-1.1549	1.5735
Real Assets[(3)]	(1.4534)	0.7677	(0.7761)	(0.3319)	(1.6607)	(0.2108)	-1.12911	1.2508
US Treasuries 1-3M	(0.0421)	0.0501	(0.0491)	0.0455	(0.1105)	0.0642	-0.00238	(0.0011)
USD Swap Rate 10Y	(0.0250)	(1.1246)	0.6459	(0.2612)	(0.3087)	(2.0594)	-0.88137	(2.3004)
USD 3M LIBOR	1.4489	(1.4202)	(0.4282)	1.1052	13.0858	(11.3593)	0.340458	(2.1367)
US CPI	0.0334	(0.0409)	0.1100	(0.1122)	(0.1723)	0.0019	-0.02436	0.0324
MSCI World	(1.1237)	1.1173	(0.7512)	0.8371	(2.2754)	0.1453	-0.92186	1.5496
S&P 500	(1.0950)	1.1552	(1.0774)	1.1681	(1.6754)	(0.0986)	-0.56844	1.2650
Broad Debt	0.0660	0.9394	0.4025	1.0322	(0.6370)	0.8066	-0.17138	0.5143
Note:	[1] Starts from 6.2006			[2] Starts from 12.2001			[3] Starts from 5.2005	

Table 6.6 (continued)

Index 3 - All Widebody Aircraft MoM MV vs. Comparative Metrics - Covariance Statistics

Time Scenario	*1* [4]		*2* [4]		*3*		*4*	
Time	*6.1996-6.2017*		*6.1996-6.2007*		*6.2007-6.2010*		*6.2010-6.2017*	
	Weighted Ave	*Simple Ave*	*Weighted Ave*	*Simple Ave*	*Weighted Ave*	*Simple Ave*	*Weighted Ave*	*Simple Ave*
Aircraft Lessors[1]	(0.8827)	1.7776	(0.1567)	(0.7409)	(0.7410)	(1.1257)	(1.9671)	3.1601
Real Estate	(0.1297)	0.2790	(0.0607)	0.2938	(0.3270)	0.6122	(0.1017)	0.0292
Shipping	5.4239	1.6665	6.4115	(2.6178)	(0.6253)	1.8151	6.5959	7.6721
Agr. Land	0.3296	0.2113	0.0944	0.7120	0.2382	0.1082	0.6892	(0.4493)
Timberland	0.1932	0.0722	(0.1525)	0.2004	0.3219	(0.2598)	0.6813	(0.0746)
Commodities	(0.5648)	0.5977	(0.4117)	0.0133	1.3458	1.2209	(1.3987)	1.0144
Crude Oil	(2.1969)	1.6625	(3.4952)	1.4943	2.8911	1.3172	(2.1326)	1.7268
Gold	(1.0481)	1.7643	0.4511	(0.1184)	2.2720	0.0991	(4.5431)	5.2014
Precious Metals	(1.2414)	2.0342	0.0253	0.1832	2.2953	0.1405	(4.4500)	5.4910
Airlines	2.7238	(0.4340)	4.4852	1.2861	(2.2757)	1.2146	1.8253	(3.4395)
Transportation	1.4538	1.2318	1.8639	2.2985	0.9604	(1.3915)	1.0657	0.5992
Infrastructure[2]	(0.1527)	0.9620	(0.0546)	0.9230	0.7871	0.2192	(0.5405)	1.1491
Real Assets[3]	(0.4975)	0.7512	(0.5561)	0.0978	0.7206	0.1383	(0.7684)	0.9995
US Treasuries 1-3M	(0.0494)	0.0497	(0.0377)	0.0379	(0.0296)	0.0678	(0.0031)	(0.0012)
USD Swap Rate 10Y	0.7269	(0.2682)	0.3992	(0.1433)	(3.4763)	2.2717	2.9642	(1.5785)
USD 3M LIBOR	1.2152	(0.5839)	0.0956	0.1458	3.4672	0.5010	1.6075	(1.9356)
US CPI	(0.0015)	(0.0048)	0.0032	(0.0875)	(0.0679)	0.2638	0.0388	(0.0134)
MSCI World	(0.2629)	1.0864	(0.7944)	1.6818	1.3147	(0.7864)	(0.1693)	0.8893
S&P 500	(0.1163)	1.1041	(0.7844)	1.8227	1.3381	(1.0743)	0.1793	0.8860
Broad Debt	(0.1721)	0.8996	0.1807	1.7449	0.3175	(0.8243)	(0.8282)	0.2928
Note:	[1] Starts from 6.2006 [4] Starts from 2.1997			[2] Starts from 12.2001			[3] Starts from 5.2005	

Index 4 - All Narrowbody Classic Aircraft MoM MV vs. Comparative Metrics - Covariance Statistics

Time Scenario	*1*		*2*		*3*		*4*	
Time	*6.1996-6.2017*		*6.1996-6.2007*		*6.2007-6.2010*		*6.2010-6.2017*	
	Weighted Ave	*Simple Ave*	*Weighted Ave*	*Simple Ave*	*Weighted Ave*	*Simple Ave*	*Weighted Ave*	*Simple Ave*
Aircraft Lessors[1]	0.3233	1.1502	0.2624	(1.3604)	5.5156	(1.1990)	(1.5466)	2.0133
Real Estate	(0.4971)	0.5248	(0.2172)	0.4537	0.0604	0.4230	(0.5956)	0.1986
Shipping	5.7420	(2.4679)	11.4642	(11.2328)	(6.2780)	4.5171	0.8864	7.7888
Agr. Land	(0.0794)	0.5107	(0.2119)	0.8882	(1.0211)	0.5285	0.5175	(0.0966)
Timberland	(0.0197)	(0.4077)	0.3440	(1.0927)	0.6154	(3.3446)	(0.5204)	1.5874
Commodities	(0.5685)	0.1685	1.9506	(2.4217)	(2.7531)	1.9182	(3.0647)	3.1693
Crude Oil	(0.8893)	0.4916	0.3139	(1.9466)	(1.3311)	2.9347	(2.2312)	3.1120
Gold	(0.4450)	0.5019	0.7699	(0.8972)	(3.1170)	1.0796	(1.8054)	2.8605
Precious Metals	(0.6239)	0.6308	0.6894	(0.9718)	(2.9791)	0.9841	(2.1847)	3.3437
Airlines	1.5578	(1.0328)	3.0713	1.1645	(1.4999)	(4.2421)	0.1395	(2.9766)
Transportation	0.7281	0.3457	1.2299	0.6966	(0.1291)	(2.2855)	0.8355	0.5259
Infrastructure[2]	(1.3338)	1.8337	(0.5019)	1.0069	(1.8281)	1.1110	(0.8632)	2.1941
Real Assets[3]	(1.4936)	1.2123	(1.1683)	(0.5337)	(0.8437)	(0.0144)	(1.1244)	1.9024
US Treasuries 1-3M	(0.0912)	0.0635	(0.0730)	0.0652	(0.1468)	0.0819	(0.0059)	(0.0031)
USD Swap Rate 10Y	(0.0235)	(0.6807)	1.1300	(0.5159)	1.9645	(1.5988)	(2.4572)	(0.7564)
USD 3M LIBOR	0.9348	(2.3853)	(0.7382)	1.5748	17.0878	(19.1063)	(1.3468)	(2.7107)
US CPI	0.0321	(0.0683)	0.1368	(0.1753)	(0.2405)	0.0662	(0.0197)	0.0198
MSCI World	(0.8078)	1.6079	(0.5844)	1.1717	(0.3594)	(0.4673)	(0.7298)	2.6648
S&P 500	(0.8790)	1.5697	(1.0196)	1.6434	(0.1786)	(1.1585)	(0.3569)	2.1420
Broad Debt	0.1599	1.5399	0.5224	1.4441	(0.6851)	0.6850	(0.0375)	1.6612
Note:	[1] Starts from 6.2006			[2] Starts from 12.2001			[3] Starts from 5.2005	

Table 6.6 (continued)

Index 5 - All Narrowbody NG Aircraft MoM MV vs. Comparative Metrics - Covariance Statistics								
Time Scenario	1 [4]		2 [4]		3		4	
Time	6.1996-6.2017		6.1996-6.2007		6.2007-6.2010		6.2010-6.2017	
	Weighted Ave[5]	Simple Ave[5]	Weighted Ave[5]	Simple Ave[5]	Weighted Ave	Simple Ave	Weighted Ave	Simple Ave
Aircraft Lessors[(1)]	(0.7395)	(0.7537)	0.5291	(1.5573)	(3.1647)	(2.1845)	(1.1186)	(0.3573)
Real Estate	(0.4486)	0.2632	(0.0576)	0.2121	(1.0894)	0.2062	(0.5411)	0.0811
Shipping	0.9376	0.9366	5.6516	(5.4197)	(7.0430)	(2.5799)	(1.2077)	8.9232
Agr. Land	(0.1631)	0.2908	(0.2099)	0.6205	(0.9222)	0.3707	0.2162	(0.1482)
Timberland	(0.8820)	0.1737	(0.2457)	(0.4212)	(2.9210)	0.2068	(0.6784)	0.5259
Commodities	(0.9610)	0.1926	0.6523	(0.6328)	(2.7445)	(0.8360)	(2.0587)	1.3805
Crude Oil	(1.0524)	(0.2514)	0.0957	0.3249	(5.2116)	(1.6427)	(0.7793)	(0.3356)
Gold	(0.0186)	(0.6698)	0.1408	(0.3910)	(1.1612)	(1.9922)	0.1328	(0.1790)
Precious Metals	(0.1718)	(0.5602)	0.0794	(0.3744)	(0.9237)	(2.6306)	(0.2788)	0.3367
Airlines	0.8863	0.0303	2.2952	1.3217	2.2459	(2.2639)	(1.0954)	(0.5150)
Transportation	(0.3174)	1.0080	0.1505	1.3618	(3.8380)	1.6873	0.7377	(0.0595)
Infrastructure[(2)]	(1.5798)	0.8606	(0.5900)	0.5055	(3.7228)	0.2442	(1.2149)	1.0553
Real Assets[(3)]	(1.3670)	0.3845	(0.3194)	(0.1065)	(2.1673)	(0.7201)	(1.1192)	0.7964
US Treasuries 1-3M	0.0065	0.0306	0.0203	0.0362	(0.0998)	0.0330	(0.0015)	0.0006
USD Swap Rate 10Y	(0.3458)	(1.8335)	0.0906	(0.0377)	(1.9148)	(1.4455)	(0.1946)	(4.2335)
USD 3M LIBOR	2.1337	(0.9766)	(0.9778)	0.9028	13.1300	(6.3021)	1.2132	(1.5182)
US CPI	0.0336	(0.0140)	0.1273	(0.0534)	(0.0963)	(0.0801)	(0.0262)	0.0415
MSCI World	(1.6620)	0.7488	(1.4114)	0.9227	(3.5404)	0.0861	(0.9685)	0.5437
S&P 500	(1.4151)	0.7438	(1.6151)	1.1104	(2.3960)	(0.0304)	(0.5913)	0.3874
Broad Debt	(0.1176)	0.4835	0.2293	1.0609	(0.7921)	0.6514	(0.2318)	(0.4610)
Note:	[1] Starts from 6.2006 [4] Starts from 10.1999			[2] Starts from 12.2001			[3] Starts from 5.2005	

4.7 *Correlation and Covariance Observations: Time 1 (6.1996–6.2017)*

In the Time 1, the 25 comparative datasets' correlation and covariance are calculated. Within the Indexes, Index 1 compared with Index 3 had the highest positive correlation at 0.9517 overall while similarly Index 2 had strong positive correlations with Indexes 4 and 5 with figures 0.9514 and 0.8708, respectively. The other combinations are in the moderately positive range of 0.3435 to 0.6985.

Compared to the other data sets, Index 1–5 had mostly slightly negative correlation at most -0.1618 between Index 5 and Infrastructure index. The highest positive correlation is 0.1375 between Index 3 and Shipping index. Surprisingly, Index 1–5 all had low negative or close to 0 correlation with Aircraft Lessors and also the Real Assets index. Between the 20 comparative sets, Aircraft Lessors had moderate positive correlations above 0.35 with Timberland, Transportation, Infrastructure, Real Assets, MSCI World, and S&P 500. There are strong positive correlations above 0.75 between Real Assets and Real Estate, Timberland,

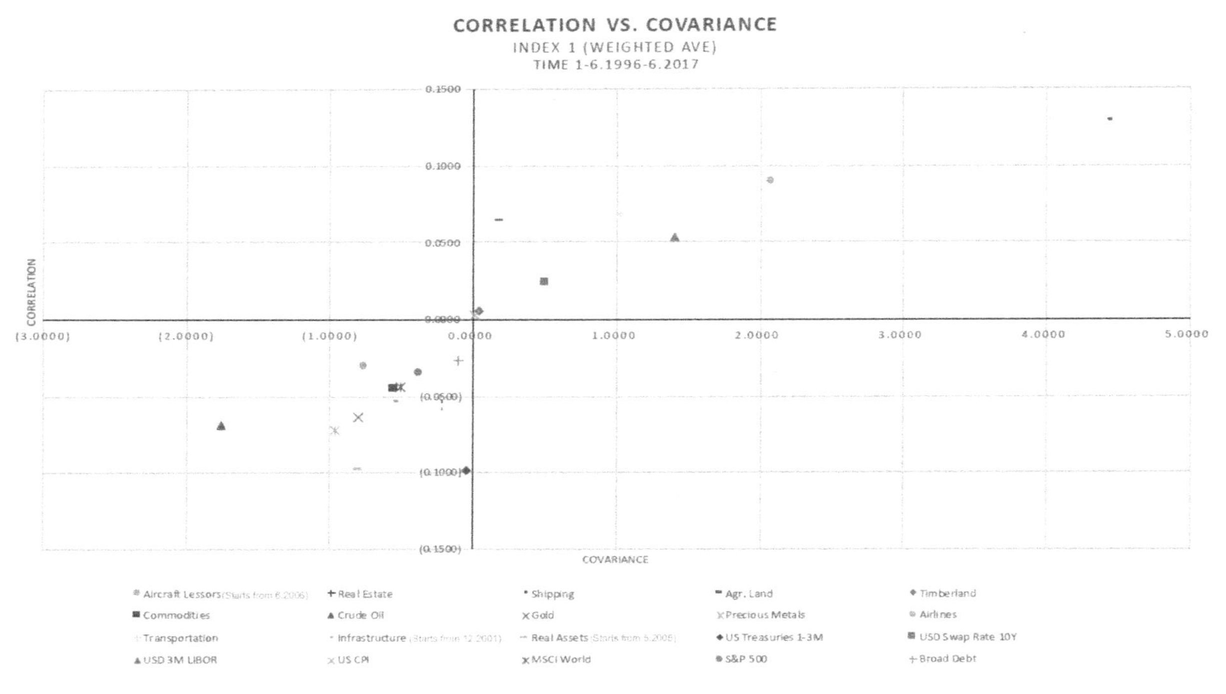

Fig. 6.8 Aircraft correlation versus covariance with comparative asset classes time 1—6.1996–6.2017

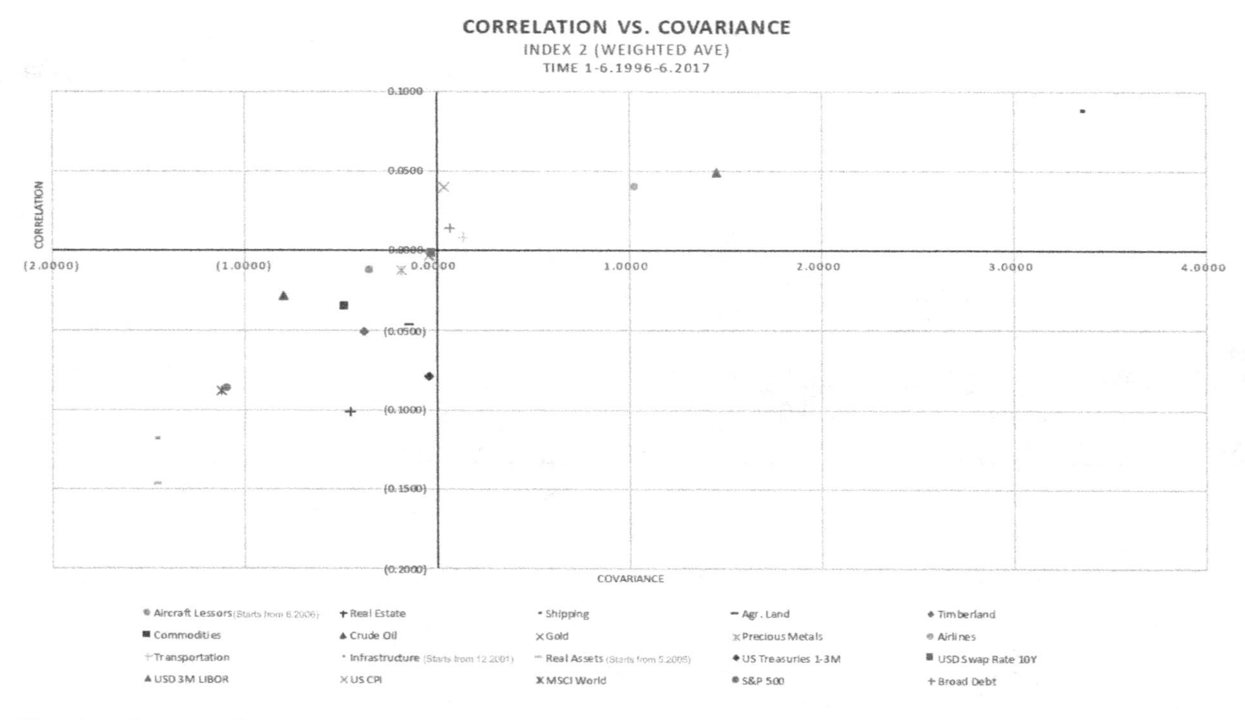

Fig. 6.8 (continued)

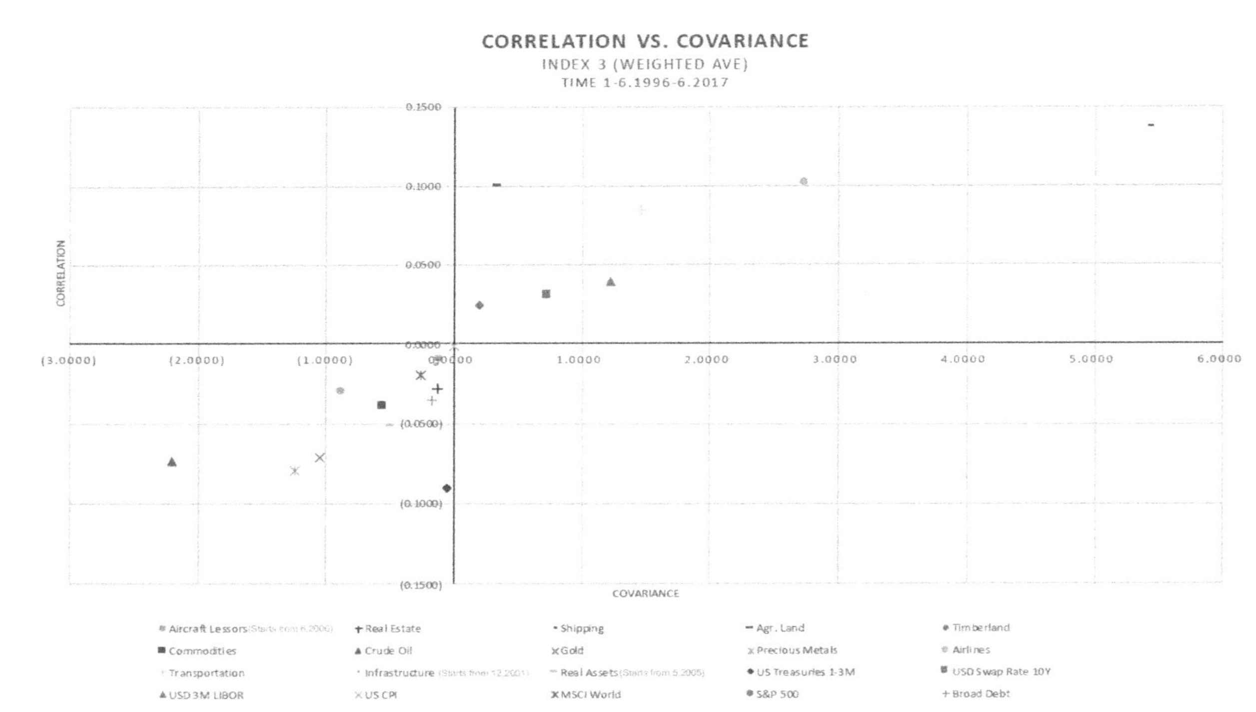

Fig. 6.8 (continued)

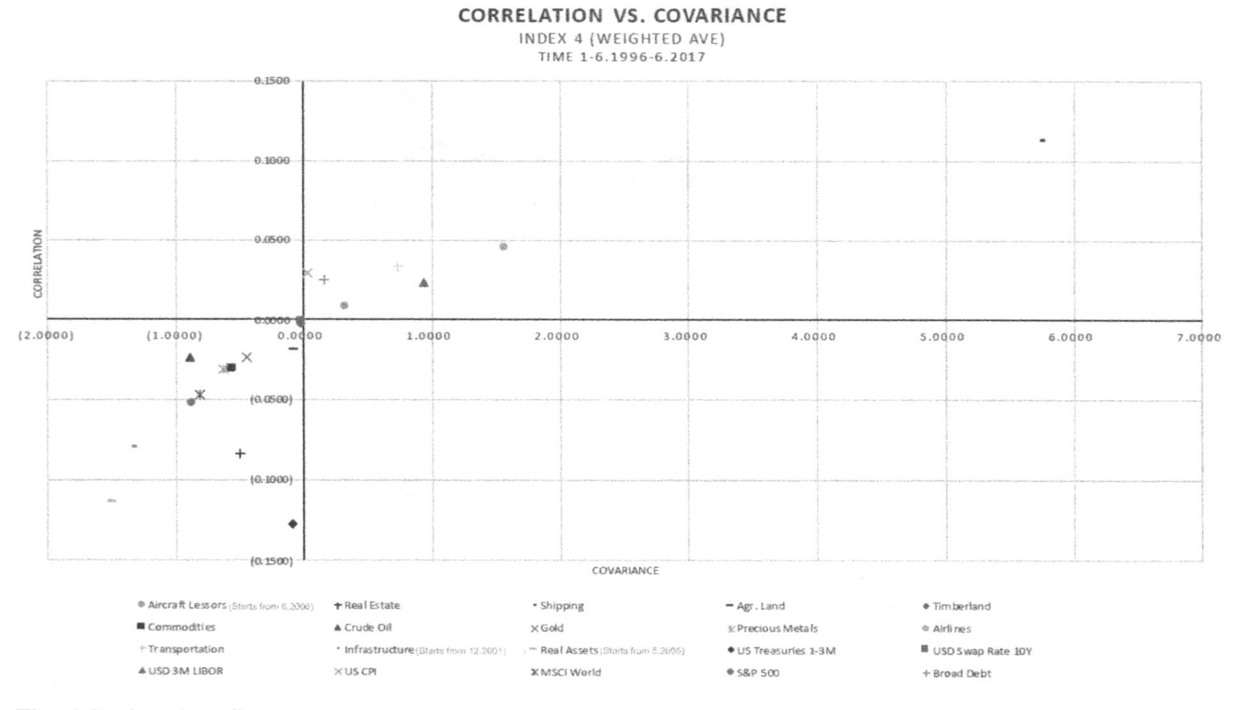

Fig. 6.8 (continued)

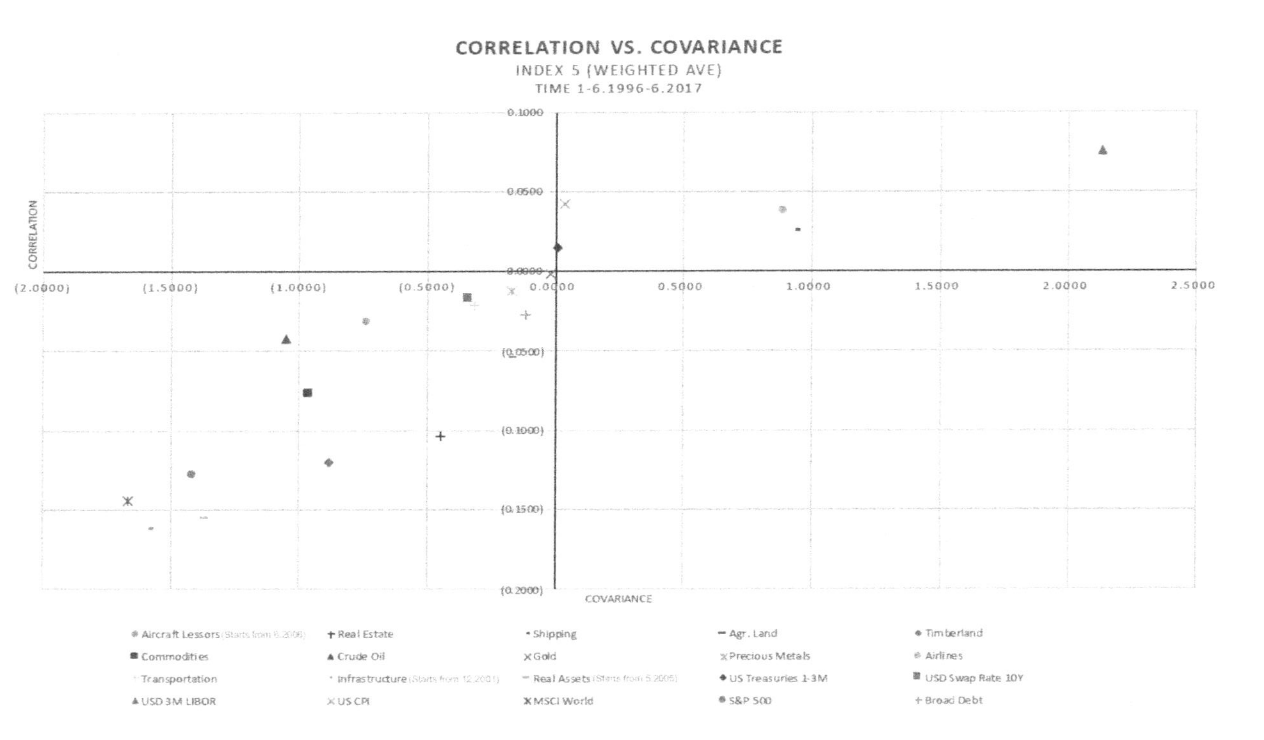

Fig. 6.8 (continued)

Infrastructure, MSCI World, S&P 500; Commodities and Crude Oil; Gold and Precious Metals; Infrastructure and MSCI World, S&P 500; MSCI World and S&P 500. The only moderately negative correlation is between Broad Debt and USD Swap Rate 10Y.

4.8 *Correlation and Covariance Observations: Time 2 (6.1996–6.2007)*

In the Time 2, the 25 comparative datasets' correlation and covariance are similarly calculated. Within the Indexes, all of the combinations exhibited moderate to strong positive correlation with the stronger range consisting of 0.8343 and higher with the highest correlation at 0.9859 between Index 2 and 4. The other combinations are in the moderately positive correlation range of 0.6235 to 0.7343.

Compared to the other data sets, Index 1–5 had a correlation distribution around 0. Aircraft Lessors exhibited a very low positive and negative correlation with Index 1–5. Real Estate showed the lowest negative correlation with the Indexes with Index 4 exhibiting a low of -0.2255. The other lowest asset class correlation is surprisingly the Real Assets, MSCI World, and S&P 500 indexes. Shipping showed the highest positive moderate correlation with the Indexes in the 0.2 range with Airlines also having a slight positive correlation.

Between the 20 comparative sets, Aircraft Lessors had the strongest correlation, a positive 0.3528, between Real Assets. The other moderately positive correlation is with Transportation at 0.2736, MSCI World at 0.2141, and S&P 500 at 0.2006. Others in conjunction with the other asset classes exhibited very weak positive and negative correlations. There are strong positive correlations above 0.75 between Commodities and Crude Oil; Gold and Precious Metals; Real Assets and Infrastructure; and MSCI World and S&P 500. The only moderately negative correlation is between Broad Debt and USD Swap Rate 10Y like in Time 1.

4.9 *Correlation and Covariance Observations: Time 3 (6.2007–6.2010)*

In the Time 3, the 25 comparative datasets' correlation and covariance are similarly calculated. Within the Indexes, the moderate to strong positive correlations in Time 1 and 2 had diverged in Time 3. Index 1 had a positive correlation with all the other indexes with Index 3 correlation of 0.8694 the highest. Index 2, while exhibiting a high correlation with Index 4 and 5 at 0.8221 and 0.8184, respectively, had a slight negative correlation with Index 3. Index 3 also had a slightly positive correlation with Index 4 and a slightly negative correlation with Index 5. Indexes 4 and 5 also had a moderate positive correlation of 0.3607.

Compared to the other data sets, Index 1–5 like previous time periods did not exhibit strong correlation in general. The highest moderate negative correlations are between Agricultural Land and Indexes 2, 4, and 5 being -0.5246, -0.4204, and -0.4621, respectively. USD 3M LIBOR had the strongest positive correlations with Index 1–5 with a range of 0.983 to 0.3065. Aircraft Lessors like other Times 1–2 did not exhibit any moderate positive or negative correlations with the range being -0.1011 to 0.1447.

Aircraft Lessors had generally moderate positive correlations with the other 20 comparative sets. The moderately positive correlations are between Real Estate, Timberland, Commodities, Crude Oil, Airlines, Transportation, Infrastructure, Real Assets, USD Swap Rate 10Y, MSCI World Index, and S&P 500 indexes with the highest correlation at 0.7350 with S&P500. The other moderate negative correlations are with Agricultural Land, US Treasuries 1–3M, and USD 3M LIBOR while the others are low positive correlations. There are strong positive correlations above 0.75 between Real Assets and Real Estate, Timberland, Infrastructure, MSCI World, S&P 500; Commodities and Crude Oil; Timberland and Transportation, Infrastructure, MSCI World, S&P 500; Gold and Precious Metals; Infrastructure and Transportation, MSCI World, S&P 500; Transportation and MSCI World, S&P 500; MSCI World and S&P 500. The moderately negative correlations are between Broad Debt and USD Swap Rate 10Y, USD 3M LIBOR; Airlines and Agricultural Land.

4.10 Correlation and Covariance Observations: Time 4 (6.2010–6.2017)

In the Time 4, the 25 comparative datasets' correlation and covariance are similarly calculated. Within the Indexes, all of the correlations are positive with several very strong correlations like Time 1 and 2. Index 1 exhibited a strong positive correlation of 0.9588 with Index 3 while moderately positive correlation is shown with the other Indexes. While Index 2 correlation with both Index 4 and 5 are strong at 0.9416 and 0.9745, respectively, it had a slight positive correlation with Index 3. Index 3 also had a slightly positive correlation with Indexes 4 and 5. Indexes 4 and 5 also had a strong positive correlation of 0.8505.

Compared to the other data sets, Index 1–5 like previous time periods did not exhibit strong correlation in general and are mostly weak negative correlation. Some of the weak positive correlations are with Shipping, Agricultural Land, Timberland, Airlines, Transportation, USD Swap Rate 10Y, USD 3M LIBOR, and US CPI. Aircraft Lessors exhibited generally a weak negative correlation. The highest weak negative correlation is with Gold, Precious Metals, Commodities and Broad Debt. The highest weak positive correlation is with Agricultural Land.

Between the 20 comparative sets, Aircraft Lessors had all weak to moderate positive correlations. The highest moderately positive correlations are between Timberland, Transportation, MSCI World, and S&P 500 indexes with the highest correlation at 0.7292 with MSCI World.

There are strong positive correlations above 0.75 between Real Estate and Real Assets, Infrastructure, MSCI World; Timberland and MSCI World, S&P 500; Commodities and Crude Oil; Gold and Precious Metals; Transportation and MSCI World, S&P 500; Infrastructure and Real Assets, MSCI World, S&P 500; Real Assets and MSCI World, S&P 500, Broad Debt; MSCI World, and S&P 500. The only moderately negative correlation is between Broad Debt and USD Swap Rate 10Y.

5 Conclusion

Following the main question, how do the aircraft asset-class segments compare in terms of various returns and volatility metrics against other real asset-backed classes and benchmarks in the four time segments, various returns, as well as standalone and relative volatility metrics, are investigated.

The main hypothesis is that aircraft asset class as represented by Index 1–5 MoM WA MV has a lower standalone risk in terms of standard deviation and variance over the historical time periods compared with the comparative sets. In terms of standalone volatility, standard deviation, and variance, Index 1–5 is ranked in the bottom half or third and fourth quartiles in all four time segments. Aircraft Lessors in terms of standalone volatility, standard deviation, and variance, are found at or near the top in Time 1–3 but moved down to eight out of 25 in the post-GFC period Time 4. The Shipping asset class is higher in the volatility metric for the periods that Aircraft Lessors are not at the top. The data suggests that the main hypothesis is not correct while the statement Aircraft Lessors asset class has a lower standalone risk in terms of standard deviation and variance over other comparable asset classes is mostly supported by the data.

Making an extension of the main research question, the second hypothesis is that the aircraft asset classes have a lower relative volatility in terms of beta or β, covariance and correlation compared with the other comparative asset classes including publicly listed aircraft lessors in the four time segments. In terms of relative volatility or beta with respect to Real Assets and MSCI World, Index 1–5 are mostly in the bottom half or third and fourth quartiles of the rankings in all four time segments. The betas are all negative with the exception of Index 3 in Time 3 for both cases.

Aircraft Lessors are found to be within the top six for beta to Real Asset and MSCI World in all four time segments. Crude Oil is the highest or higher ranked beta to Real Asset in Time 1, 2, and 4 while Time 3 Shipping is the highest and Crude Oil is slightly lower ranked. Similarly, the Airlines index is the highest or higher in Time 1, 2, and 4 while Time 3 is lower at four under top ranked Aircraft Lessor index under beta to MSCI World. Other than Time 3 in the beta to Real Assets case, Shipping is ranked lower than Aircraft Lessors asset class.

The data suggests that the second hypothesis is mainly correct. While not the lowest in each case, the aircraft asset class is at the lower end of the relative comparison. The statement Aircraft Lessors asset class has a higher relative volatility in terms of beta, covariance, and correlation compared with the other comparative asset classes is mostly supported by the data.

Continuing in the same line of questioning, the third hypothesis is that the publicly listed aircraft leasing companies generate higher excess returns or α relative to the risk-free rate, Real Assets and MSCI World compared to the aircraft asset class in the four time segments. The relative rankings of the asset classes are the same for the various return metrics. The US Treasuries 1–3M is set as the risk-free benchmark and Real Assets and MSCI World are assumed to be the other two market proxy benchmarks.

For Time 1 by these return measures, Index 1–5 is ranked 12, 13, 7, 19, and 20, respectively, out of 25. Aircraft Lessors index is ranked in second. The top ranked is the Shipping asset class. The mean for all datasets is positive except for the USD 3M LIBOR and USD Swap Rate 10 Yr. Excess return over the risk-free rate are all positive for all the comparative set except for Commodities and rates data US CPI, USD 3M LIBOR, and USD Swap Rate 10 Yr. For alpha over MSCI World, MSCI World is ranked 11 so more than half of the asset sets are negative including all Index except for Index 3. Real Assets is ranked below MSCI World at 17 so Indexes 1 and 4 are positive alpha along with Index 3.

For Time 2, the aircraft asset classes have moved down slightly. Index 3 is ranked 12 while Indexes 1, 5, 4,and 2 are in the third and fourth quartiles ranked 16, 18, 19, and 21, respectively. Like Time 1 by return measures, Aircraft Lessors is ranked second and the Shipping asset class is top ranked. The mean return for all datasets is positive except for the USD Swap Rate 10 Yr. Excess return over the risk-free rate are all positive for all the Indexes except for Index 2 MoM WA MV, Airlines, US CPI, USD 3M LIBOR, and USD Swap Rate 10 Yr. MSCI World is ranked 11 and for alpha over MSCI World, one more than half the asset classes are negative including Index 1–5. With Real Assets benchmark ranked fifth, all five Index 1–5 are still negative.

For Time 3, Index 1–5 are ranked generally in the upper quartile with Index 4 ranked fourth out of 25 while Index 3, 2, 1, and 5 ranking seven–nine and 11, respectively. In Time 3, Aircraft Lessors is ranked right under the median at 13 below Index 1–5 while Shipping, Gold, and Precious Metals are top-ranked asset classes. A little less than half of the datasets for the mean return are negative. Excess return over the risk-free rate, again a little less than half the group is negative but not Index 1–5 nor Aircraft Lessors. During Time 3, MSCI World performed poorly and had a mean of -0.7412% and ranked 23. That means for alpha or excess returns over MSCI World the only negative excess returns are USD Swap Rate 10Y and USD 3M LIBOR. All other indexes exhibit positive alpha to MSCI World. Real Assets, ranked 16, while still negative, performed much better the MSCI World. Given Real Assets is slightly under the median, a bit less than half have negative alpha.

For Time 4, Index 3 is ranked sixth out of 25 while Index 1, 4, 2, and 5 are ranked at 11, 13, 16, and 21 respectively in the comparative set.

Time 4 Index 1–5 rankings are similar to Time 1. Aircraft Lessors is ranked near the top at three. The top ranked is the Airlines asset class. For the mean return for all datasets are positive except for USD Swap Rate 10Y, Crude Oil, and Commodities. Excess return over the risk-free rate is all positive except for USD Swap Rate 10Y, Crude Oil, and Commodities. For alpha over MSCI World, more than half of the comparative set is negative as the benchmark is ranked nine including all Index but Index 3 and Aircraft Lessors. Real Assets is ranked 14 and is higher than MSCI World which is similar as Times 1–3. Negative alpha over Real Assets occurs in slightly less than half the comparative sets including Indexes 2 and 5.

The data suggests that the third hypothesis is mostly correct. Aircraft Lessors has higher excess returns or α relative to the risk-free rate, Real Assets and MSCI World in all of the time segments except for Time 3.

Combining return and volatility concepts, the fourth hypothesis is that the publicly listed aircraft leasing companies generate higher Sharpe ratios than the aircraft asset market in the four time segments. Sharpe ratios adjust the excess return to the risk-free rate to take into account for the individual volatility and standard deviation. The rankings varied with the four time scenarios.

In Time 1, Indexes 3, 1, 4, 2, and 5 Sharpe ratios are ranked three, six, 11, 19, and 20, respectively. Ahead of Index 3 in the ranking are Agricultural Land and Real Estate asset classes. Aircraft Lessors is ranked 12 after Index 4 and ahead of Index 2 and 5.

In Time 2, Index 1–4 moved down compared to Time 1. Indexes 3 and 1 moved down to nine and 14, respectively, while Index 4 and 2 are in the lower quartile at 18 and 22. Index 5 increased two rankings to 18. The highest Sharpe ratios are Real Estate, Real Assets, and Agricultural Land asset classes. Aircraft Lessors stayed at the same rank at 12 as Time 1 but ranked after Index 3 ahead of the other 4 Indexes.

In Time 3, Index 1–4 in the top two–five rank while Index 5 is at ten. Index 1–5 all moved significantly higher compared to Time 2. The highest Sharpe ratio is the Agricultural Land asset class. Aircraft Lessors decrease one rank to 13 from Time 2 but below Index 1–5.

In Time 4, most of the comparative sets are negative except for the top three ranked Commodities, Crude Oil, and USD Swap Rate 10Y. Index 5 moved to rank five on top of Index 1–4 which are spread out with a bigger range from eight to 21. Aircraft Lessors decreased slightly to rank 15 from Time 3 and ranked after Index 4 and ahead of Index 1 and 3.

The data suggests that the fourth hypothesis is partially correct. Aircraft Lessors Sharpe ratios are only higher compared to some Indexes and in some time segments.

5.1 *Correlation and Covariance Observations*

Within Index 1–5, there are mostly moderate to high positive correlations in the four time segments. Compared to the other data sets, Index 1–5 did not exhibit strong correlations with other benchmarks and the figures are around 0 and slightly negative in the four time segments. Shipping was the highest positive correlation with Index 1–5 in Time 2 in the 0.2 range and otherwise showed a low positive correlation in the other time segments. Otherwise, the lowest negative correlation with Index 1–5 was Agricultural Land in Time 3. Aircraft Lessors in general exhibited a very low positive and negative correlation with Index 1–5 in each of the time segments. Airlines also exhibited mostly weak positive correlations.

Comparing between the 20 comparative sets, the themes that were found across the time segments reinforced intuitive thoughts such as the negative correlation between Broad Debt and USD Swap Rate 10Y and less so with USD 3M LIBOR. Other strong correlations are between Crude Oil and Commodities and Gold and Precious Metals given the subcomponent is a subset of the larger group. Aircraft Lessors had the strongest correlation with the two equity indexes, MSCI World and S&P 500, Transportation, Real Assets, Infrastructure, and Timberland for three time segments except for Time 2 where it only had a weak correlation with all the other asset classes.

Bibliography

Acquila Capital. (2015). *Real Assets—Investments in Timberland.* Retrieved from www.aquila-capital.de/en/research/stellungnahme-zum-presseartikel-aquila-waldinvest-iii

Bloomberg. (2017). *Economic, FX, Fuel, and Funding Data.* Retrieved from: www.bloomberg.com

Brookfield Asset Management. (2017). *Real Assets, Real Diversification.* Retrieved from www.brookfield.com

Cambridge Associates. (2017). *Infrastructure Index*. Retrieved from: www.CambridgeAssociates.com

MSCI. (2017). *MSCI World Index*. Retrieved from: www.msci.com/indexes

NCREIF. (2017). *Indices*. Retrieved from: www.NCREIF.org

S&P Dow Jones Indexes. (2017). *S&P Dow Jones Indexes*. Retrieved from: https://eu.spindices.com

CHAPTER 7

Conclusion

This research establishes the status of the aircraft asset class and its segments. Chapters 3 and 4 show the dynamics and drivers of the aviation finance and leasing landscape with respect to both the global market and more specifically the China market affecting aircraft pricing. The demand, supply, and business model change drivers are laid out. Demand drivers include economic factors, business cycles, exogenous shocks, fuel prices, and traffic flows along with population demographics. Supply drivers include aircraft manufacturers, parked or retired aircraft, operating leases, secondary trading of aircraft, the financing environment along with current trends and segments such as commercial banks, capital markets, and export credit financing. In addition, an analysis of the drivers affecting cross-border mergers and acquisitions in the industry is developed along with the characteristics of the increased use of leasing, specifically aircraft operating leasing.

A brief history of the aircraft leasing industry's development is overlaid on top of the progress in both China and globally. Interlaced with this history, the drivers affecting cross-border mergers and acquisitions in the industry and specifically sidecars are analyzed along with the development and characteristics of the major global jurisdictions including Ireland, Singapore, Hong Kong, and China. Tax and government incentives are major drivers but there are other non-tax factors for leasing and these differences and similarities between China and the other global jurisdictions are discussed in depth.

D. Yu, *Aircraft Valuation*,
https://doi.org/10.1007/978-981-15-6743-8_7

Chapter 4 focuses on the empirical data and the historical market characteristics and analysis of aircraft asset pricing in terms of returns and volatility characteristics over different time segments and across multiple cycles. The main question addressed is determining the market characteristics of the aircraft asset class in terms of its returns, volatility, and trends. This chapter fills in the gap in the academic literature by examining empirical analysis and looks into the economic shocks through aviation asset pricing. This unique dataset is hand collected including time series of specific aircraft type valuation data from a collection of major aircraft appraisers over a 21-year period from 1996 to 2017 representing the entire range of the aircraft asset class.

The aviation asset class is segmented into five different major aircraft type groupings with different weighting construction effects including all aircraft types, all narrowbody aircraft, all widebody aircraft, all narrowbody classic aircraft, and all narrowbody next generation aircraft. These groups are analyzed under four time segments that look at the overall and subhistorical periods and the effects of the great financial crisis.

There are many takeaways from the observations from both cross-index and cross-time analyses. The results suggest that the main hypothesis is correct that the aircraft market represented by All Aircraft Types Index 1 MoM WA MV has higher value depreciation compared to the accounting depreciation of 0.283% per month. In terms of the second hypothesis, the data also suggests this is correct as the standard deviation is less than 3% throughout the four time segments. The data suggests some mixed results for the second hypothesis. Looking at the MoM WA MV case, narrowbody aircraft Index 2 has lower value depreciation throughout the four time segments than widebody aircraft. Narrowbody aircraft standard deviation is lower for all the time segments except for the Time 3 case. While there is one deviation in Time 3 standard deviation, the data suggests that most of the second hypothesis is correct.

The data suggests mixed results for the third hypothesis for both value depreciation and standard deviation. The data suggests that most of the third hypothesis is correct that the narrowbody NG aircraft represented by All Narrowbody NG Aircraft Index 5 will have lower value depreciation and standard deviation throughout the four time segments than narrowbody classic aircraft represented by All Narrowbody Classic Aircraft Index 4 as well as narrowbody aircraft represented by Index 2. The data suggests mixed results for the fourth hypothesis. The data partially supports the first part of the fourth hypothesis that the five aircraft market segments will have lower value depreciation and standard deviation in the pre-GFC period Time 2 than during and post the GFC Time 3 and Time 4, respectively. Looking at the Index 1 results, the overall aircraft market has lower value depreciation in Time 2 than the other times while for most indexes in Time 3 the standard deviation is the lowest. For the second part of the fourth hypothesis, the data does not support this. The overall aircraft market has a higher value depreciation in Time 4 than in the other times while for all indexes Time 4 also has the highest standard deviation.

Also, based on the correlation between BV and MV the data suggests mixed results for the fifth hypothesis. The first part of the hypothesis that there is slight to moderate correlation between MV and BV for the aircraft types throughout the four time segments is found to be supported for all the index and time scenarios except for the high correlation found in Index 5's Time 2 and 3. The data mostly supports the first part of the fifth hypothesis. The data suggests that Time 2 is the highest relative correlation for all indexes except for Index 3, where Time 3 is the highest relative correlation and thus the data mostly supports the second part of the fifth hypothesis. The data suggests that Time 4 is the lowest relative correlation for all indexes except for Index 4 where Time 3 is the lowest relative correlation and thus the data does not support the third part of the fifth hypothesis. The lowest correlation between MV and BV is after the GFC Time 4.

Chapter 5 focuses on the comparison of the aircraft asset pricing characteristics with other real asset-backed classes and benchmarks in the four time segments with investigative focus on various returns as well as stand-alone and relative volatility metrics. The main hypothesis that aircraft asset class as represented by Index 1–5 MoM WA MV has a lower stand-alone risk in terms of standard deviation and variance over the historical time periods compared with the comparative sets is not supported by the data. The data mostly supports the statement that the aircraft lessors asset class has a lower stand-alone risk in terms of standard deviation and variance over other comparable asset classes.

In terms of stand-alone volatility, standard deviation, and variance, Index 1–5 is ranked in the bottom half or third and fourth quartiles in all four time segments. Aircraft lessors in terms of stand-alone volatility, standard deviation, and variance are found at or near the top in Time 1–3 but are down to 8 out of 25 in the post-GFC period Time 4. The shipping asset class is higher in the volatility metric for the periods that aircraft lessors are not at the top.

Making an extension of the main research question, the second hypothesis, aircraft asset classes has a lower relative volatility in terms of beta or β, covariance, and correlation compared with the other comparative asset classes including publicly listed aircraft lessors in the four time segments, is mostly supported by the data. While the aircraft asset class is not the lowest in each case, it is at the lower end of the relative comparison. The statement aircraft lessors asset class has a higher relative volatility in terms of beta, covariance, and correlation compared with the other comparative asset classes is mostly supported by the data.

Continuing in the same line of questioning, the third hypothesis, publicly listed aircraft leasing companies generate higher excess returns or α relative to the risk-free rate, real assets, and MSCI World compared to the aircraft asset class in the four time segments, is mostly supported by the data. Aircraft lessors have higher excess returns or α relative to the risk-free rate, real assets, and MSCI World compared to the aircraft asset class in all of the time segments except for Time 3.

Combining return and volatility concepts, the fourth hypothesis, publicly listed aircraft leasing companies generate higher Sharpe ratios than the aircraft asset market in the four time segments, is partially supported by the data. Aircraft lessors Sharpe ratios are only higher compared to some indexes and in some time segments. The rankings varied with the four time scenarios.

In summary, the correlation and covariance of Index 1–5 both within itself and with the other comparative datasets showed that within Index 1–5 there are mostly moderate to high positive correlations in the four time segments. Compared to the other datasets, Index 1–5 did not exhibit strong correlations with other benchmarks and the figures are around 0 and slightly negative in the four time segments. Shipping was the highest positive correlation with Index 1–5 in Time 2 in the 0.2 range and otherwise showed a low positive correlation in the other time segments. Otherwise, the lowest negative correlation with Index 1–5 was agricultural land in Time 3. Aircraft lessors in general exhibited a very low positive and negative correlation with Index 1–5 in each of the time segments. Airlines also exhibited mostly weak positive correlations.

After the establishment of the aircraft asset class and the market segmentations, further comparative analyses are conducted in Chap. 5 to deduce the implications to the other real asset classes and major investable benchmarks. The main hypothesis is that aircraft asset class as represented by Index 1–5 MoM WA MV has a lower stand-alone risk in terms of standard deviation and variance over the historical time periods compared with the comparative sets. The data suggests that the main hypothesis is not correct while the statement that aircraft lessors asset class has a lower stand-alone risk in terms of standard deviation and variance over other comparable asset classes is mostly supported by the data.

In case of the second hypothesis, an extension of the main question is that the aircraft asset classes have a lower relative volatility in terms of beta or β, covariance, and correlation compared with the other comparative asset classes including publicly listed aircraft lessors in the four time segments. The data suggests that the second hypothesis is mainly correct. While not the lowest in each case, the aircraft asset class is at the lower end of the relative comparison. The statement aircraft lessors asset class has a higher relative volatility in terms of beta, covariance, and correlation compared with the other comparative asset classes is mostly supported by the data.

The third hypothesis is that the publicly listed aircraft leasing companies generate higher excess returns or α relative to the risk-free rate, real assets, and MSCI World compared to the aircraft asset class in the four time segments. The data suggests that the third hypothesis is mostly correct. Aircraft lessors have higher excess returns or α relative to the risk-free rate, real assets, and MSCI World compared to the aircraft asset class in all of the time segments except for Time 3.

Combining return and volatility concepts, the fourth hypothesis is that the publicly listed aircraft leasing companies generate higher Sharpe ratios than the aircraft asset market in the four time segments. The data suggests that the fourth hypothesis is partially correct. Aircraft lessors Sharpe ratios are only higher compared to some indexes and in some time segments.

Within Index 1–5, there are mostly moderate to high positive correlations in the four time segments. Compared to the other datasets, Index 1–5 did not exhibit strong correlations with other benchmarks and the figures are around 0 and slightly negative in the four time segments. Shipping had the highest positive correlation with Index 1–5 in Time 2 in the 0.2 range and otherwise showed a low positive correlation in the other time segments. Otherwise, the lowest negative correlation with Index 1–5 was agricultural land in Time 3. Aircraft lessors in general exhibited a very low positive and negative correlation with Index 1–5 in each of the time segments. Airlines also exhibited mostly weak positive correlations.

Comparing between the 20 comparative sets, the themes that were found across the time segments reinforced intuitive thoughts such as the negative correlation between Broad Debt and USD Swap Rate 10Y and less so with USD 3M LIBOR. Other strong correlations are between crude oil and commodities and gold and precious metals given the subcomponent is a subset of the larger group. Aircraft lessors had the strongest correlation with the two equity indexes, MSCI World and S&P 500, transportation, real assets, infrastructure, and timberland for three time segments except for Time 2 where it had only a weak correlation with all the other asset classes.

The unrestricted model of independent variables in the regression and statistical testing resulted in low R^2 statistics under 0.100 Index 1–5. For the main Index 1–5 regressions, the null hypothesis that all the coefficients are not significant with a significance level of 5%, no model rejects the null hypothesis. There are only a couple of independent variables that were significant at the 5% significance level, namely, real estate in Index 1 and 2, while labor and log (agricultural) were significant for Index 2.

The model was restricted for higher correlated variables, gold, general commodities, and S&P 500, in order to reduce multicollinearity and VIF figures and improve the model; the R^2 figure remained very low. All of Index 1–5 regressions, except Index 2, again did not reject the null hypothesis that all the coefficients are not significant with a significance level of 5%. For Index 2, the model accepts the null hypothesis with a p value of 0.217. Similar to the unrestricted case, the specific independent variables are significant at the 5% significance level and include real estate for Index 1 and 2 and labor and log (agricultural) for Index 2.

These results support that there is very little explanatory power of the comparative asset classes for the aircraft asset classes represented by Index 1–5. Individually, there are a few independent variables at a 5% significance level. This supports the notation found earlier in other results that the other comparable classes do not have much explanatory power for the aircraft asset class and its subsegments and should be considered for inclusion in portfolios for asset allocation.

Further subsequent studies with additional data are helpful to expand the time horizon of the commercial aircraft asset class characteristics as well as its segments by aircraft types. While the time segmentation data provides a view for the responses of this market and segments toward a large financial crisis, exogenous shock, namely, the great financial crises, additional data would show the effects of other previous exogenous shocks to the market. There can also be additional research that can be done to expand the analysis in addition to the time horizon. The expansion of the aircraft types to include regional and turboprops aircraft can also extend the analysis as can the creation of prediction models for future values to possibly improve the forecasts of the data based on historical data. In addition, if specific matching historical lease rate data to the aircraft types and times can be potentially gathered, this would enable enhancement of the return analysis and comparative analysis.

APPENDIX

D. Yu, *Aircraft Valuation*,
https://doi.org/10.1007/978-981-15-6743-8

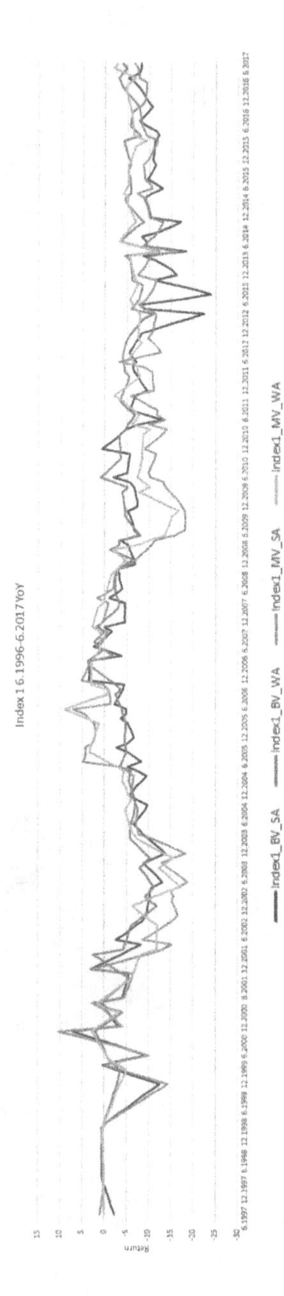

Fig. A.1 Index 1 YoY—6.1996–6.2017

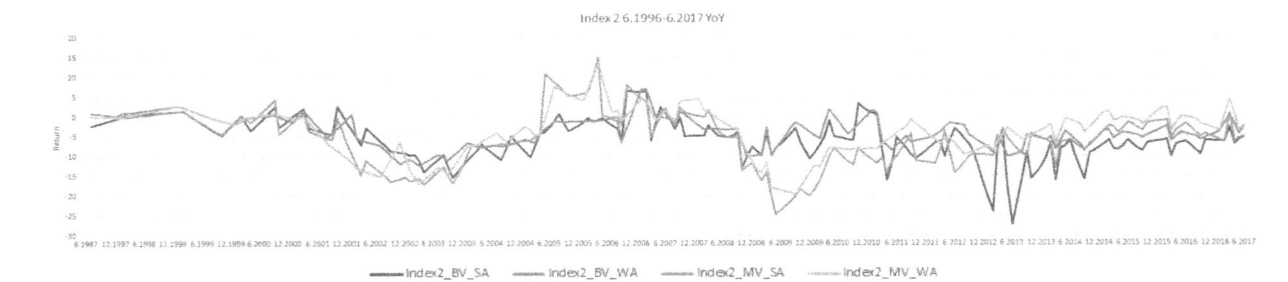

Fig. A.2 Index 2 YoY—6.1996–6.2017

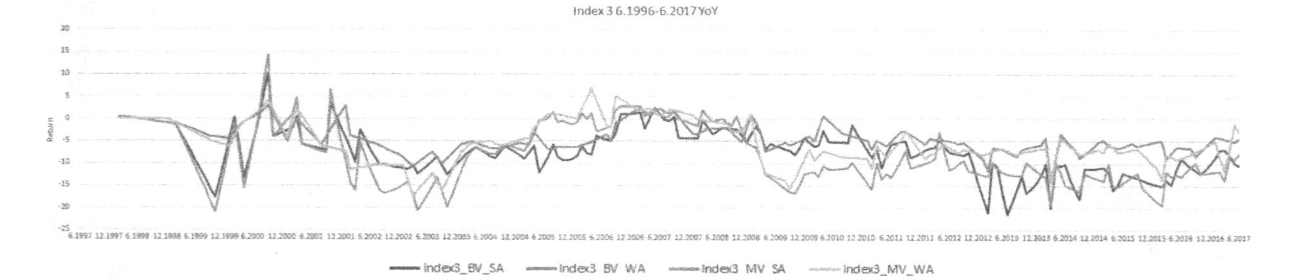

Fig. A.3 Index 3 YoY—6.1996–6.2017. (Note: Index 3 data starts from 2.1997)

Fig. A.4 Index 4 YoY—6.1996–6.2017

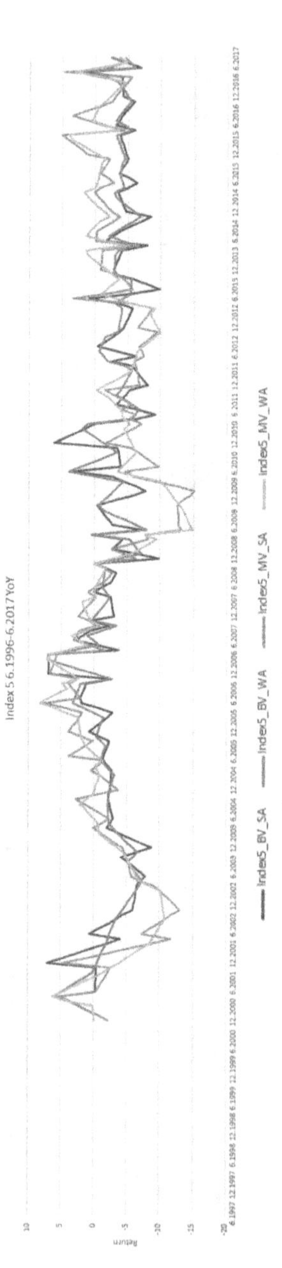

Fig. A.5 Index 5 YoY—6.1996–6.2017. (Note: Index 5 data starts from 10.1999)

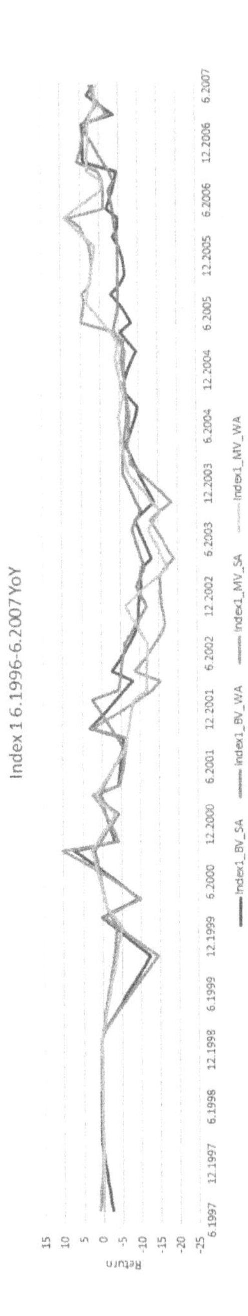

Fig. A.6 Index 1 YoY—6.1996–6.2007

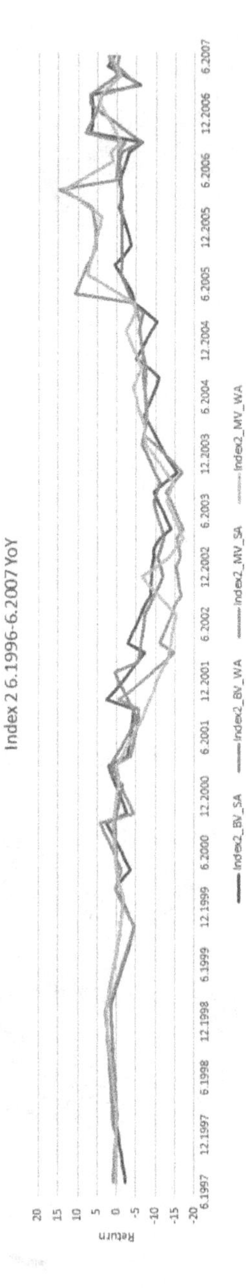

Fig. A.7 Index 2 YoY—6.1996–6.2007

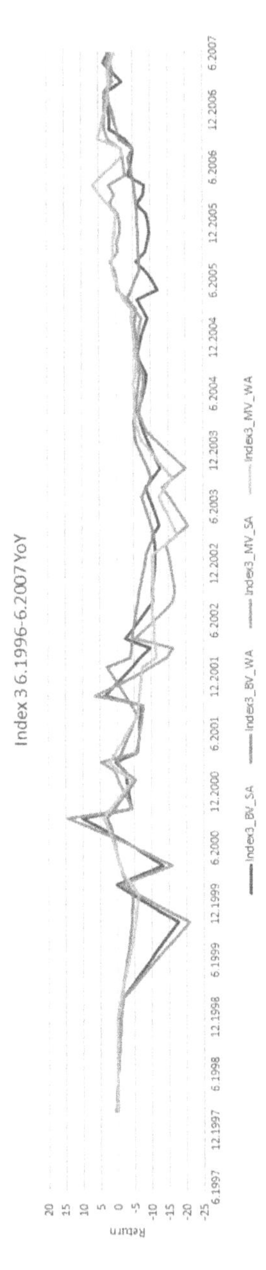

Fig. A.8 Index 3 YoY—6.1996–6.2007. (Note: Index 3 data starts from 2.1997)

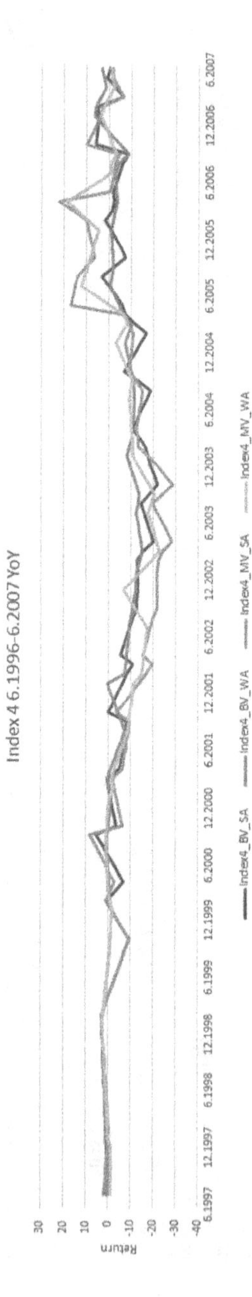

Fig. A.9 Index 4 YoY—6.1996–6.2007

Fig. A.10 Index 5 YoY—6.1996–6.2007. (Note: Index 5 data starts from 10.1999)

Fig. A.11 Index 1 YoY—6.2007–6.2010

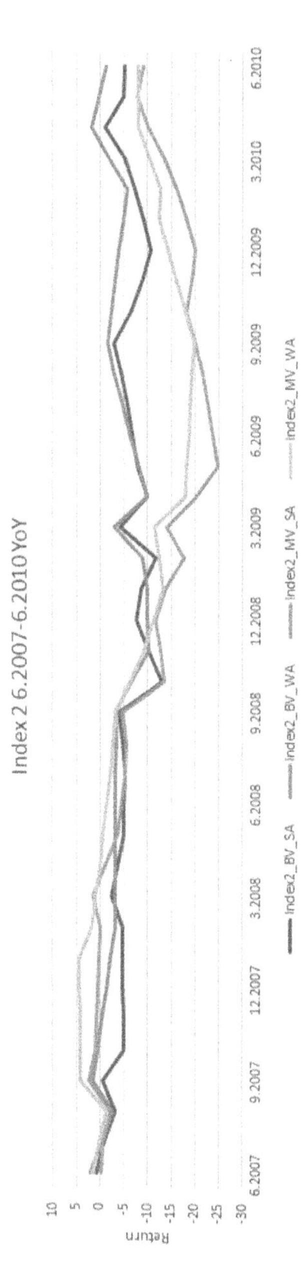

Fig. A.12 Index 2 YoY—6.2007–6.2010

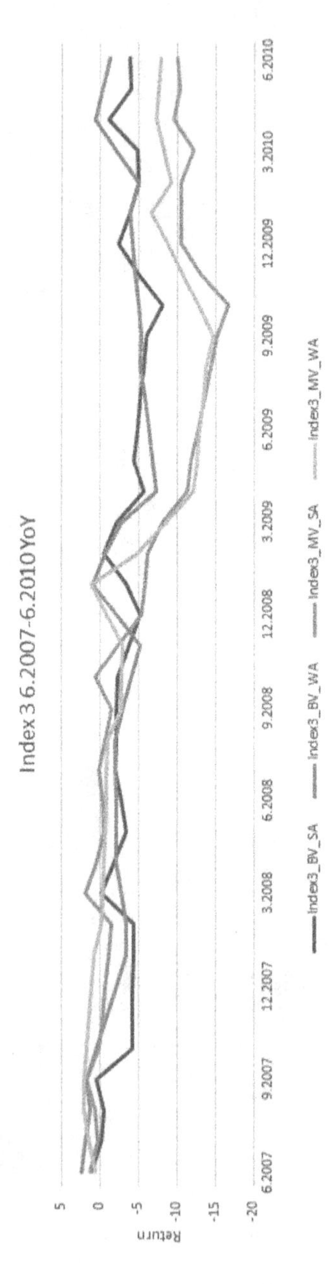

Fig. A.13 Index 3 YoY—6.2007–6.2010

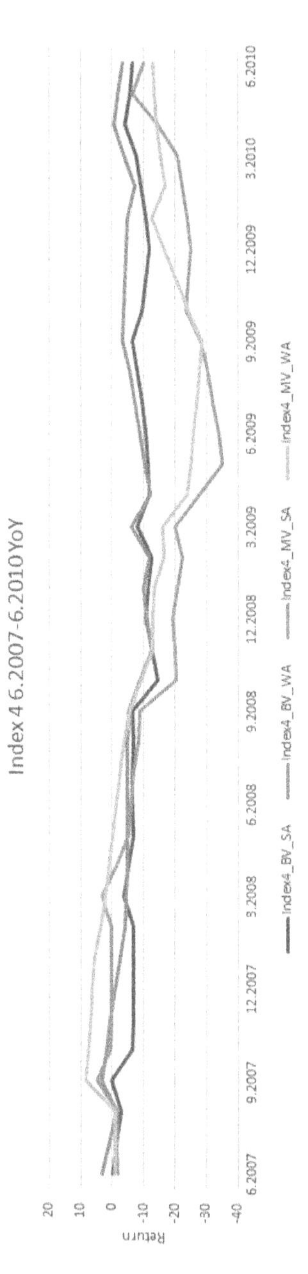

Fig. A.14 Index 4 YoY—6.2007–6.2010

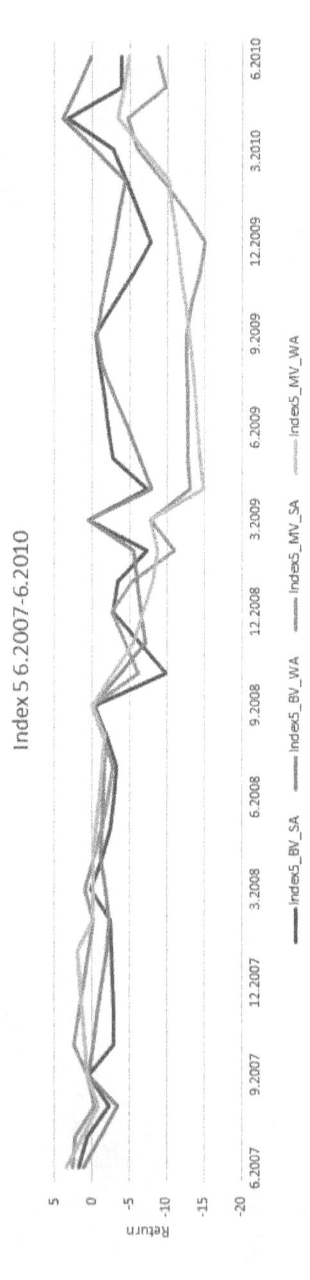

Fig. A.15 Index 5 YoY—6.2007–6.2010

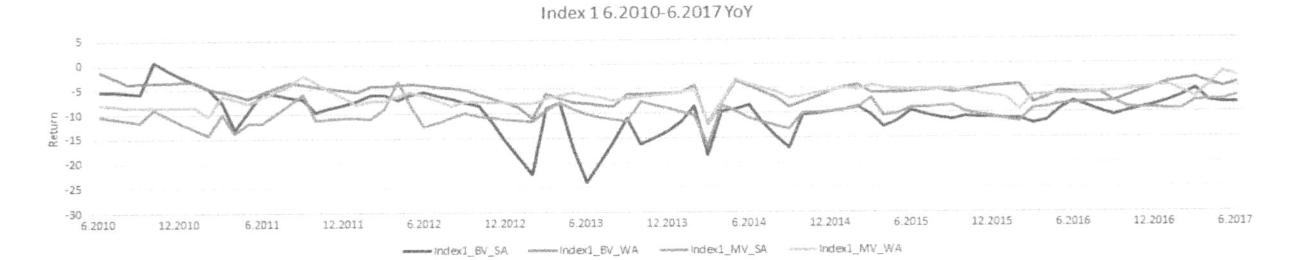

Fig. A.16 Index 1 YoY—6.2010–6.2017

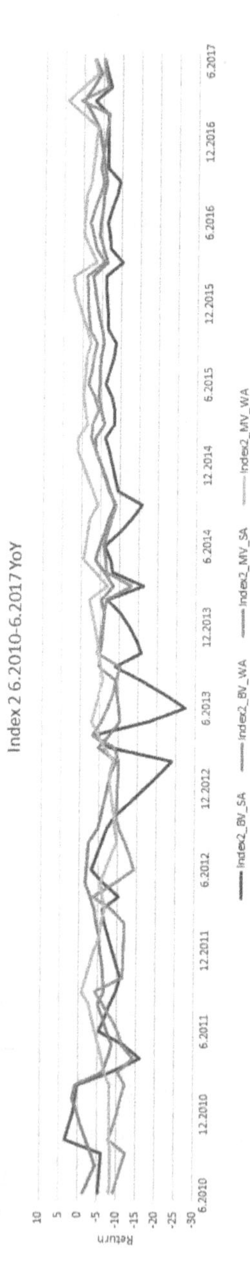

Fig. A.17 Index 2 YoY—6.2010–6.2017

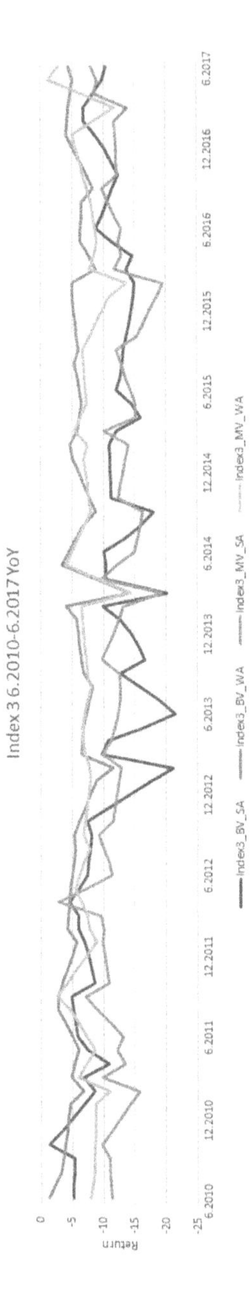

Fig. A.18 Index 3 YoY—6.2010–6.2017

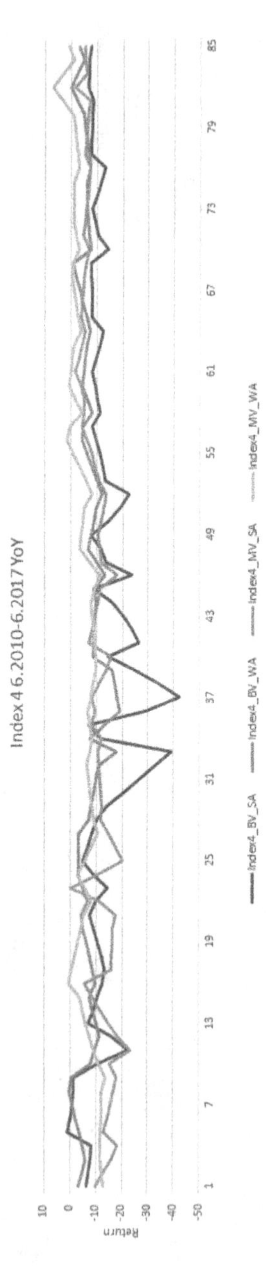

Fig. A.19 Index 4 YoY—6.2010–6.2017

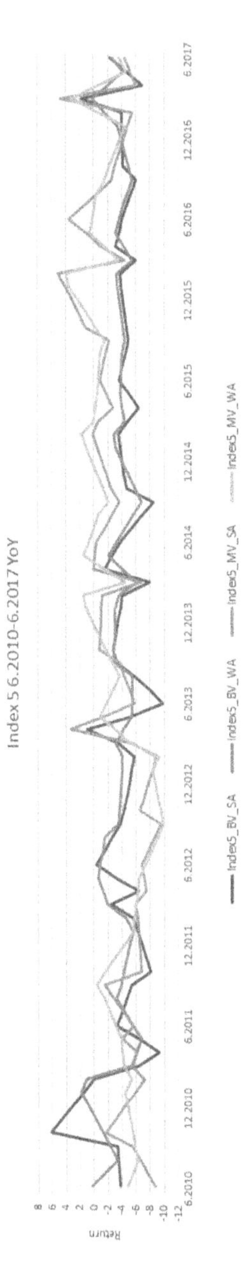

Fig. A.20 Index 5 YoY—6.2010–6.2017

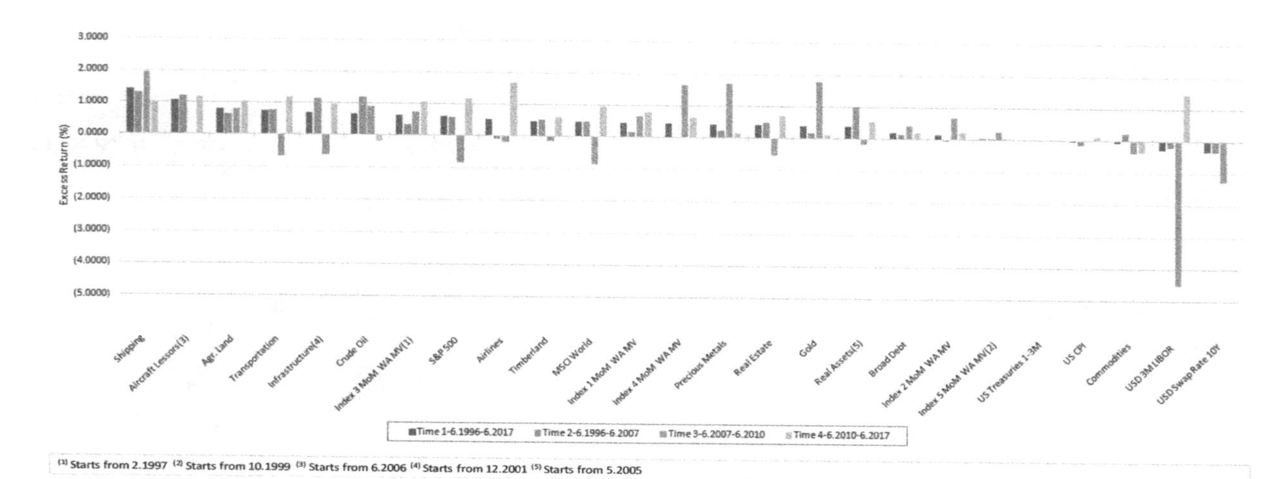

Fig. A.21 Asset class excess return over risk free rate (US Treasuries 1-3M) over 4 time Segments

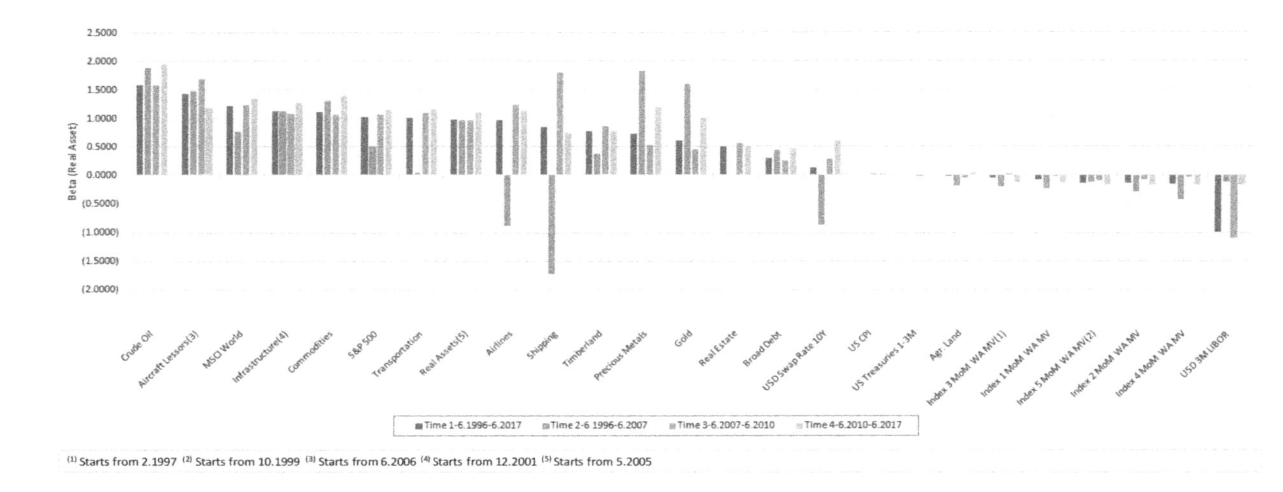

Fig. A.22 Asset class rankings by Beta (Real Asset) over 4 time segments

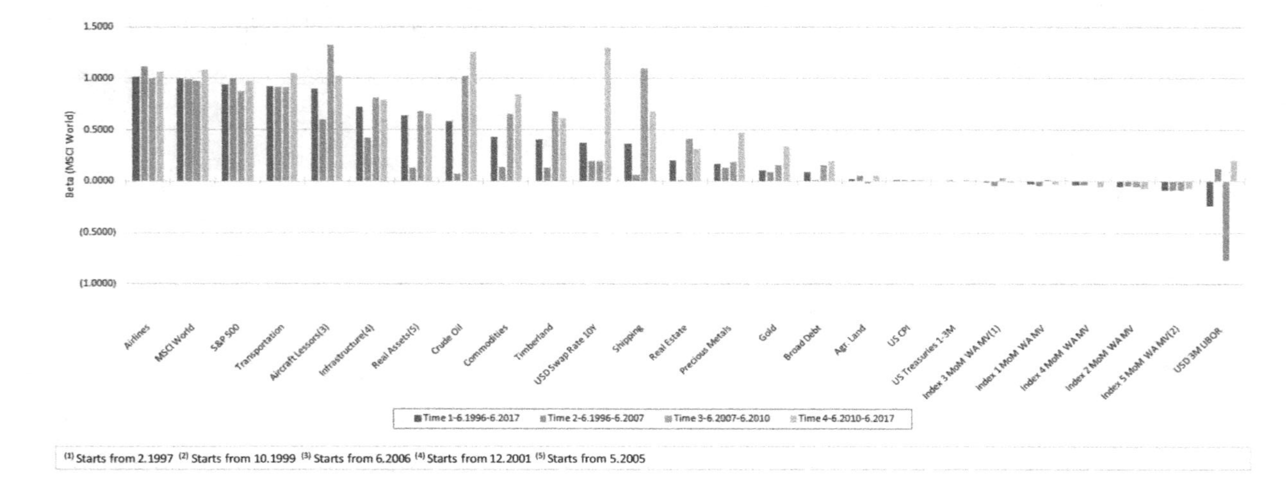

Fig. A.23 Asset class rankings by Beta (MSCI World) over 4 time segments

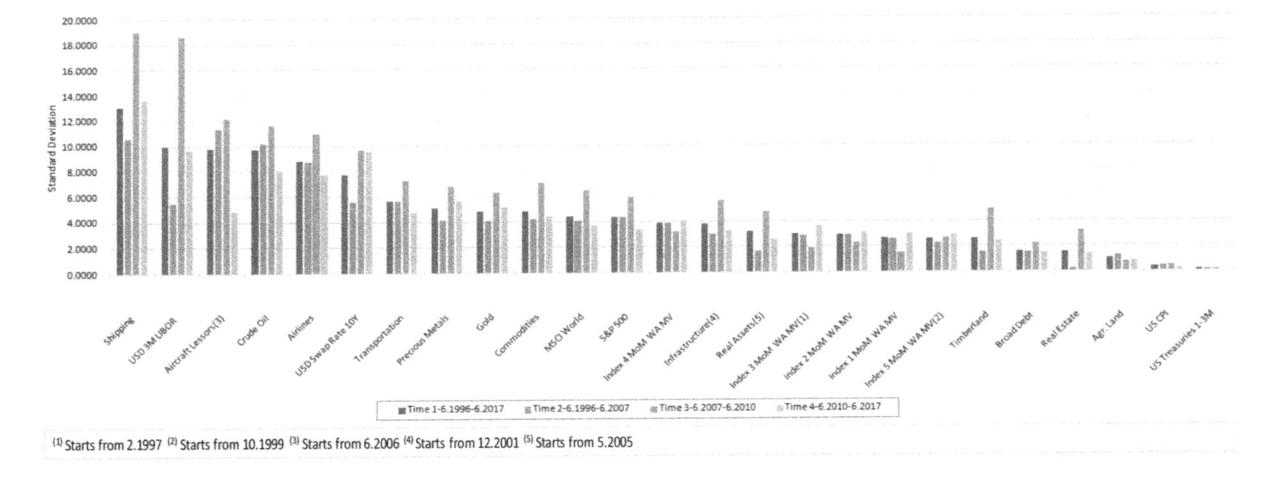

Fig. A.24 Asset class rankings by standard deviation over 4 time segments

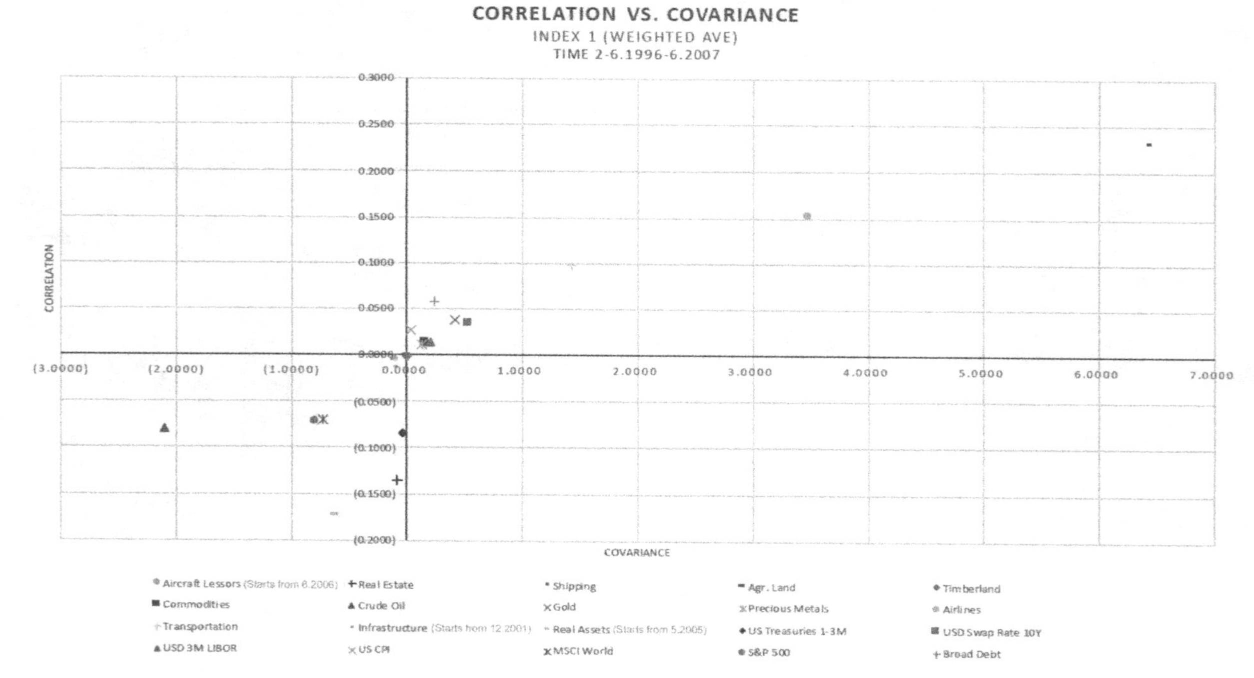

Fig. A.25 Aircraft correlation vs. covariance with comparative asset classes time 2—6.1996–6.2007

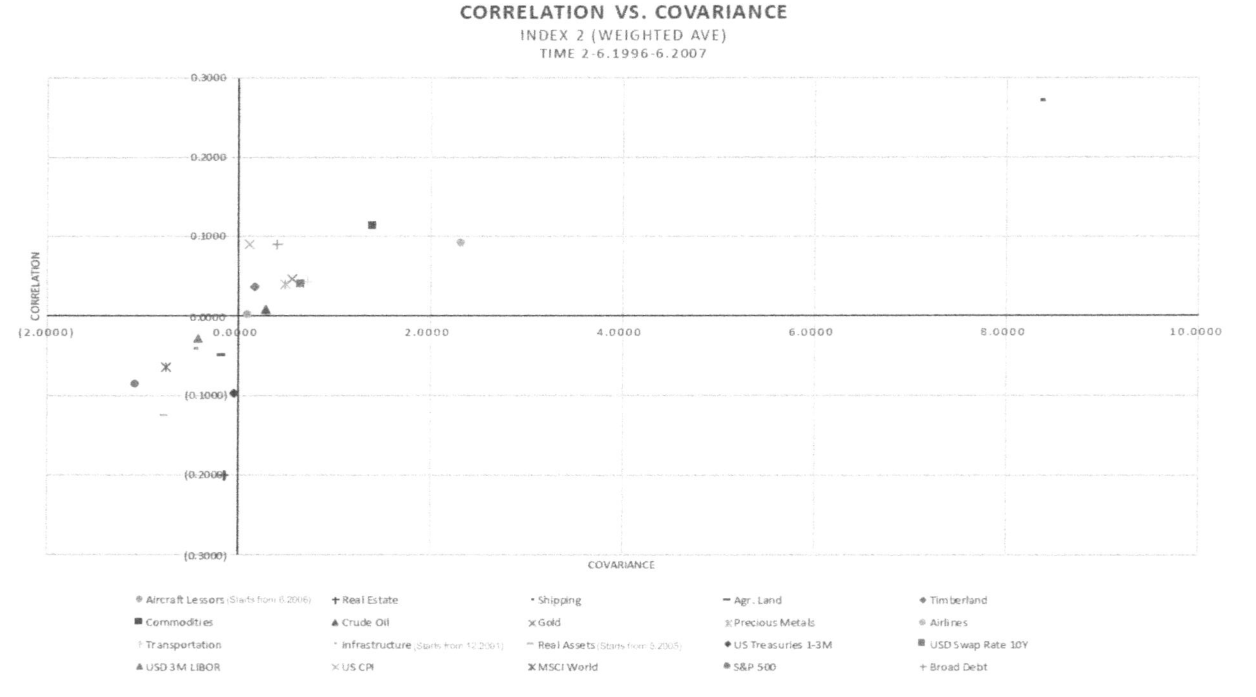

Fig. A.26 Aircraft correlation vs. covariance with comparative asset classes time 3—6.2007–6.2010

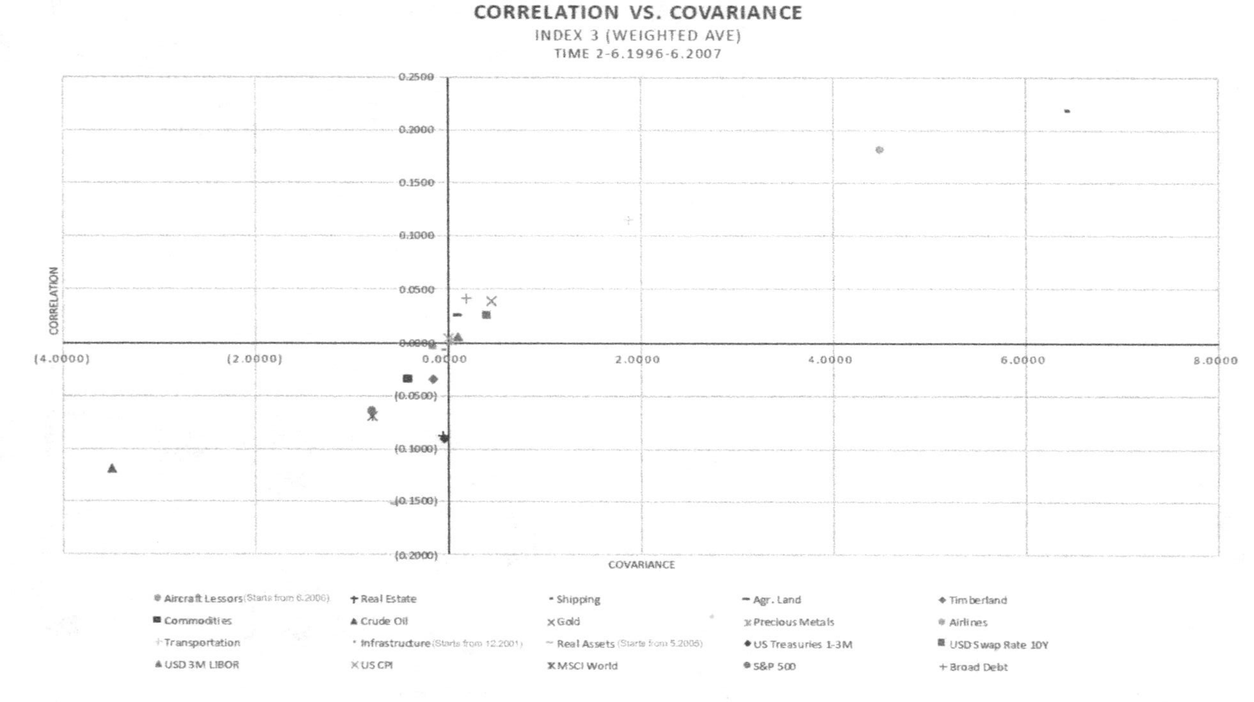

Fig. A.27 Aircraft correlation vs. covariance with comparative asset classes time 4—6.2010–6.2017

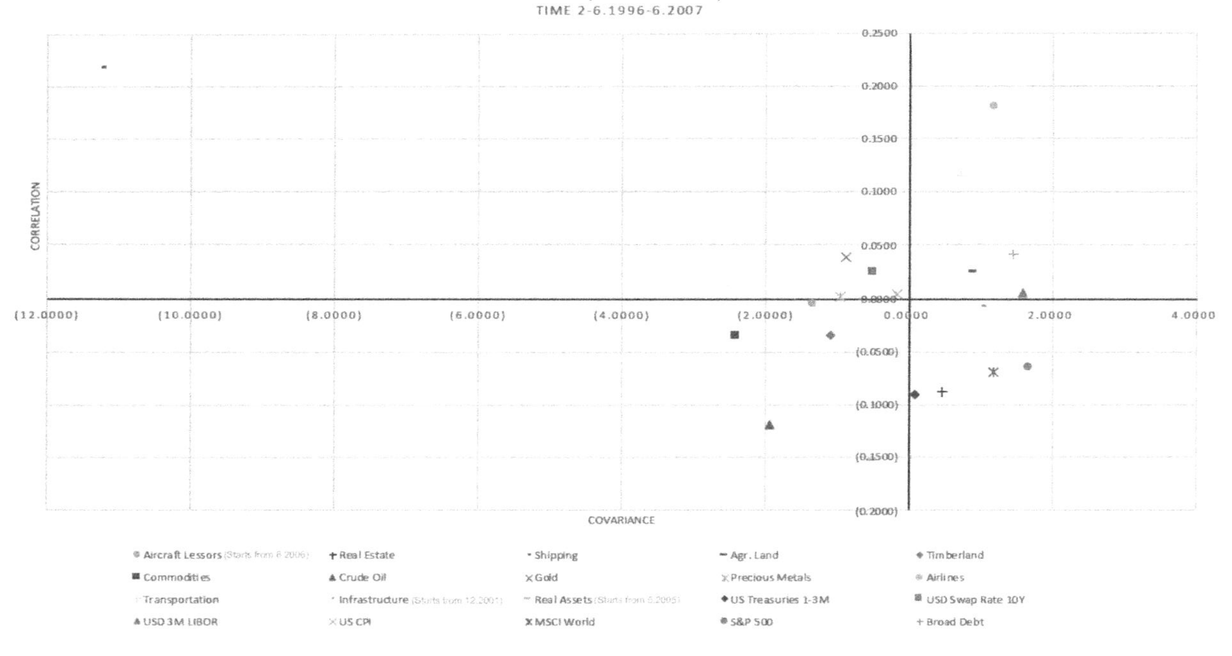
CORRELATION VS. COVARIANCE
INDEX 4 (WEIGHTED AVE)
TIME 2-6.1996-6.2007
CORRELATION
0.2500
0.2000
0.1500
0.1000
0.0500
0.0000
(0.0500)
(0.1000)
(0.1500)
(0.2000)
(12.0000)
(10.0000)
(8.0000)
(6.0000)
(4.0000)
(2.0000)
0.0000
2.0000
4.0000
COVARIANCE
Aircraft Lessors (Starts from 6.2006)
Real Estate
Shipping
Agr. Land
Timberland
Commodities
Crude Oil
Gold
Precious Metals
Airlines
Transportation
Infrastructure (Starts from 12.2001)
Real Assets (Starts from 6.2005)
US Treasuries 1-3M
USD Swap Rate 10Y
USD 3M LIBOR
US CPI
MSCI World
S&P 500
Broad Debt

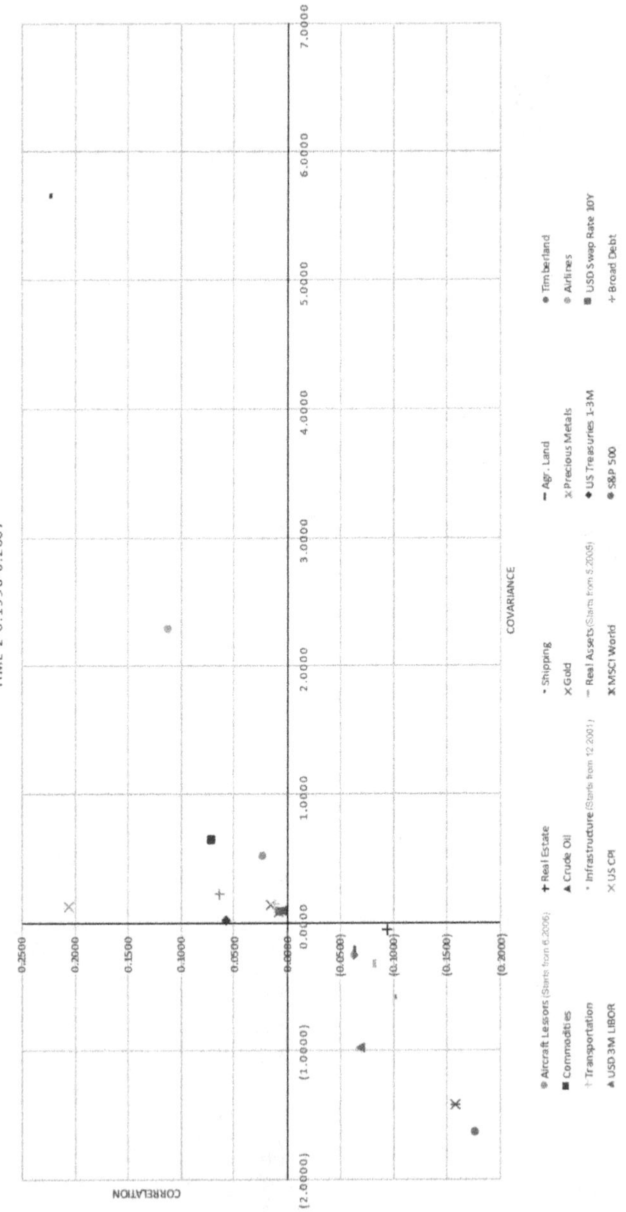
CORRELATION VS. COVARIANCE
INDEX 5 (WEIGHTED AVE)
TIME 2-6.1996-6.2007
CORRELATION
COVARIANCE
0.2500
0.2000
0.1500
0.1000
0.0500
0.0000
(0.0500)
(0.1000)
(0.1500)
(0.2000)
(2.0000)
(1.0000)
0.0000
1.0000
2.0000
3.0000
4.0000
5.0000
6.0000
7.0000
Aircraft Lessors (Starts from 6.2006)
Real Estate
Shipping
Agr. Land
Timberland
Commodities
Crude Oil
Gold
Precious Metals
Airlines
Transportation
Infrastructure (Starts from 12.2001)
Real Assets (Starts from 5.2005)
US Treasuries 1-3M
USD Swap Rate 10Y
USD 3M LIBOR
US CPI
MSCI World
S&P 500
Broad Debt

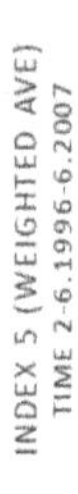

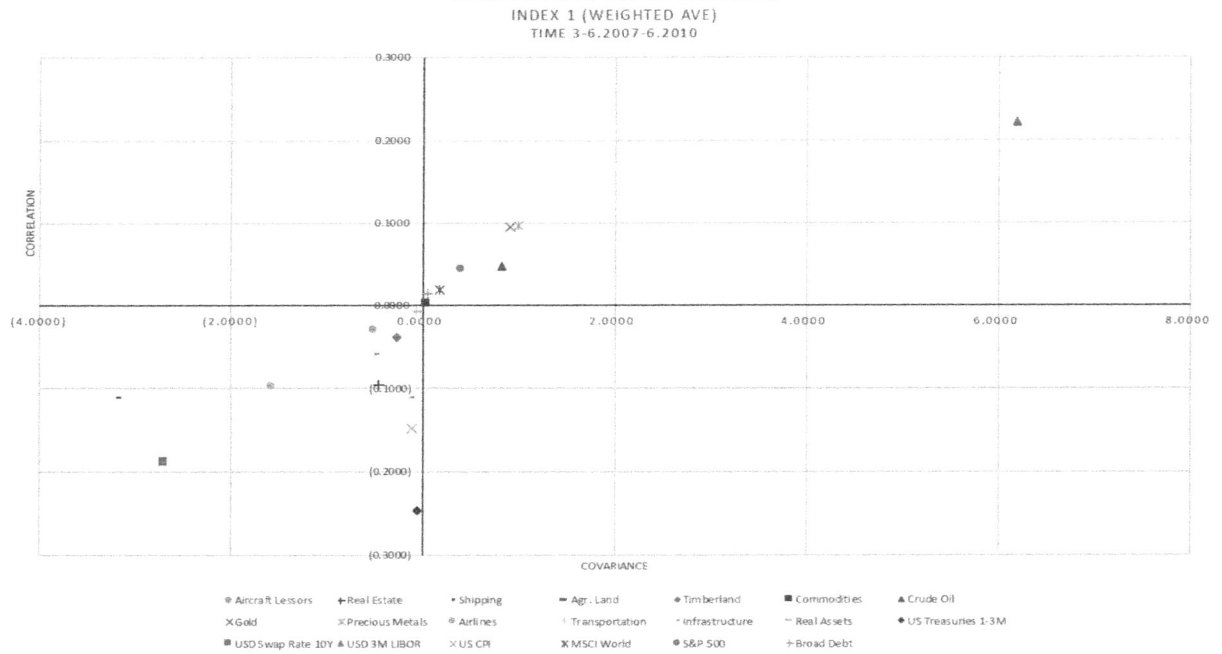
CORRELATION VS. COVARIANCE
INDEX 1 (WEIGHTED AVE)
TIME 3-6.2007-6.2010
CORRELATION
0.3000
0.2000
0.1000
0.0000
(0.1000)
(0.2000)
(0.3000)
(4.0000)
(2.0000)
0.0000
2.0000
4.0000
6.0000
8.0000
COVARIANCE
Aircraft Lessors
Real Estate
Shipping
Agr. Land
Timberland
Commodities
Crude Oil
Gold
Precious Metals
Airlines
Transportation
Infrastructure
Real Assets
US Treasuries 1-3M
USD Swap Rate 10Y
USD 3M LIBOR
US CPI
MSCI World
S&P 500
Broad Debt

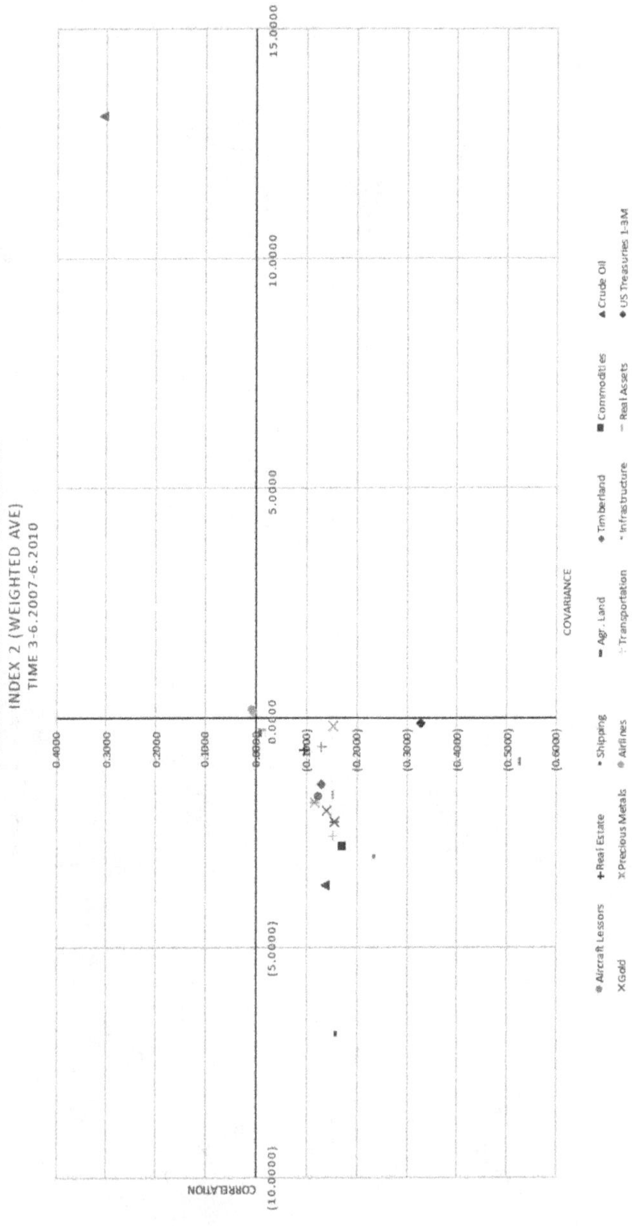
CORRELATION VS. COVARIANCE
INDEX 2 (WEIGHTED AVE)
TIME 3-6.2007-6.2010
CORRELATION
0.4000
0.3000
0.2000
0.1000
0.0000
(0.1000)
(0.2000)
(0.3000)
(0.4000)
(0.5000)
(0.6000)
(10.0000)
(5.0000)
0.0000
5.0000
10.0000
15.0000
COVARIANCE
Aircraft Lessors
Real Estate
Shipping
Agr. Land
Timberland
Commodities
Crude Oil
Gold
Precious Metals
Airlines
Transportation
Infrastructure
Real Assets
US Treasuries 1-3M
USD Swap Rate 10Y
USD 3M LIBOR
US CPI
MSCI World
S&P 500
Broad Debt

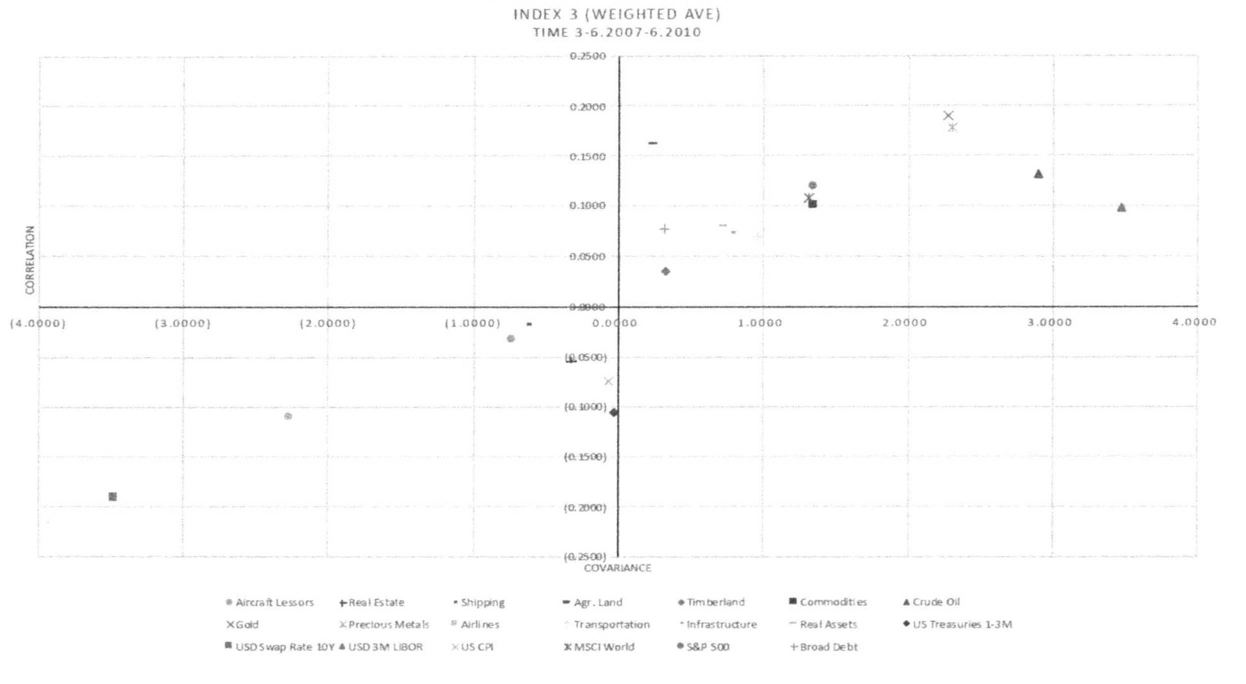
CORRELATION VS. COVARIANCE
INDEX 3 (WEIGHTED AVE)
TIME 3-6.2007-6.2010
0.2500
0.2000
0.1500
0.1000
0.0500
0.0000
(0.0500)
(0.1000)
(0.1500)
(0.2000)
(0.2500)
CORRELATION
(4.0000)
(3.0000)
(2.0000)
(1.0000)
0.0000
1.0000
2.0000
3.0000
4.0000
COVARIANCE
Aircraft Lessors
Real Estate
Shipping
Agr. Land
Timberland
Commodities
Crude Oil
Gold
Precious Metals
Airlines
Transportation
Infrastructure
Real Assets
US Treasuries 1-3M
USD Swap Rate 10Y
USD 3M LIBOR
US CPI
MSCI World
S&P 500
Broad Debt

CORRELATION VS. COVARIANCE
INDEX 4 (WEIGHTED AVE)
TIME 4-6.2007-6.2010

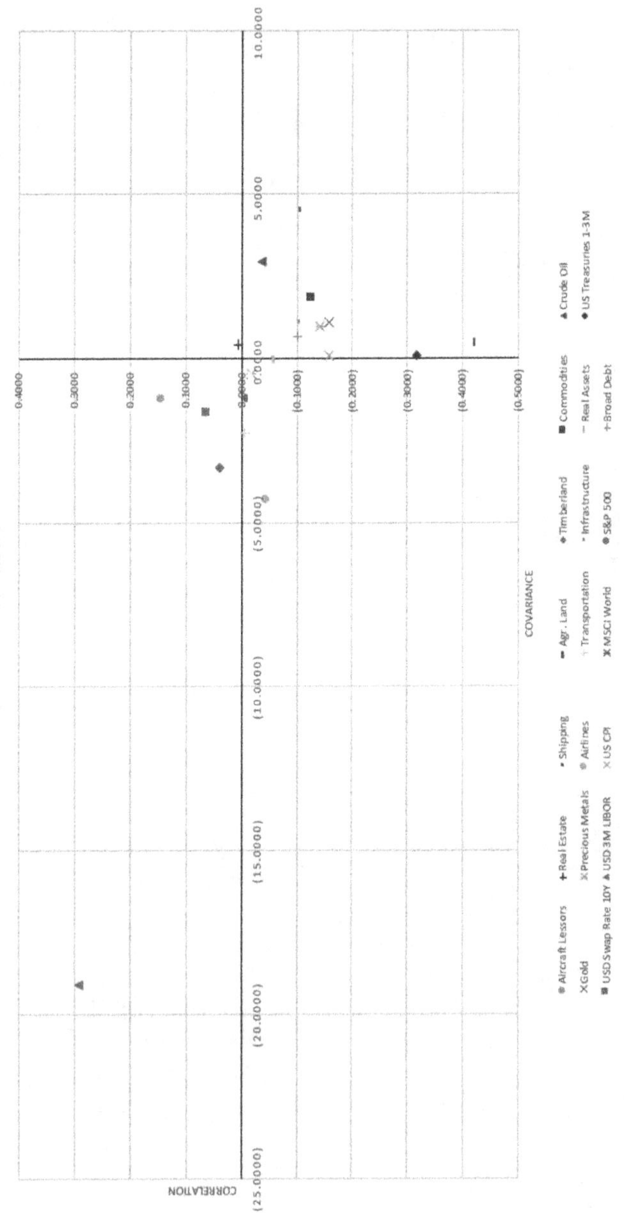
CORRELATION
COVARIANCE
(25.0000)
(20.0000)
(15.0000)
(10.0000)
(5.0000)
0.0000
5.0000
10.0000
0.4000
0.3000
0.2000
0.1000
0.0000
(0.1000)
(0.2000)
(0.3000)
(0.4000)
(0.5000)
Aircraft Lessors
Real Estate
Shipping
Agr. Land
Timberland
Commodities
Crude Oil
Gold
Precious Metals
Airlines
Transportation
Infrastructure
Real Assets
US Treasuries 1-3M
USD Swap Rate 10Y
USD 3M LIBOR
US CPI
MSCI World
S&P 500
Broad Debt

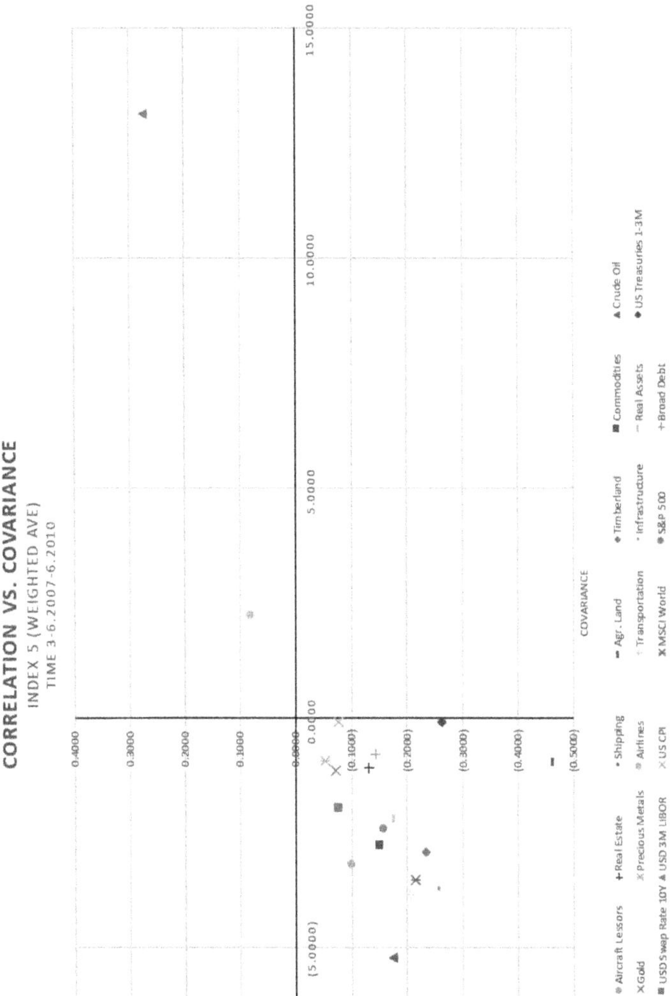

CORRELATION VS. COVARIANCE
INDEX 5 (WEIGHTED AVE)
TIME 3-6.2007-6.2010

CORRELATION
0.4000
0.3000
0.2000
0.1000
0.0000
(0.1000)
(0.2000)
(0.3000)
(0.4000)
(0.5000)

(10.0000)
(5.0000)
0.0000
5.0000
10.0000
15.0000
COVARIANCE

Aircraft Lessors
Real Estate
Shipping
Agr. Land
Timberland
Commodities
Crude Oil
Gold
Precious Metals
Airlines
Transportation
Infrastructure
Real Assets
US Treasuries 1-3M
USD Swap Rate 10Y
USD 3M LIBOR
US CPI
MSCI World
S&P 500
Broad Debt

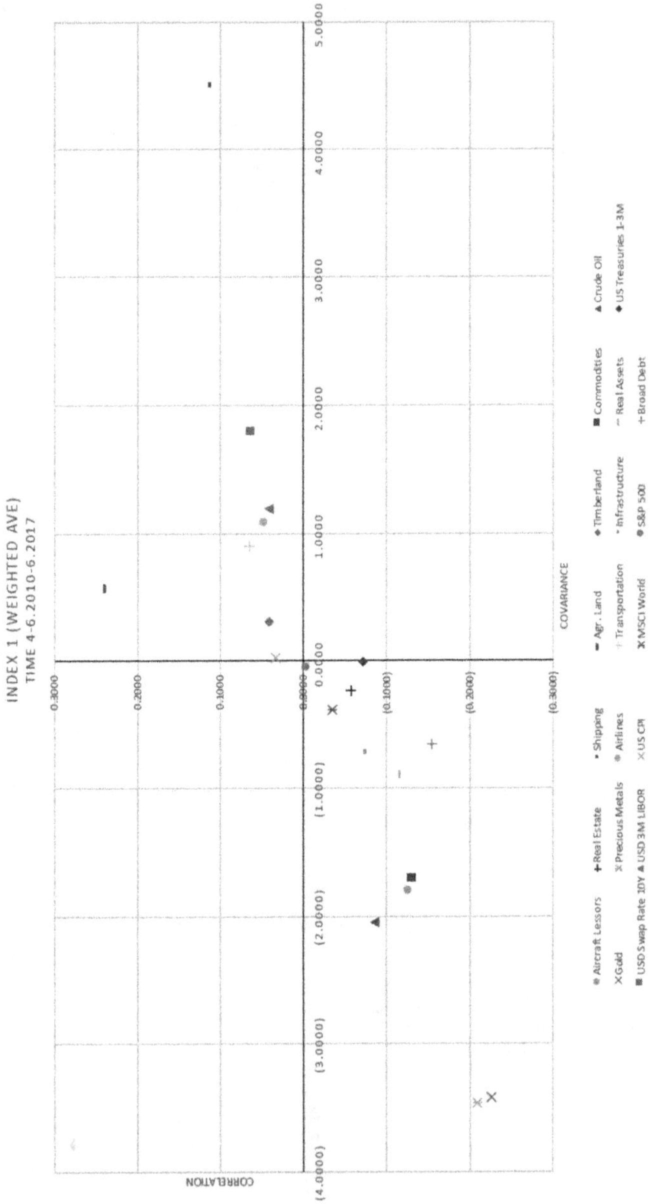
CORRELATION VS. COVARIANCE
INDEX 1 (WEIGHTED AVE)
TIME 4-6.2010-6.2017
CORRELATION
0.3000
0.2000
0.1000
0.0000
(0.1000)
(0.2000)
(0.3000)
COVARIANCE
(4.0000)
(3.0000)
(2.0000)
(1.0000)
0.0000
1.0000
2.0000
3.0000
4.0000
5.0000
Aircraft Lessors
Real Estate
Shipping
Agr. Land
Timberland
Commodities
Crude Oil
Gold
Precious Metals
Airlines
Transportation
Infrastructure
Real Assets
US Treasuries 1-3M
USD Swap Rate 10Y
USD 3M LIBOR
US CPI
MSCI World
S&P 500
Broad Debt

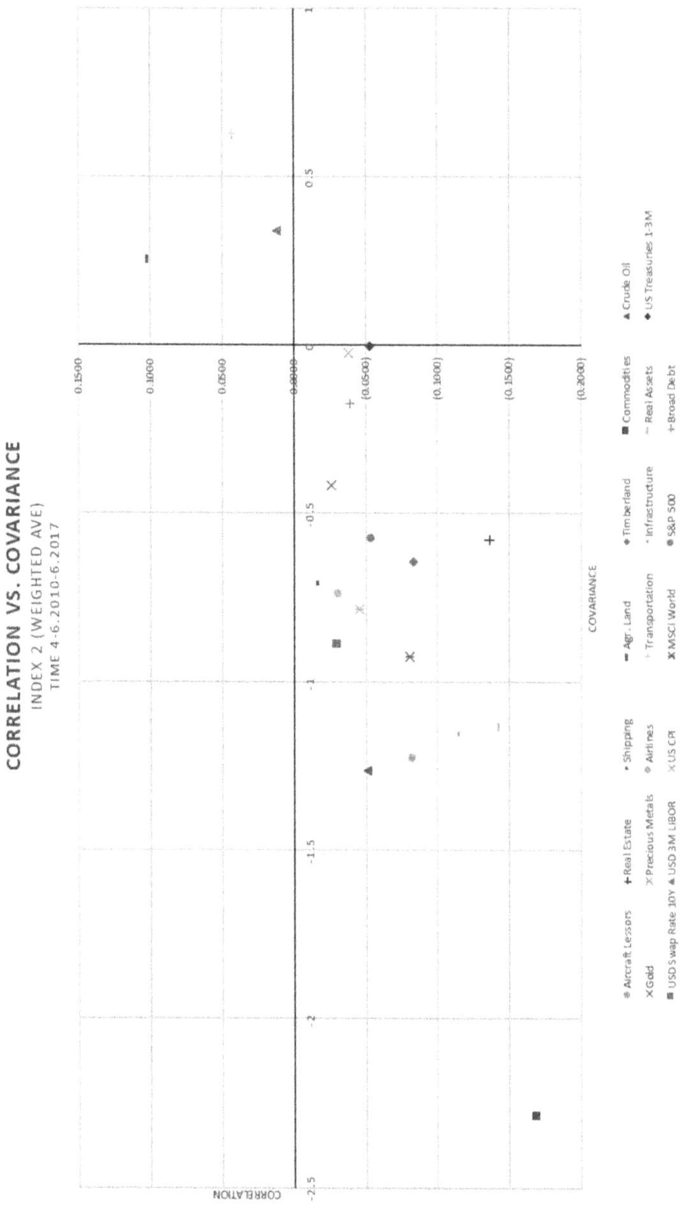
CORRELATION VS. COVARIANCE
INDEX 2 (WEIGHTED AVE)
TIME 4-6.2010-6.2017
CORRELATION
COVARIANCE
0.1500
0.1000
0.0500
0.0000
(0.0500)
(0.1000)
(0.1500)
(0.2000)
-2.5
-2
-1.5
-1
-0.5
0
0.5
1
Aircraft Lessors
Real Estate
Shipping
Agr. Land
Timberland
Commodities
Crude Oil
Gold
Precious Metals
Airlines
Transportation
Infrastructure
Real Assets
US Treasuries 1-3M
USD Swap Rate 10Y
USD 3M LIBOR
US CPI
MSCI World
S&P 500
Broad Debt

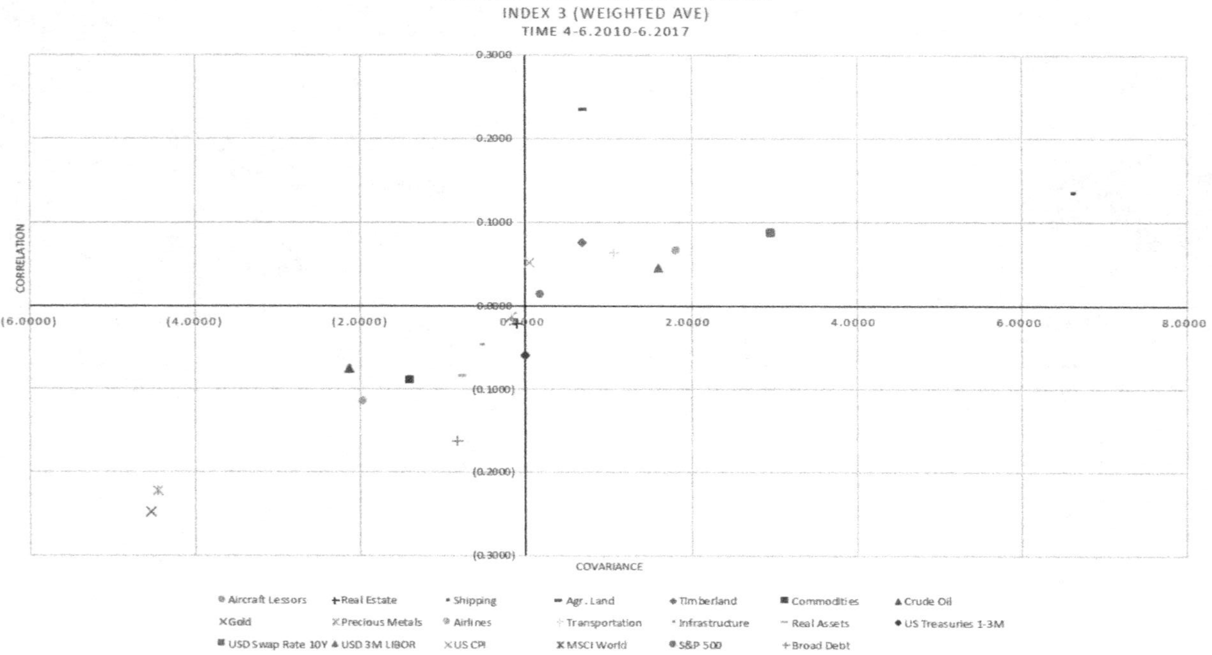

CORRELATION VS. COVARIANCE
INDEX 3 (WEIGHTED AVE)
TIME 4-6.2010-6.2017
0.3000
0.2000
0.1000
0.0000
(0.1000)
(0.2000)
(0.3000)
CORRELATION
(6.0000)
(4.0000)
(2.0000)
2.0000
4.0000
6.0000
8.0000
COVARIANCE
Aircraft Lessors
Real Estate
Shipping
Agr. Land
Timberland
Commodities
Crude Oil
Gold
Precious Metals
Airlines
Transportation
Infrastructure
Real Assets
US Treasuries 1-3M
USD Swap Rate 10Y
USD 3M LIBOR
US CPI
MSCI World
S&P 500
Broad Debt

CORRELATION VS. COVARIANCE
INDEX 4 (WEIGHTED AVE)
TIME 4-6.2010-6.2017

CORRELATION
0.2000
0.1500
0.1000
0.0500
0.0000
(0.0500)
(0.1000)
(0.1500)
(0.2000)

(3.5000) (3.0000) (2.5000) (2.0000) (1.5000) (1.0000) (0.5000) 0.0000 0.5000 1.0000 1.5000
COVARIANCE

Aircraft Lessors, Real Estate, Shipping, Agr. Land, Timberland, Commodities, Crude Oil, Gold, Precious Metals, Airlines, Transportation, Infrastructure, Real Assets, US Treasuries 1-3M, USD Swap Rate 10Y, USD 3M LIBOR, US CPI, MSCI World, S&P 500, Broad Debt

CORRELATION VS. COVARIANCE
INDEX 5 (WEIGHTED AVE)
TIME 4-6.2010-6.2017

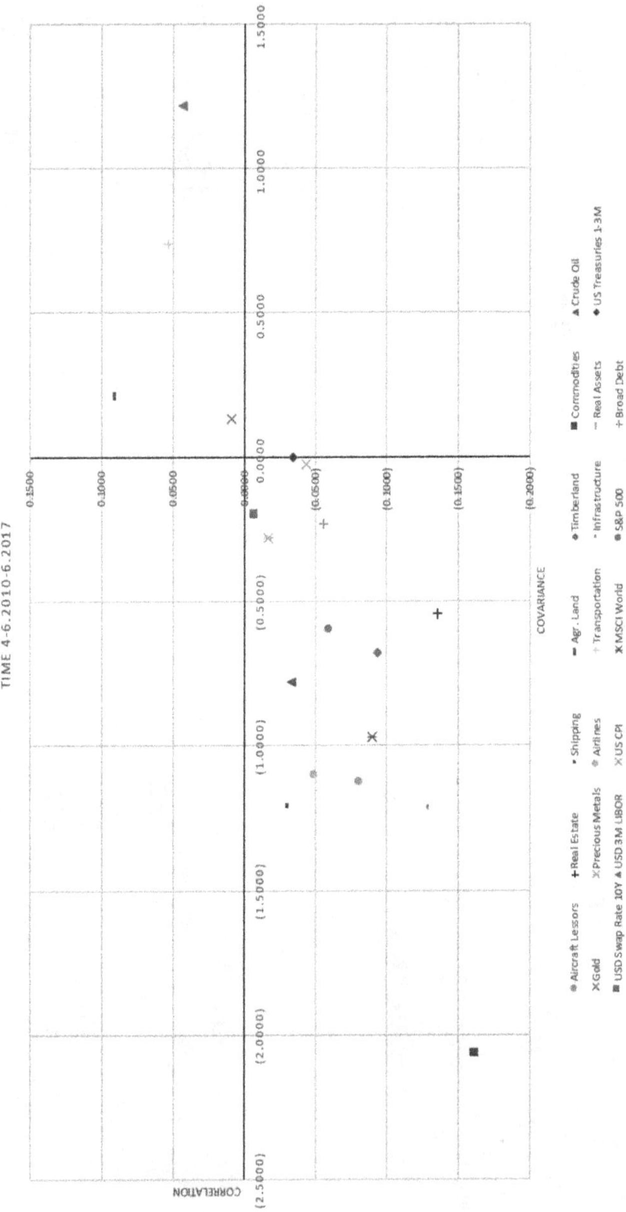
CORRELATION
COVARIANCE
1.5000
1.0000
0.5000
0.0000
(0.5000)
(1.0000)
(1.5000)
(2.0000)
(2.5000)
0.1500
0.1000
0.0500
0.0000
(0.0500)
(0.1000)
(0.1500)
(0.2000)
Aircraft Lessors
Real Estate
Shipping
Agr. Land
Timberland
Commodities
Crude Oil
Gold
Precious Metals
Airlines
Transportation
Infrastructure
Real Assets
US Treasuries 1-3M
USD Swap Rate 10Y
USD 3M LIBOR
US CPI
MSCI World
S&P 500
Broad Debt

Bibliography

Xian Aircraft wins order for 60 MA60 twin-turboprops. (2000, November 14). *Flight International.*

Chinese Civil Aviation Set to Take Off. (2002, September 10). *Xinhua News Agency.* Retrieved from http://china.org.cn/english/BAT/42435.htm

Private Sector's Flying Dream. (2002, November 28). *China Daily.*

High-flying Irishman who became global entrepreneur in aviation sector. (2007, October 6). *The Irish Times.*

5 million Chinese citizens hold ordinary passports. (2012, May 15). *CCTV.* Retrieved from http://news.cntv.cn/20120515/118225.shtml

China to start construction on 35 railway projects: report. (2017, February 19). *Xinhua News Agency.*

A4A. (2017). *Passenger Airline Cost Index.* Retrieved from: http://airlines.org/dataset/a4a-quarterly-passenger-airline-cost-index-u-s-passenger-airlines/

Acquila Capital. (2015). *Real Assets—Investments in Timberland.* Retrieved from www.aquila-capital.de/en/research/stellungnahme-zum-presseartikel-aquila-waldinvest-iii

Airbus. (2016). *Global Market Forecast 2016–2035.* Retrieved from www.airbus.com/aircraft/market/global-market-forecast.html

Airbus. (2017). *Orders and Deliveries.* Retrieved from https://www.airbus.com/aircraft/market/orders-deliveries.html

Airfinance Journal. (2016). Leasing Top 50 2016. *Airfinance Journal.*

Airline Monitor. (2016). *Airline and Aircraft Data.* Retrieved from: www.airline-monitor.com

D. Yu, *Aircraft Valuation*,
https://doi.org/10.1007/978-981-15-6743-8

Alcock, C. (2020). Airbus Restarts A320 Assembly in China as Virus Hits Airlines. *Aviation International News.* Retrieved from: https://www.ainonline.com/aviation-news/air-transport/2020-02-12/airbus-restarts-a320-assembly-china-virus-hits-airlines

An, B. (2013, December 18). Hukou reforms target 2020: official. *China Daily.*

Ang, J., & Peterson, P. P. (1984). The leasing puzzle. *The Journal of Finance, 39*(4), 1055–1065.

Ascend. (2017). *Yearly Operating Lease Percentage as of September 14.* Retrieved from: www.flightglobal.com

Asian Development Bank. (2020). The Economic Impact of the COVID-19 Outbreak on Developing Asia. *ADB Briefs.* Retrieved from: https://www.adb.org/sites/default/files/publication/571536/adb-brief-128-economic-impact-covid19-developing-asia.pdf

Ballantyne, T. (2017, May 1). The long and winding road. *Orient Aviation.*

Baltic Exchange Information Services Ltd. (2018a). *Guide to Market Benchmarks Version 3.4.* Retrieved from www.balticexchange.com/dyn/_assets/_forms/guide-to-market-benchmarks.shtml

Baltic Exchange Information Services Ltd. (2018b). *Shipping Indexes.* Retrieved from: www.balticexchange.com

Barclays. (2016). *Barclays Global Aggregate Bond Index.* Retrieved from: https://indicies.barclays

Barton, D., Chen, Y., & Jin, A. (2013, June). Mapping China's middle class. *McKinsey Quarterly.*

Bazargan, M., & Hartman, J. (2012). Aircraft replacement strategy: Model and analysis. *Journal of Air Transport Management, 25*, 26–29.

Beechy, T. H. (1969). Quasi-Debt Analysis of Financial Leases. *The Accounting Review, 44*(2), 375–381.

Bellalah, M., THEMA, & CEREG. (2002). Valuing lease contracts under incomplete information: A real-options approach. *The Engineering Economist, 47*(2), 194–212.

Bertomeu, J., & Magee, R. P. (2011). From low-quality reporting to financial crises: Politics of disclosure regulation along the economic cycle. *Journal of Accounting and Economics, 52*(2), 209–227.

Bird, R., Liem, H., & Thorp, S. (2014). Infrastructure: Real assets and real returns. *European Financial Management, 20*(4), 802–824.

Bjelicic, B. (2012). Financing airlines in the wake of the financial markets crisis. *Journal of Air Transport Management, 21*, 10–16.

Bloomberg. (2017). *Economic, FX, Fuel, and Funding Data.* Retrieved from: www.bloomberg.com

Bloomberg. (2020). *Economic, FX, Fuel, and Funding Data.* Retrieved from: www.bloomberg.com

Boeing Capital Corporation. (2010). *Current Market Finance Outlook 2010.* Retrieved from Seattle, WA: www.boeing.com/resources/boeingdotcom/company/capital/pdf/2010_BCC_market_report.pdf

Boeing Capital Corporation. (2017). *Current Market Finance Outlook 2017.* Retrieved from Seattle, WA: www.boeing.com/resources/boeingdotcom/company/capital/pdf/2017_BCC_market_report.pdf

Boeing Corporation. (2016). *Current Market Outlook 2016–2035.* Retrieved from Seattle, WA: www.boeing.com/resources/boeingdotcom/commercial/about-our-market/assets/downloads/cmo_print_2016_final_updated.pdf

Boeing Corporation. (2017). *Orders and Deliveries.* Retrieved from Seattle, WA: http://www.boeing.com/commercial/#/orders-deliveries

Bourjade, S., Huc, R., & Muller-Vibes, C. (2017). Leasing and profitability: Empirical evidence from the airline industry. *Transportation Research Part A: Policy and Practice, 97*, 30–46.

Bower, R. S. (1973). Issues in lease financing. *Financial Management*, 25–34.

Bowers, W. C. (1998). Aircraft Lease Securitization: ALPS to EETCs. Retrieved from http://pages.stern.nyu.edu/~igiddy/ABS/bowers2.html

Bowman, R. G. (1980). The debt equivalence of leases: An empirical investigation. *Accounting Review, 55*, 237–253.

Brander, J. A., & Lewis, T. R. (1986). Oligopoly and financial structure: The limited liability effect. *The American Economic Review*, 956–970.

Brookfield Asset Management. (2017). *Real Assets, Real Diversification.* Retrieved from www.brookfield.com

Bunker, D. H. (2005). *International Aircraft Financing* (2nd ed., Vol. 1 General Principles). Montreal: IATA.

Burgess, K., Agnew, H., & Daneshkhu, S. (2016, January 13). Reporting rule adds $3tn of leases to balance sheets globally. *Financial Times.*

Caixin. (2020). Beijing, Shanghai and Other Cities Enact Highest Level Emergency in Response to Coronavirus. Retrieved from: https://www.caixinglobal.com/2020-01-25/beijing-shanghai-and-other-cities-enact-highest-level-emergency-in-response-to-coronavirus-101508199.html

Cambridge Associates. (2017). *Infrastructure Index.* Retrieved from: www.CambridgeAssociates.com

CAPA. (2016). *Airline and Aircraft Data.* Retrieved from: https://centreforaviation.com

CAPA. (2009). Global economic crisis has cost the aviation industry two years of growth: IATA. Retrieved from: https://centreforaviation.com/analysis/reports/global-economic-crisis-has-cost-the-aviation-industry-two-years-of-growth-iata-15921

Case, K. E., & Shiller, R. J. (1987). Prices of single family homes since 1970: New indexes for four cities. In: National Bureau of Economic Research. Cambridge, Mass., USA.

Center for Disease Control and Prevention. (2019). 1918 Pandemic. Retrieved from: https://www.cdc.gov/flu/pandemic-resources/1918-pandemic-h1n1.html

Chen, W.-T., Huang, K., & Ardiansyah, M. N. (2018). A mathematical programming model for aircraft leasing decisions. *Journal of Air Transport Management, 69*, 15–25.

China Banking Regulatory Commission. (2016). *Leasing Companies.* Retrieved from www.cbrc.gov.cn

China Daily. (2020). Over 300 mln train tickets sold for Spring Festival travel rush. Retrieved from: https://global.chinadaily.com.cn/a/202001/04/WS5e1039b7a310cf3e3558274b.html

Civil Aviation Administration of China. (2004). Monthly Statistics of Civil Aviation Industry. Retrieved from: http://www.caac.gov.cn/

Civil Aviation Administration of China. (2009). 2008 Civil Aviation Industry Annual Report. Retrieved from: http://www.caac.gov.cn/

Civil Aviation Administration of China. (2016a). *China Civil Aviation Development 13th Five-Year Plan (2016–2020).* Retrieved from www.caac.gov.cn/XXGK/XXGK/ZCFBJD/201702/P020170215595539950195.pdf

Civil Aviation Administration of China. (2016b). *Statistical Bulletin of Civil Aviation Industry Development in 2016.* Retrieved from www.caac.gov.cn/en/HYYJ/NDBG/201706/W020170602336938087866.pdf

Civil Aviation Administration of China. (2018). *Monthly Statistics of Civil Aviation Industry.* Retrieved from www.caac.gov.cn/

Civil Aviation Administration of China. (2020). China Aviation Industry February Major Production Statistics. Retrieved from: http://www.caac.gov.cn/XXGK/XXGK/TJSJ/202004/P020200420522649148513.pdf

Civil Aviation Administration of China, & The Ministry of Finance of the People's Republic of China. (2020). Announcement of Corona Crisis Financial Subsidy for Aviation Industry. Retrieved from: http://www.caac.gov.cn/XXGK/XXGK/TZTG/202003/t20200304_201269.html

Clarke, J.-P., Miller, B., & Protz, J. (2003). *An options-based analysis of the large aircraft market.* Paper presented at the AIAA/ICAS International Air and Space Symposium and Exposition, Daytona Ohio.

Cornaggia, K. J., Franzen, L. A., & Simin, T. T. (2013). Bringing leased assets onto the balance sheet. *Journal of Corporate Finance, 22*, 345–360.

Dallos, R. E. (1986, February 9). Leasing Jets Takes Off, but May Be Throttled Back : Tax Reform a Threat to Airline Suppliers Such as GE, Xerox. *Los Angeles Times.*

Damodaran, A. (2017). Operating versus Capital Leases. Retrieved from http://pages.stern.nyu.edu/~adamodar/New_Home_Page/AccPrimer/lease.htm

Dechant, T., & Finkenzeller, K. (2010). Real estate or infrastructure? Evidence from conditional asset allocation.

Deegan, G. (2015, October 22). Aircraft firms pay just €23m in tax. *Irish Examiner.*

Deloitte. (2015). *China Outbound Momentum Unabated Despite Economic Uncertainty*. Retrieved from www2.deloitte.com/cn/en/pages/about-deloitte/articles/pr-china-outbound-investment-momentum-unabated-despite-economic-uncertainty.html

Deloitte. (2017). Taxation and Investment in China 2017. Retrieved from www2.deloitte.com/content/dam/Deloitte/global/Documents/Tax/dttl-tax-chinaguide-2017.pdf

DIIO. (2017). *Airline Data*. Retrieved from: www.diio.net

Ding, Y., Ge, X., & Lv, F. (2014). Requirement Analysis and Prediction of Aviation Finance in China. *Information Technology Journal, 13*(1), 140–146.

Doganis, R. (1985). *Flying off course: the economics of international airlines*: Allen & Unwin.

Eaton, S. (2016). *The advance of the state in contemporary China : State-market relations in the reform era*. Cambridge: Cambridge University Press.

The Economist. (2020). Coronavirus is grounding the world's airlines. Retrieved from: https://www.economist.com/business/2020/03/15/coronavirus-is-grounding-the-worlds-airlines

EDB Singapore. (2012). Aircraft Leasing Scheme (ALS) 2012 Circular. Retrieved from www.edb.gov.sg/content/dam/edb/ja/resources/pdfs/financing-and-incentives/Aircraft%20Leasing%20Scheme%20(ALS)%202012%20Circular.pdf

Edwards, J. S., & Mayer, C. P. (1991). Leasing, taxes, and the cost of capital. *Journal of Public Economics, 44*(2), 173–197.

Eisfeldt, A. L., & Rampini, A. A. (2008). Leasing, ability to repossess, and debt capacity. *The Review of Financial Studies, 22*(4), 1621–1657.

Exchange-rates.org. (2015). USD-CNY Exchange Rate 12.31.15. Retrieved from www.exchange-rates.org/Rate/USD/CNY/12-31-2015

Export–Import Bank of the US. (2018). Historical Timeline. Retrieved from www.exim.gov/about/history-exim/historical-timeline/full-historical-timeline

EY. (2020). Overview of China Outbound Investment in 2019. Retrieved from: https://www.ey.com/Publication/vwLUAssets/ey-overview-of-china-outbound-investment-in-2019-en/$FILE/ey-overview-of-china-outbound-investment-in-2019-en.pdf

Ezzell, J. R., & Vora, P. P. (2001). Leasing versus purchasing: Direct evidence on a corporation's motivations for leasing and consequences of leasing. *The Quarterly Review of Economics and Finance, 41*(1), 33-47.

Federal Aviation Administration. (2018). Glossy of Terms. Retrieved from www.faa.gov/air_traffic/flight_info/avn/maintenanceoperations/programstandards/webbasedtraining/CBIChg29/AVN300WEB/Glossarytems.htm

Financial Accounting Standards Board. (1976). Statement of Financial Accounting Standards No. 13. Retrieved from www.fasb.org/jsp/FASB/Document_C/DocumentPage?cid=1218220124481&acceptedDisclaimer=true

Financial Accounting Standards Board. (2016). Accounting Standards Update: Leases (Topic 842). Retrieved from www.fasb.org/jsp/FASB/Document_C/DocumentPage?cid=1176167901010&acceptedDisclaimer=true

Finkenzeller, K., Dechant, T., & Schäfers, W. (2010). Infrastructure: a new dimension of real estate? An asset allocation analysis. *Journal of Property Investment & Finance, 28*(4), 263–274.

Finucane, T. J. (1988). Some empirical evidence on the use of financial leases. *Journal of Financial Research, 11*(4), 321–333.

Franke, M., & John, F. (2011). What comes next after recession?–Airline industry scenarios and potential end games. *Journal of Air Transport Management, 17*(1), 19–26.

Freightos. (2018). *Freightos Baltic Container Route Indexes.* Retrieved from: www.Freightos.com

Froot, K. A. (1995). Hedging portfolios with real assets. *Journal of Portfolio Management, 21*(4), 60.

Gavazza, A. (2010). Asset liquidity and financial contracts: Evidence from aircraft leases. *Journal of Financial Economics, 95*(1), 62–84.

Gavazza, A. (2011). Leasing and secondary markets: Theory and evidence from commercial aircraft. *Journal of Political Economy, 119*(2), 325–377.

Ge, L. (2017, February 21). China to Build 74 New Civil Airports by 2020: CAAC. *China Aviation Daily.* Retrieved from www.chinaaviationdaily.com/news/60/60893.html

Gibson, W., & Morrell, P. (2004). Theory and practice in aircraft financial evaluation. *Journal of Air Transport Management, 10*(6), 427–433.

Gibson, W., & Morrell, P. (2005). *Airline finance and aircraft financial evaluation: evidence from the field.* Paper presented at the Conference conducted at the 2005 ATRS World Conference.

Goldman Sachs. (2015). The Chinese Tourist Boom. *The Asian Consumer.* Retrieved from: https://www.goldmansachs.com/insights/pages/macroeconomic-insights-folder/chinese-tourist-boom/report.pdf

Goolsbee, A. (1998). The business cycle, financial performance, and the retirement of capital goods. *Review of Economic Dynamics, 1*(2), 474-496.

Gorjidooz, J., & Vasigh, B. (2010). Aircraft valuation in dynamic air transport industry. *Journal of Business & Economics Research (JBER), 8*(7).

Graham, J. R., Lemmon, M. L., & Schallheim, J. S. (1998). Debt, leases, taxes, and the endogeneity of corporate tax status. *The Journal of Finance, 53*(1), 131–162.

Gritta, R., & Lynagh, P. (1973). Aircraft Leasing-Panacea or Problem. *Transp. LJ, 5*, 9.

Gritta, R. D. (1974). The Impact of the Capitalization of Leases on Financial Analysis. *Financial Analysts Journal, 30*(2), 47–52.

Gritta, R. D., Lippman, E., & Chow, G. (1994). The impact of the capitalization of leases on airline financial analysis: An issue revisited. *Logistics and Transportation Review, 30*(2), 189.

Habib, M. A., & Johnsen, D. B. (1999). The financing and redeployment of specific assets. *The Journal of Finance, 54*(2), 693–720.

Hallerstrom, N. (2013). *Modeling Aircraft Loan & Lease Portfolios.* Retrieved from www.pkair.com/common/docs/modeling-aircraft-loan-and-lease-portfolios.pdf

Handa, P. (1991). An economic analysis of leasebacks. *Review of Quantitative Finance and Accounting, 1*(2), 177–189.

Harris, R. S., & Pringle, J. J. (1985). Risk adjusted discount rates extensions from the average risk case. *Journal of Financial Research, 8*(3), 237–244.

Haver Analytics. (2012). *Economic Data.* Retrieved from: www.haver.com

Haver Analytics. (2017). *China Transportation Data.* Retrieved from: www.haver.com

Heng, M. & Yong, C. (2020). Singapore Airlines cuts more flights; Changi sees 33% drop in passengers. *The Jakarta Post.* Retrieved from: https://www.thejakartapost.com/travel/2020/03/14/sia-cuts-more-flights-changi-sees-33-drop-in-passengers.html

HNA. (2020). *HNA Group Announcement.* Retrieved from: http://www.hnagroup.com/news/%E6%B5%B7%E8%88%AA%E9%9B%86%E5%9B%A2%E6%9C%89%E9%99%90%E5%85%AC%E5%8F%B8%E5%85%AC%E5%91%8A-1/

Holloway, S. (1997). *Straight and level: practical airline economics* (2nd ed.): Avebury.

Hong Kong Inland Revenue Department. (2017). Comprehensive Double Taxation Agreements Concluded. Retrieved from www.ird.gov.hk/eng/tax/dta_inc.htm

Hong Kong Legislative Council. (2016). Key drivers for developing an aircraft leasing centre. Retrieved from www.legco.gov.hk/research-publications/english/essentials-1516ise17-key-drivers-for-developing-an-aircraft-leasing-centre.htm

Hong Kong Legislative Council. (2017). Proposed Dedicated Tax Regime to Develop Aircraft Leasing Business in Hong Kong. Retrieved from www.legco.gov.hk/yr16-17/english/panels/edev/papers/edev20170123cb4-410-8-e.pdf

Hsu, C.-I., Li, H.-C., Liu, S.-M., & Chao, C.-C. (2011). Aircraft replacement scheduling: a dynamic programming approach. *Transportation research part E: logistics and transportation review, 47*(1), 41–60.

Hsu, K. (2014). *China's Airspace Management Challenge.* Retrieved from www.uscc.gov/sites/default/files/Research/China%27s%20Airspace%20Management%20Challenge.pdf

Humphries, C., & Hepher, T. (2014). Insight: Revenge of the Irish as mega-merger restores air finance crown. *Reuters.*

IBA Group. (2017). *Traffic and Aircraft Data.* Retrieved from: www.iba.aero

IDA Financial Services. (2017). International Financial Services. Retrieved from www.idaireland.com/business-in-ireland/industry-sectors/financial-services

IFRS. (2003). IAS 17. Retrieved from www.ifrs.org/Documents/IAS17.pdf

IFRS. (2012). IFRS Foundation Staff Analysis of SEC Final Staff Report on IFRS. Retrieved from www.ifrs.org/Alerts/PressRelease/Pages/IFRS-Foundation-Staff-Analysis-of-SEC-Final-Staff-Report-on-IFRS.aspx

IHS Economics. (2016). *Economics Data.* Retrieved from: www.ihsmarkit.com

Inderst, G. (2009). Pension fund investment in infrastructure.

Inderst, G. (2010). Infrastructure as an asset class. *EIB papers, 15*(1), 70–105.

Innovata. (2016). *LCC Data.* Retrieved from: www.innovata-llc.com

International Air Transport Association. (2003). Industry Recovery Starts June Shows Signs of Improvement. Retrieved from: https://www.iata.org/en/pressroom/pr/2003-08-04-01/

International Air Transport Association. (2009). Deeper Losses Forecast—Falling Yields, Rising Fuel Costs. Retrieved from: https://www.iata.org/en/pressroom/pr/2009-09-15-01/

International Air Transport Association. (2010). Annual Report 2010. Retrieved from: https://www.iata.org/contentassets/c81222d96c9a4e0bb4ff6ced0126f0bb/iata-annualreport2010.pdf

International Air Transport Association. (2012). *Financial Monitor for Jan/Feb-2012 released on 01-Mar-2012.* Retrieved from www.iata.org

International Air Transport Association. (2016a). *Air Passenger Forecasts, October 2016.* Retrieved from www.iata.org

International Air Transport Association. (2016b). *Airline Industry Profitability.* Retrieved from www.iata.org

International Air Transport Association. (2020a). COVID-19 Hits January Passenger Demand. Retrieved from: https://www.iata.org/en/pressroom/pr/2020-03-04-03/

International Air Transport Association. (2020b). Initial impact* assessment of the novel Coronavirus. IATA Economics. Retrieved from: https://www.iata.org/en/iata-repository/publications/economic-reports/coronavirus-initial-impact-assessment/

International Air Transport Association. (2020c). COVID-19 Cuts Demand and Revenues. Retrieved from: https://www.iata.org/en/pressroom/pr/2020-02-20-01/

International Air Transport Association. (2020d). IATA Updates COVID-19 Financial Impacts—Relief Measures Needed. Retrieved from: https://www.iata.org/en/pressroom/pr/2020-03-05-01/

International Air Transport Association. (2020e). Updated impact* assessment of the novel Coronavirus. *IATA Economics.* Retrieved from: https://www.iata.org/en/iata-repository/publications/economic-reports/coronavirus-updated-impact-assessment/

International Air Transport Association. (2020f). COVID-19 Updated Impact Assessment. *IATA Economics.* Retrieved from: https://www.iata.org/en/iata-repository/publications/economic-reports/covid-fourth-impact-assessment/

International Civil Aviation Organization. (2012a). *Airline Industry Profitability.* Retrieved from www.icao.int

International Civil Aviation Organization. (2012b). *WASA database.* Retrieved from: www.icao.intl

International Civil Aviation Organization. (2016). *Airline Demand.* Retrieved from: www.icao.intl

International Civil Aviation Organization. (2017). *Airline Operating Costs and Productivity.* Retrieved from: www.icao.intl

International Monetary Fund. (2016). *Country Economic Data.* Retrieved from: www.imf.org

International Monetary Fund. (2017). *Country Economic Data.* Retrieved from: www.imf.org

International Society of Transport Aircraft Trading. (2009). *The Principles Of Appraisal Practice And Code Of Ethics.* Retrieved from https://members.istat.org/p/do/sd/sid=140&fid=168&req=direct

International Society of Transport Aircraft Trading. (2013). *Appraisers' Program Handbook Revision 7.* Retrieved from www.istat.org/d/do/382

Jiang, H., & Hansman, R. (2004). *An analysis of profit cycles in the airline industry.* Thesis (SM), Department of Aeronautics and Astronautics, Massachusetts.

Johns Hopkins University CSSE. (2020). *COVID-19 Map.* Retrieved from: https://coronavirus.jhu.edu/map.html

Johnson, R. W., & Lewellen, W. G. (1972). Analysis of the Lease or Buy Decision. *The Journal of Finance, 27*(4), 815-823.

Kasper, D. M. (1988). *Deregulation and globalization: Liberalizing international trade in air services.* Washington, DC: American Enterprise Institute

Kothari, S., & Lester, R. (2012). The role of accounting in the financial crisis: Lessons for the future. *Accounting Horizons, 26*(2), 335–351.

Krishnan, V. S., & Moyer, R. C. (1994). Bankruptcy costs and the financial leasing decision. *Financial Management*, 31–42.

Laprade, D. B. (1981). *The Basic Economics of Air Carrier Operations.* Ottawa: Canada Transport Commission.

Lease, R. C., McConnell, J. J., & Schallheim, J. S. (1990). Realized returns and the default and prepayment experience of financial leasing contracts. *Financial Management*, 11–20.

Lee, A. (2020). Coronavirus sends China's aviation industry into free fall, damaging hopes of becoming global hub. *South China Morning Post*. Retrieved from: https://www.scmp.com/economy/china-economy/article/3065236/coronavirus-sends-chinas-aviation-industry-free-fall-damaging

Lenoir, N. (1998). *Cycles in the air transportation industry.* Paper presented at the WCTR 1998, 8th World Conference on Transportation Research.

Lewellen, W. G., Long, M. S., & McConnell, J. J. (1976). Asset leasing in competitive capital markets. *The Journal of Finance, 31*(3), 787–798.

Lewis, C. M., & Schallheim, J. S. (1992). Are debt and leases substitutes? *Journal of Financial and Quantitative Analysis, 27*(4), 497–511.

Li, X. (2010, December 13). China aircraft leasing makes new steps. *China Economic News.*

Liehr, M., Größler, A., Klein, M., & Milling, P. M. (2001). Cycles in the sky: understanding and managing business cycles in the airline market. *System Dynamics Review, 17*(4), 311–332.

Lim, S. C., Mann, S. C., & Mihov, V. T. (2017). Do operating leases expand credit capacity? Evidence from borrowing costs and credit ratings. *Journal of Corporate Finance, 42*, 100–114.

Lyneis, J. M. (2000). System dynamics for market forecasting and structural analysis. *System Dynamics Review: The Journal of the System Dynamics Society, 16*(1), 3–25.

Mann, E. D. (2009). Aviation finance: An overview. *Journal of Structured Finance, 15*(1), 109.

Martin, G. A. (2010). The long-horizon benefits of traditional and new real assets in the institutional portfolio. *The Journal of Alternative Investments, 13*(1), 6.

McKinsey & Company. (2019). China and the world: Inside the dynamics of a changing relationship. *McKinsey Global Institute*. Retrieved from: https://www.mckinsey.com/featured-insights/china/china-and-the-world-inside-the-dynamics-of-a-changing-relationship

Meadows, D. L. (1971). *Dynamics of commodity production cycles.*

MergerMarket, & ICF. (2016). *Aviation and Aerospace M&A Quarterly Q2 2016.* Retrieved from www.mergermarket.com/assets/ICF_Q2_2016_Newsletter_Final_LR.pdf

Ministry of Commerce of the People's Republic of China. (2016). *Leasing Companies.* Retrieved from www.mofcom.gov.cn

Ministry of Public Security of the People's Republic of China. (2016). *Twenty-fourth Meeting of the Standing Committee of the Twelfth National People's Congress.* Retrieved from www.mps.gov.cn/n2253534/n2253535/n2253536/c5538068/content.html

MSCI. (2017). *MSCI World Index.* Retrieved from: www.msci.com/indexes

Myers, S. C. (1974). Interactions of corporate financing and investment decisions—implications for capital budgeting. *The Journal of Finance, 29*(1), 1–25.

Myers, S. C., Dill, D. A., & Bautista, A. J. (1976). Valuation of financial lease contracts. *The Journal of Finance, 31*(3), 799–819.
National Bureau of Statistics of China. (2011). *Basic Statistics on National Population Census in 1953, 1964, 1982, 1990, 2000 and 2010.* Retrieved from: www.stats.gov.cn/tjsj/Ndsj/2011/html/D0305e.htm
National Bureau of Statistics of China. (2016). *China Statistical Yearbook 2016.* Retrieved from: www.stats.gov.cn/tjsj/ndsj/2016/indexeh.htm
National Development and Reform Commission of China. (2016). *Leasing Companies.* Retrieved from www.ndrc.gov.cn
NCREIF. (2017). *Indices.* Retrieved from: www.NCREIF.org
Newell, G., & Peng, H. W. (2007). *The significance of infrastructure in investment portfolios.* Paper presented at the Pacific Rim Real Estate Conference, Fremantle.
Newell, G., & Peng, H. W. (2008). The role of US infrastructure in investment portfolios. *Journal of Real Estate Portfolio Management, 14*(1), 21–34.
OAG. (2020). Schedules Analyser. Retrieved from: https://www.oag.com/
Oum, T. H., Zhang, A., & Zhang, Y. (2000a). Optimal demand for operating lease of aircraft. *Transportation Research Part B: Methodological, 34*(1), 17–29.
Oum, T. H., Zhang, A., & Zhang, Y. (2000b). Socially optimal capacity and capital structure in oligopoly: the case of the airline industry. *Journal of Transport Economics and Policy,* 55–68.
Oxford Economics. (2016). *World Population.* Retrieved from www.oxfordeconomics.com
The Paper. (2020). Aviation Supply Chain Back to Normal: Airbus Tianjin Facilities Fully Resume Production. Retrieved from: https://www.thepaper.cn/newsDetail_forward_6683802
Park, Y. (2020). China's Aviation Market Shrinks to Smaller Than Portugal's. *Bloomberg Business.* Retrieved from: https://www.ainonline.com/aviation-news/air-transport/2020-02-12/airbus-restarts-a320-assembly-china-virus-hits-airlines
Parrish, R. (1970). Aircraft Leasing. *Airline Management and Marketing, II (June 1970), 50.*
Pearce, B. (2012). The state of air transport markets and the airline industry after the great recession. *Journal of Air Transport Management, 21,* 3–9.
Pearson, R. J. (1976). Airline managerial efficiency. *The Aeronautical Journal, 80*(791), 475–482.
People's Bank of China. (2017). Foreign Exchange Reserve Figures. Retrieved from www.pbc.gov.cn
Pierson, K., & Sterman, J. D. (2013). Cyclical dynamics of airline industry earnings. *System Dynamics Review, 29*(3), 129–156.
Pryke, R. (1987). *Competition among international airlines.*
Pulvino, T. C. (1998). Do asset fire sales exist? An empirical investigation of commercial aircraft transactions. *The Journal of Finance, 53*(3), 939–978.

Pulvino, T. C. (1999). Effects of bankruptcy court protection on asset sales. *Journal of Financial Economics, 52*(2), 151–186.

PWC. (2012). *Overview of China's Aircraft Leasing Industry Promoting Industry Development.* Retrieved from www.pwccn.com

Quigley, J. M. (1995). A simple hybrid model for estimating real estate price indexes. *Journal of Housing Economics, 4*(1), 1–12.

Relbanks.com. (2016a). *The Largest Banks in China.* Retrieved from www.relbanks.com/asia/china

Relbanks.com. (2016b). *World's Top Insurance Companies.* Retrieved from www.relbanks.com/top-insurance-companies/world

Reuters. (1990). Company News; Guinness Peat In Joint Venture. Retrieved from www.nytimes.com/1990/06/15/business/company-news-guinness-peat-in-joint-venture.html

Robertson, D. H. (1915). *A study of industrial fluctuation: an enquiry into the character and causes of the so-called cyclical movements of trade*: London: PS King.

Roenfeldt, R. L., & Osteryoung, J. S. (1973). Analysis of financial leases. *Financial Management*, 74–87.

Rothballer, C., & Kaserer, C. (2012). The risk profile of infrastructure investments: Challenging conventional wisdom. *Journal of Structured Finance, 18*(2), 95.

RTE. (2020). Chinese tourism—the main engine of global travel. Retrieved from: https://www.rte.ie/news/business/2020/0203/1112823-chinese-tourism/

Russell, D. W. (2014). *Good risks : discovering the secrets to ORIX's 50 years of success.* Singapore: John Wiley.

S&P Dow Jones Indexes. (2017). *S&P Dow Jones Indexes.* Retrieved from: https://eu.spindices.com

Schallheim, J., Wells, K., & Whitby, R. J. (2013). Do leases expand debt capacity? *Journal of Corporate Finance, 23*, 368–381.

Schallheim, J. S., & McConnell, J. (1985). A model for the determination of "fair" premiums on lease cancellation insurance policies. *The Journal of Finance, 40*(5), 1439–1457.

Sharpe, S. A., & Nguyen, H. H. (1995). Capital market imperfections and the incentive to lease. *Journal of Financial Economics, 39*(2), 271–294.

Shearman, P. (1992). *Air Transport: Strategic Issues in Planning and Development.*

Shi, Y. (2005). *Foundations and practices of leasing*, Beijing, China: University of International Business and Economics Press.

Shleifer, A., & Vishny, R. W. (1992). Liquidation values and debt capacity: A market equilibrium approach. *The Journal of Finance, 47*(4), 1343–1366.

Slovin, M. B., Sushka, M. E., & Polonchek, J. A. (1990). Corporate Sale and Leasebacks and Shareholder Wealth. *The Journal of Finance, 45*(1), 289–299.

Smith, A. (1776). *The wealth of nations.*

Smith, C. W., & Wakeman, L. (1985). Determinants of corporate leasing policy. *The Journal of Finance, 40*(3), 895–908.
Smith, M. A. (1968, Dec. 19). New Douglas Finance Company. *Flight International, 81,* 1020.
Smith, O. (2019). The unstoppable rise of the Chinese traveller—where are they going and what does it mean for overtourism? *The Telegraph.* Retrieved from: https://www.telegraph.co.uk/travel/comment/rise-of-the-chinese-tourist/
State Administration of Foreign Reserves of China. (2017). *Foreign Reserves.* Retrieved from www.safe.gov.cn
Statista. (2020). Number of visitors to the United States from China from 2003 to 2024. Retrieved from: https://www.statista.com/statistics/214813/number-of-visitors-to-the-us-from-china/
Stonier, J. (1999). Airline long-term planning under uncertainty. *Real Options and Business Strategy, Lenos Trigeorgis ed., Risk Books.*
Stonier, J. (2001). Airline fleet planning, financing, and hedging decisions under conditions of uncertainty. *Handbook of Airline Strategy*, McGraw-Hill.
Stonier, J. (1998). Marketing From A Manufacturer's Perspective: Issues In Quantifying The Economic Benefits Of New Aircraft, And Making The Correct Financing Decision. *Handbook Of Airline Marketing.*
Straszheim, M. R. (1969). *The international airline industry*: Brookings Institution, Transport Research Program.
Thomson Reuters Datastream. (2017). *Economic Data.* Retrieved from: www.thomsonreuters.com
United States Senate. (2020). S.3578—COVID-19 Funding Accountability Act of 2020. Retrieved from: https://www.congress.gov/bill/116th-congress/senate-bill/3578?s=2&r=10
UN Population Division. (2016). *Population Databases.* Retrieved from: www.un.org/en/development/desa/population/publications/database/index.shtml
Unsworth, E. (1990, June 20). GPA Leases 1st Aircraft To China. *JOC.com.* Retrieved from www.joc.com/gpa-leases-1st-aircraft-china_19900620.html
US Census Bureau. (2018). Population Clock. Retrieved from www.census.gov/popclock
US Securities And Exchange Commission. (2012). *Work Plan for the Consideration of Incorporating International Financial Reporting Standards into the Financial Reporting System for U.S. Issuers.* Retrieved from www.sec.gov/spotlight/globalaccountingstandards/ifrs-work-plan-final-report.pdf
US State Department. (2017). Passport Statistics. Retrieved from https://travel.state.gov/content/travel/en/passports/after/passport-statistics.html
Vasigh, B., & Erfani, G. (2004). Aircraft value, global economy and volatility. *Airlines Magazine, 28.*

Vasigh, B., Fleming, K., & Humphreys, B. (2014). *Foundations of airline finance: Methodology and practice*: Routledge.

Wang, W. (2010, March 10). Airline boss in plea for more airspace. *Shanghai Daily.*

Wayne, L. (2007, May 10). The Real Owner of All Those Planes. *New York Times.*

Williamson, O. E. (1988). Corporate finance and corporate governance. *The Journal of Finance, 43*(3), 567–591.

Wojahn, O. W. (2012). Why does the airline industry over-invest? *Journal of Air Transport Management, 19*, 1–8.

World Bank. (2017). *Country Economic Data*. Retrieved from: www.worldbank.org

Wyman, H. E. (1973). Financial lease evaluation under conditions of uncertainty. *The Accounting Review, 48*(3), 489–493.

Yan, A. (2006). Leasing and debt financing: substitutes or complements? *Journal of Financial and Quantitative Analysis, 41*(3), 709–731.

Yang, H. (2016). *China financial leasing industry development report for 2015.* Tianjin: Nankai University Press.

Yu, D. (2017). *Aviation Joint Ventures, Sidecars & M&A*. Unpublished manuscript, Johns Hopkins University.

Yu, M. (2013, January 31). Inside China: No airspace for holiday travel. *Washington Times.* Retrieved from www.washingtontimes.com/news/2013/jan/31/inside-china-no-airspace-for-holiday-travel/.

Index

D. Yu, *Aircraft Valuation*,
https://doi.org/10.1007/978-981-15-6743-8

CPSIA information can be obtained
at www.ICGtesting.com
Printed in the USA
LVHW041523221120
672385LV00004B/117